ARCHAEOLOGY & GEOMATICS

Sidestone Press

ARCHAEOLOGY & GEOMATICS

HARVESTING THE BENEFITS OF 10 YEARS OF TRAINING IN THE IBERIAN PENINSULA (2006-2015)

edited by
Victorino Mayoral Herrera
César Parcero-Oubiña
Pastor Fábrega-Álvarez

Published by Sidestone Press, Leiden
www.sidestone.com

Lay-out & cover design: Sidestone Press
Photograph cover: Visualization of LIDAR data,
by João Fonte and José Manuel Costa.

ISBN 978-90-8890-451-6 (softback)
ISBN 978-90-8890-452-3 (hardcover)
ISBN 978-90-8890-453-0 (PDF e-book)

Contents

Unboxing the black box. Some reflections after ten years of teaching geospatial technologies to archaeologists

Victorino Mayoral Herrera,[1] *César Parcero-Oubiña*[2] *and Pastor Fábrega-Álvarez*[2]

Introduction

Over the course of recent decades, the use of Geospatial Technologies in archaeology has grown exponentially. At the time of their inception, they were seen as complex, expensive and not very user-friendly tools the use of which was limited to just a few highly specialised experts. Furthermore, geographic data were limited, hard to acquire and manage and, quite often, too expensive for the ordinary archaeologist.

All this has changed substantially in the last few years. Today, software tools are usually quite powerful and relatively easy to use, even at a low cost or completely for free. Geographic data are widely accessible, including a range of large scale datasets which are extremely useful for archaeologists, such as high resolution aerial and satellite imagery, LiDAR data, etc. (Bennett *et al.* 2014). Geospatial approaches are extremely common in archaeology nowadays, as is proven by a steady stream of papers, conferences and books (e.g. Bevan and Lake 2013; Comer and Harrower 2013; Opitz and Cowley 2013; Kamermans *et al.* 2014; Polla and Verhagen 2014; Posluschny 2015; Forte and Campana 2016, to mention only a few recent syntheses).

---

1 Instituto de Arqueología de Mérida, Consejo Superior de Investigaciones Científicas. Plaza de España, 15. 06800 Mérida (Spain) vmayoral@iam.csic.es.

2 Instituto de Ciencias del Patrimonio, Consejo Superior de Investigaciones Científicas. Avenida de Vigo, s/n. 15705 Santiago de Compostela (Spain) cesar.parcero-oubina@incipit.csic.es; pastor.fabrega-alvarez@incipit.csic.es.

Such a combination of user-friendly tools and extensive free datasets has its obvious benefits, although a certain degree of risk may also be entailed, which can be summarised in what J. Huggett labelled more than 10 years ago as a "technological fetishism" (Huggett 2004). Geospatial technologies seem to be especially prone to such a risk, since nowadays highly complex analytical processes are just a click of the mouse away. As a consequence, specialists in these fields are very much in demand and training has become an urgent need in recent years. However, at least in the case of Spain, training in geospatial technologies for archaeologists has been notable by its absence in most university departments. This was indeed the case back in 2006, when we first decided to organise the course around which this book has been written. At that time, not one university in Spain offered, on a regular basis, any kind of formal training in the field of geomatics, geographic information, remote sensing, etc., for archaeology. Back then, and even still today to a great extent, such training was only possible within other academic fields, such as Natural Sciences, Geography or Topography. In stark contrast to this fact, a growing number of younger archaeologists were eager to acquire such knowledge. The combination of these two factors (an interest from practitioners and the lack of an adequate supply) was what triggered the organisation, back in 2006, of the first edition of a course then titled "La aplicación de las tecnologías de información geográfica en arqueología: de la reflexion a la práctica" ("The application of geospatial technologies in archaeology: from reflection to practice").

Quite unexpectedly, the course has been held annually ever since and has trained more than 300 young archaeologists in the critical use of popular tools such as GIS, GPS, remote sensing or LiDAR for the documentation and analysis of the archaeological record. To commemorate the 10th anniversary of the course, a conference was organised in Mérida (Spain) in October 2015, in which former students were invited to present and discuss pieces of work in which they had made comprehensive use of these technologies. This book brings together some of these contributions, the selection of which was based solely on the voluntary desire of their authors to prepare a written version of their contribution to the conference. Our aim, as editors of this volume, is mainly to give international visibility to what has been, and is being, done in the Spanish-speaking world (with particular emphasis on the case of Spain) as far as the use of geospatial technologies in archaeology is concerned. There is now a well-established body of archaeological literature in this field, with a relatively long history (Blasco Bosqued *et al.* 1997; Grau 2006; Mayoral Herrera and Celestino Pérez 2011). However, most of the literature in this field has been published in Spanish so far and it remains largely unknown to an international audience. The same applies to the growing body of literature produced in Latin America (e.g. Pastor *et al.* 2013; Figuerero Torres and Izeta 2013).

Unboxing the black box

Archaeology as a discipline has been increasingly branching into a multitude of sub-fields which, along with the development of specialised tools and techniques, are leading us towards a kind of hyper-specialisation. In this context, the acquisition of a comprehensive overview seems to be imperative in order to take efficient, informed decisions regarding the development of a coherent research process. Nevertheless, quite often, the process of achieving our desired objectives leads us into a true labyrinth of procedures with a complex mathematical basis. Overwhelmed by such a huge quantity of specialised skills, we tend more and more to rely, sometimes uncritically or inadvertently, on what computers offer us through their user-friendly interfaces. Rather than making calculations simpler and more flexible, all of this prevents us from revealing the internal mechanics with which the computer works. We are continually at risk of falling into the "push button syndrome", which saves us the task of figuring out what is really going on inside the "black box". When studying the past, we can add to this the opportunistic way in which we usually apply these tools. Many misunderstandings in our efforts to use them originate from a mismatch between our expectations and the way geospatial technologies, especially GIS, work. From their inception they were designed to deal with very specific problems, which were completely disconnected from our need to explore the peculiarities of cultural phenomena across time. We must therefore face the challenge of acquiring a deep and critical knowledge of geospatial technologies, as it is the only means of efficiently using and tailoring this wide spectrum of possibilities to our needs.

Ten years of training

Ten years in a nutshell

As has already been mentioned, in 2006 what was, at that time, a pioneering initiative of specialised training in the field of geospatial technologies for archaeology was first organised by the Mérida Institute of Archaeology (IAM). At that time, the course was a single 8-day event, held in Mérida, the contents of which were designed to give a general introduction to the field to an audience which was expected to have little or no previous experience. Encouraged by the positive feedback from the more than 20 students from Spain and Portugal who attended the course, a second edition was organised in 2007 following a similar pattern. Again, around 20 students attended the course, and it seemed that there was, indeed, a consistent demand for specialised training in the field, mainly taking into consideration the fact that most of these students were coming from far-flung places in Spain and Portugal, as Mérida is a town which is not especially easy to reach from most regions in the Iberian Peninsula. The organisation of the International Symposium on GIS applications in Archaeology later in the year (published in Mayoral Herrera and Celestino Pérez 2011) consolidated the vocation of the IAM to develop a research programme focused on this topic.

Following those first two years, the course began to be co-organised by the IAM and the institution which is now known as the Institute of Heritage Sciences (Incipit), located in Santiago de Compostela (Spain). This institute shared the considerable interest of the IAM in the use of geospatial technologies for archaeology. Another common characteristic of both institutions was their peripheral geographic position in Spain, which we considered as a challenging factor to prove that the interest of students in this field was explicit and durable.

Year	2006	2007	2008	2009	2010	2011	2012	2013	2014	2015	2016
Structure	Single course, 56 hours			Introductory module, 56 hours						Introductory online module, 90 hours	
				Advanced module, 56 hours						Advanced module, 45 hours	
Nº of students	25	22	21	18+17	17+10	17+10	20+20	20+9	15+12	19+14	25+15

Table 1: Changes in length of course and number of students per year.

In 2006 and 2007, the course retained its initial structure as an introductory, 8-day event, and was held alternately in Mérida and Santiago de Compostela.

In those early years we noticed that the students attending the course had a rather uneven degree of prior expertise in the field, which made it advisable to refine the programme and contents of the course in order to offer "something for all tastes". Therefore, from 2009 onwards the course adopted a two-tiered structure with an 8-day introductory module followed by an 8-day advanced module, with the students being free to choose to participate in one or both modules. The introductory module is aimed at providing the students with only the basic skills in order to understand digital geographic information and tools, and their potential application in the field of archaeology. As far as the advanced module is concerned, it is aimed at those students seeking to deepen their knowledge of some specific issues relating to the application of digital tools in archaeology. The programme of this advanced module changes every year, depending on the need to focus on some specific subjects. For instance, in 2016 it was focused on the detection of archaeological evidence, whereas in 2017 the subject was the digital analysis of archaeological landscapes.

The course has maintained this two-tiered structure ever since, although in 2015 one final change was made: the introductory module began to be taught completely online, thereby allowing, among other things, for the expansion of its contents up to 90 hours of training (plus 45 more for the advanced module). The advanced module remains a face-to-face event, held alternately in Mérida and Santiago de Compostela.

A quick overview of the contents

A basic principle of the course has been to focus on the relationship between theory and practice. We have always made it clear that our aim is not to teach the students how to use a couple of specific computer programs or what buttons to click, but to explore what these tools can do (and what they cannot do) and to critically analyse their application in archaeology. Questioning, reflection and discussion has always been a major concern, especially in the advanced module. For us, exploring the relationship between technology and the diversity of theoretical trends in archaeological research has always been especially important, as well as addressing the influence of other disciplines, such as Geography. This leads us to evaluate the evolution of the application of GIS in Archaeology from its beginnings to the present.

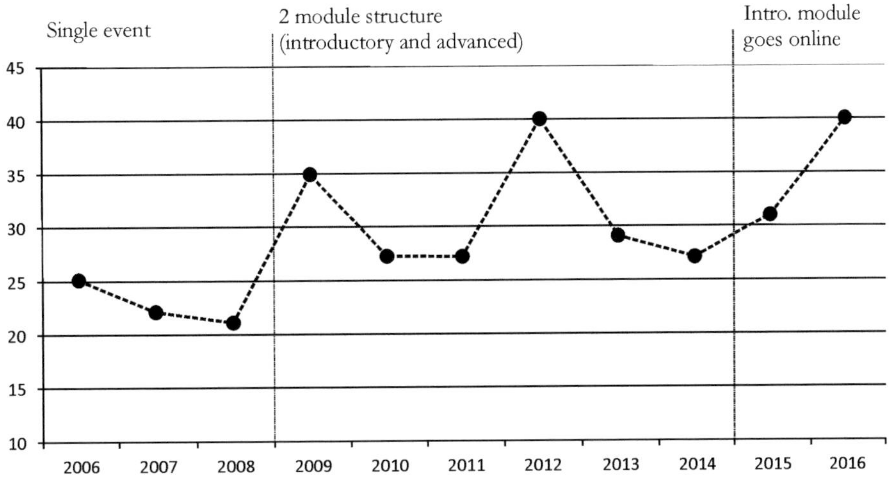

Figure 1: Number of students per year, including both modules after 2009.

On a more practical level, a strong emphasis has been placed on guiding the students towards the design and implementation of projects and research questions, explaining in detail the general stages of the process. Many of them came to Mérida and Santiago with questions relating to their own research. The relaxed environment of reduced groups with common interests in a small-sized town encouraged a high degree of feedback. Numerous personal and professional academic collaborations were born during the holding of the courses.

The core of the training programme has been designed around the application of geospatial technologies on the scale of sites, regions and landscapes. As far as data detection and capture are concerned, a strong emphasis has been placed on tools such as remote aerial sensing, surface surveys, geophysical prospection and, more recently, LiDAR. Quite often, these subjects have been addressed on both theoretical and practical levels, including the hands-on use of GPS, UAVs, helium kite balloons, magnetometers, etc. As far as the analysis of the evidence is concerned, the course has particularly focused on those branches of spatial analysis which deal with the modelling of human action on the landscape, with a major emphasis on highly popular approaches in archaeology, such as visibility or mobility analyses. The rapid growth of geospatial tools and datasets in recent years has encouraged us to incorporate new contents in the course which simply did not exist 10 years ago, such as LiDAR and structure from motion.

The students

In the period from 2006 to 2015, more than 300 students attended the course. On average, the first editions had 20-25 students. Following the division into two modules in 2009, the figures increased to an average of ca. 30-35 students per year. Although most of them came from different parts of Spain and, to a lesser extent, Portugal, the change to an online format for the introductory module in 2015 opened up the possibility of participation to a larger number of Latin

American students (whose involvement had been lower up to that moment). A list of countries of origin up to 2016 includes students from: Spain, Portugal, Italy, France, Chile, Uruguay, Argentina, Ecuador, Peru, Mexico, Venezuela and the USA.

In terms of professional and academic background, obviously the vast majority of the students are archaeologists, although the occasional presence of people from different backgrounds, such as geographers, architects, historians, art historians or computer specialists is worthy of mention. Most of them had recently graduated and were typically involved in master's or doctorate programmes, with some even having recently finished their PhD studies. In addition, the regular attendance of professional archaeologists from private companies and consultants must also be mentioned, along with the occasional presence of archaeologists working in public CRM agencies. The profound impact of the crisis on contract archaeology across Europe, which was especially noticeable in Spain with a loss of 66% of work positions in archaeology between 2009 and 2013 (Parga-Dans *et al.* 2014), was also visible here, with some professional archaeologists seeking to improve their employment possibilities through the acquisition of new skills in the field of digital archaeology.

Finally, the progressive development of free and open-source software tools and an increase in the availability of free digital geospatial data has also had an effect on the course, both in terms of a growing number of young archaeologists interested in the field and of a steady increase in their foundational level of knowledge.

About this book

As mentioned previously, the 10th edition of the course took place in 2015. To commemorate the occasion, a conference was organised in Mérida in which former students were invited to present and discuss case studies making use of geospatial technologies after the completion of the course. This book brings together some of these contributions, which are mainly excerpts from the master's or doctorate theses of the authors, parts of ongoing research projects or even, in some cases, a description and discussion of the daily professional activity of the authors.

The book has been divided into four sections. Three of them focus on the application of geospatial tools for documenting (section 1), analysing (section 3) and disseminating (section 4) archaeological data respectively. The fourth (section 2) mainly deals with methodological approaches. Although some of the papers deal with a combination of these fields, we have tried to keep them within the section which best encompasses their contents.

Section 1. Shapes and locations. Documenting and characterising the archaeological record. This section includes four papers. Two of them, written by Luis Antonio Sevillano Perea and Jesús García Sánchez, illustrate the use of geospatial tools for the documentation and analysis of surface artefact distributions at a detailed spatial resolution. This is a field with a substantial background, in which digital tools such as GIS have proved their usefulness to manage and inject meaning into large and complex datasets in which uncertainty is usually a major issue. In a rather different domain, the paper by José Manuel Costa García and João Fonte deals with LiDAR data and their usefulness and limitations in the detection of earthworks, in their case Roman military settlements in previously poorly-known areas (following well known methods such as those proposed in Hesse 2010; Štular *et al.* 2012). Finally, Miguel Carrero-Pazos, Benito Vilas-Estévez and Alia Vázquez-Martínez focus on the usefulness of photogrammetry and computer vision techniques in order to improve the geometric documentation of archaeological carvings, both in terms of efficiency and clarity, in a similar way to other recent proposals (Santos and Pires 2014; Pires *et al.* 2015).

Section 2 is composed of three papers which propose and illustrate some methodological and procedural approaches for the analysis of archaeological elements in spatial terms. Pablo Paniego Díaz discusses and compares the different methods which exist for measuring the visual prominence of sites, an aspect which has previously been the focus of some interesting proposals (e.g. Llobera 2001; De Reu *et al.* 2011; Bernardini *et al.* 2013), and advocates a comparative approach. José M. Señorán Martín explores the distribution of megalithic monuments in relation to mobility patterns across the landscape. Again, this is something which has previously been explored by many authors (e.g. Wheatley *et al.* 2010; Murrieta-Flores *et al.* 2011; Llobera 2015). His contribution includes not only the use of the well-known least cost paths, but also of more sophisticated approaches to mobility, such as focal mobility networks (Llobera *et al.* 2011; Fábrega-Álvarez 2006). Marcos García García starts with a rather straightforward question ('Were prehistoric settlements preferably located on areas protected from prevailing winds?') and proposes a method of evaluation which is illustrated with a case study in the Northern Plateau of Spain.

Section 3. Patterns, behaviour, decisions. Analysing the archaeological evidence. This section includes the largest number of contributions. They all have in common a strong emphasis on spatial analysis on different scales and in different archaeological and geographic contexts. Maite García-Rojas, Alejandro Prieto, Aitor Sánchez, Cristina Camarero and the late Lydia Zapata (who very sadly passed away in 2015) combine different techniques (landscape characterisation, cost analysis and statistical analysis) to explore the possible sources of raw material for Pleistocene groups in the Northern Iberian Peninsula. María Sebatián López relies on a revision of the traditional "settlement pattern distribution" approach in order to analyse the distribution of Neolithic rock art engravings and to conclude that it might essentially have been driven by the basic structure of the physical landscape.

Moving into the Iron Age, the contribution by Raquel Liceras-Garrido, Enrique Cerrillo-Cuenca and Alfredo Jimeno-Martínez is again rooted on the tradition of settlement pattern analysis. They have chosen to examine the relationship between a large number of sites from the Upper Duero river valley (more than 300) and terrain landforms, detecting some statistically significant differences between sites from different periods, from the Early Iron Age into the Early Roman period.

Leticia Tobalina Pulido, Benoît Pace and Alain Campo examine the distribution of a particular type of archaeological site: caves used in the Roman period, in an area between the Ebro River and the Pyrenees. These caves had largely been ignored by archaeologists in the past, so the analysis they present here constitutes an initial attempt at understanding them within their geographic context.

José Manuel Costa-García takes us back to the study of Roman military sites, this time in an attempt to understand them in terms of their spatial distribution, morphology and settlement pattern. A stark contrast seems to emerge between marching camps, lookout posts and more permanent settlement sites.

In a completely different geographic and archaeological context, Jordi A. López Lillo analyses the distribution of Formative (early agricultural) communities in the Southern Andes. His contribution combines a strong focus on the discussion of certain anthropological models of social complexity (with an emphasis on the work of P. Clastres) and the use of spatial analysis in order to explore what he defines as "centrifugal landscapes", which consist of multiple self-centred housing clusters distributed across a "more or less continuous landscape where no sharp-edges are noticeable".

Finally, moving yet again into a completely different context, Manuel Antonio Franco Fernández and Pedro Rodríguez Simón illustrate the usefulness of different geospatial technologies for the documentation and analysis of archaeological evidence from the Spanish Civil War (1936-1939). Among other aspects, a combined use of viewshed analysis and surface artefact distributions has allowed them to clarify the organisation and intensity of combat in the Battle of the Ebro River and around the village of Belchite.

The final section, *Archaeology and the public. Disseminating to a wider audience*, includes two contributions which show two different, complementary dimensions of this area. Gustavo Sanz Palomera illustrates how the use of GIS for the management of cultural heritage data by the regional government of Cantabria has facilitated the creation and management of protected areas and the assessment of impacts around listed properties or archaeological sites.

Cristina Charro Lobato, Eduardo Martín Agúndez and Agustín Cabria Ramos describe a very interesting cooperation project for the development of heritage-oriented tourist maps of Gondar, the capital town of the region of Amhara (Ethiopia). This was an experience reaching beyond the purely academic sphere which also tells a cautionary tale about the series of practical, mundane problems which may also influence our activity.

Finale

Back in 2006, one of the students suggested during a discussion that maybe, in the future, such courses would not be necessary anymore, as these tools and technologies would have become widespread in everyday archaeological practice. It has become quite clear to us nowadays that this is only partially true. On the one hand, it is much easier today to gain access to training materials, documents, tutorials, etc., which teach us how to use specific software or how to perform a particular piece of analysis. To a great extent, the Internet has become a major source for quick and easy "how-to learning" in this field, as in many others. But on the other hand, and precisely because of this, the need for a solid, robust and critical approach to these technologies is more necessary today than ever before. The growing capacities of these technologies, paired with their increasingly apparent simplicity for the end user, make it necessary, more than ever, to "unbox these black boxes" and approach them from a critical and informed perspective. Otherwise, instead of using the tools to match our needs, we could easily find ourselves in exactly the opposite position. To put it in the words of Marcos Llobera, "while it is highly advantageous, and ultimately necessary, that whatever methods we develop become widely available, we also need to keep in mind the difference between "simple methods" and "methods made simple". The former are desirable provided we do not fall into the routine of adjusting our questions to them, the latter require skill, time and effort to develop which often goes unrewarded or dismissed as not being part of archaeology" (Llobera 2012, 506). We could not agree more.

Acknowledgements

We would like to thank all of the students who have attended the course since it was first organized in 2006. We fully acknowledge that their comments, discussions and critical points of view have provided us with strong stimuli over the course of all these years. Special thanks go to those who have contributed to this volume with their texts.

We are especially grateful to all the colleagues who have participated as teachers in these years, some of them after having attended the course as students in the past (in alphabetical order): Emilio Abad, Javier Baena, Jesús Bermúdez, Rocío Blas, Sebastián Celestino, Teresa Chapa, Felipe Criado, Brais Currás, Ángel Felicísimo, Silvia Fernández Cacho, Carlos Fernández Freire, Alejandro Güimil-Fariña, Luis Guitiérrez Soler, João Fonte, Celeste García Paredes, Jesús García-Sánchez, Leonardo García Sanjuán, Luis Gonçalves-Seco, César González-García, César González-Pérez, Ignasi Grau, Marcos Llobera, José María Rodrigo, Elías López-Romero, Patricia Mañana, José Ángel Martínez del Pozo, Juan Francisco Murillo, Patricia Murrieta-Flores, Ana Nieto Masot, Juan Luis Pecharromán, Juan Antonio Pérez Álvarez, Patricia Ríos, José Ángel Salgado, Patricio Soriano, Pau de Soto, Arianna Traviglia, Antonio Uriarte, Frank Vermeulen, Juan Vicent, Sabah Walid, David Wheatley, María Yubero and Mar Zamora.

We would also like to thank our respective institutes, the IAM and the Incipit for their support in organising the course all these years, as well as the following supporting institutions and organisations: Departamento de Postgrado del CSIC, Universidad de Extremadura, Consorcio de la Ciudad de Mérida, Centro de Supercomputación de Galicia, the Archaeolandscapes Europe Project and Archaeolandscapes International.

Last, but by no means least, we would like to thank all the colleagues who have played an active role in the practical organisation of the course in these 10+ years (in alphabetical order): Jorge Canosa, Estíbaliz García Gómez, Beatriz Gómez Arribas, Martina González Veiga, Alejandro Güimil-Fariña, Patricia Mañana, Carlos Morán, Teresa Neo, Aurora Porto, Sofía Quiroga, Anxo Rodríguez Paz, Ernesto Salas, Luis Sevillano and Iñaki Villa.

References

Bennett, R., Cowley, D. and De Laet, V. 2014. The data explosion: tackling the taboo of automatic feature recognition in airborne survey data. *Antiquity* 88, 896-905.

Bernardini, W., Barnash, A., Kumler, M. and Wong, M. 2013. Quantifying visual prominence in social landscapes. *Journal of Archaeological Science* 40, 3946-3954.

Bevan, A. and Lake, M. 2013. *Computational Approaches to Archaeological Spaces.* Walnut Creek, CA, USA: Left Coast Press.

Blasco Bosqued, M.C., Baena Preysler, J. and Quesada Sanz, F. 1997. *Los SIG y el análisis espacial en arqueología.* Madrid: Universidad Autónoma de Madrid.

Comer, D.C. and Harrower, M.J. 2013. *Mapping Archaeological Landscapes from Space*. Springer.

De Reu, J., Bourgeois, J., De Smedt, P., Zwertvaegher, A., Antrop, M., Bats, M., De Maeyer, P., Finke, P., Van Meirvenne, M., Verniers, J. and Crombé, P. 2011. Measuring the relative topographic position of archaeological sites in the landscape, a case study on the Bronze Age barrows in northwest Belgium. *Journal of Archaeological Science* 38, 3435-3446.

Fábrega-Álvarez, P. 2006. Moving without destination. A theoretical GIS – based determination of movement from a given origin. *Archaeological Computing Newsletter* 64, 7-11.

Figuerero Torres, M.J. and Izeta, A.D. 2013. *El uso de Sistemas de Información Geográfica (SIG) en arqueología sudamericana.* BAR International Series. Oxford: Archaeopress.

Forte, M. and Campana, S. 2016. *Digital Methods and Remote Sensing in Archaeology. Archaeology in the Age of Sensing*. Quantitative Methods in the Humanities and Social Sciences, Springer.

Grau, I., 2006. *La Aplicación de los SIG en Arqueología del Paisaje.* Alicante: Universidad de Alicante.

Hesse, R. 2010. LiDAR-derived Local Relief Models – a new tool for archaeological prospection. *Archaeological Prospection* 17, 67-72.

Huggett, J. 2004. Archaeology and the New Technological Fetishism. *Archeologia e Calcolatori* 15, 81-92.

Kamermans, H., Gojda, M. and Posluschny, A.G. 2014. *A sense of the past: studies in current archaeological applications of remote sensing and non-invasive prospection methods.* BAR-Internatioal Series. Oxford: Archaeopress.

Llobera, M. 2001. Building Past Landscape Perception With GIS: Understanding Topographic Prominence. *Journal of Archaeological Science* 28, 1005-1014.

Llobera, M. 2012. Life on a Pixel: Challenges in the Development of Digital Methods Within an "Interpretive" Landscape Archaeology Framework. *Journal of Archaeological Method and Theory* 19, 495-509.

Llobera, M. 2015. Working the Digital: Some Thoughts from Landscape Archaeology, in: Chapman, R. and Wylie, A. (eds.). *Material Evidence. Learning from Archaeological Practice*. Oxon: Routledge, 173-188.

Llobera, M., Fabrega-Alvarez, P. and Parcero-Oubina, C. 2011. Order in movement: a GIS approach to accessibility. *Journal of Archaeological Science* 38, 843-851.

Mayoral Herrera, V. and Celestino Pérez, S., 2011. *Tecnologías de Información Geográfica y Análisis Arqueológico del Territorio. Actas del V Simposio Internacional de Arqueología de Mérida.* Madrid: CSIC.

Murrieta-Flores, P., Garcia Sanjuan, L. and Wheatley, D. 2011. Movilidad y vias de paso en los paisajes prehistoricos: Megalitos y vias pecuarias en Almaden de la Plata (Sevilla, Espana), in: Mayoral Herrera, V. and Celestino Perez, S. (eds.). *Tecnologías de Informacion Geografica y Anaisis Arqueologico del Territorio. Actas del V Symposio Internacional Arqueología de Merida (Merida, 7-10 de Noviembre de 2007).* Madrid: CSIC, 412-423.

Opitz, R.S. and Cowley, D.C. 2013. *Interpreting Archaeological Topography: Lasers, 3D Data, Observation, Visualisation and Applications.* Oxford: Oxbow.

Parga-Dans, E., Barreiro, D. and Varela-Pousa, R. 2016. Isomorphism and legitimacy in Spanish contract archaeology: the free-fall of an institutional model and the caveat of change. *International Journal of Heritage Studies* 22, 291-301.

Pastor, S., Murrieta Flores, P. and García Sanjuán, L. 2013. Los SIG en la arqueología de habla hispana: Temas, técnicas y perspectivas. *Comechingonia* 17, 9-29.

Pires, H., Gonçalves-Seco, L., Fonte, J., Mañana-Borrazás, P., Parcero-Oubiña, C., Fábrega-Álvarez, P. and Señorán, J.M. 2015. From point clouds to archaeological evidence: Improving visualization and spatial analysis of 3D data, in: Posluschny, A. (ed.). *Sensing the Past. Contributions from the ArcLand Conference on Remote Sensing for Archaeology.* Frankfurt: ArcLand, 52-53.

Polla, S. and Verhagen, P. 2014. *Computational Approaches to the Study of Movement in Archaeology, Theory, Practice and Interpretation of Factors and Effects of Long Term Landscape Formation and Transformation.* DeGruyter.

Posluschny, A. 2015. *Sensing the Past. Contributions from the ArcLand Conference on Remote Sensing for Archaeology.* Frankfurt: ArcLand.

Santos, M.J.C. and Pires, H. 2014. A estela funeraria de Capela, Penafiel. *Ficheiro Epigráfico (Suplemento de Conimbriga)* 119, 510.

Štular, B., Kokalj, Ž., Oštir, K. and Nuninger, L. 2012. Visualization of lidar-derived relief models for detection of archaeological features. *Journal of Archaeological Science* 39, 3354-3360.

Wheatley, D.W., García Sanjuán, L., Murrieta Flores, P.A. and Márquez Pérez, J. 2010. Approaching the Lanscape Dimension of the Megalithic Phenomenon in Southern Spain. *Oxford Journal of Archaeology* 29, 387-405.

Notes on contributors

Victorino Mayoral Herrera (Madrid, 1970; PhD 2001, Complutense University of Madrid, Spain). Staff Scientist at the Instituto de Arqueología-Mérida (IAM), Consejo Superior de Investigaciones Científicas (CSIC) in Spain. Before joining the IAM, he developed his career at the Complutense University of Madrid and in the Heritage Administration of the Autonomous Region of Extremadura. His research has been mainly concerned with an archaeological approach to the study of landscape, with two main vectors. On the one hand, a historical analysis focused on the transition between indigenous cultures and the Roman rule in the Mediterranean, without neglecting the longue durée nature of this kind of research. On the other hand, in methodological terms his efforts have been mainly devoted to the exploration of the potential of non-invasive methods like remote sensing, aerial, geophysical, and especially surface survey with the support of geospatial technologies.

César Parcero-Oubiña (Santiago de Compostela, 1969; PhD 2001, University of Santiago de Compostela, Spain). Staff Scientist at the Instituto de Ciencias del Patrimonio (Incipit), Consejo Superior de Investigaciones Científicas (CSIC) in Spain. Before joining the Incipit, he developed his career at the University of Santiago de Compostela. His research has been mainly concerned with landscape archaeology and focused on the analysis of the productive, social and political dimensions of later prehistoric contexts (the Iron Age in Western Europe, and now the later pre-Hispanic context in South America). He is also interested in the application of geospatial technologies for both archaeology and the wider field of Cultural Heritage. He has carried out fieldwork in Spain, Uruguay, Argentina and Chile, and collaborated in projects in Ethiopia or Mongolia.

Pastor Fábrega-Álvarez (Ourense, 1978; PhD 2017, University of Jaen, Spain). Technical staff member at the Instituto de Ciencias del Patrimonio (Incipit), Consejo Superior de Investigaciones Científicas (CSIC) in Spain. His fields of interest are the design of methodologies for analyzing and managing archaeological heritage from a geographic and landscape perspective. More specifically, his work is focused on the use of geospatial technologies such as GIS, Remote Sensing and 3D representation techniques for the analysis of the archaeological record. Recently, he has been involved in projects like SPATRIAL or CHARM which aim to develop conceptual and spatial models of the cultural heritage. In this regard, he has been member of the Spanish Spatial Data Infraestructures Working Group. He has also participated in projects and initiatives related to the archaeological study of ancient landscapes, such as ArchaeoLandscapes Europe or Agriculture and Empire in the High Altitude Atacama Desert (Chile).

Section 1

Shapes and locations

Documenting and characterising the archaeological record

Pursuing ancient rural life through surface survey: composition and diversity of artifact distributions

Luis Antonio Sevillano Perea[1]

Introduction: Background of the study. What is the surface archaeological record? How should we approach it?

Shott (1995) proposed the use of the following three attributes to describe the surface archaeological record: abundance, distribution and composition. Although they can be independently assessed, there are obvious benefits derived from their combined analysis. In fact, such a way of proceeding is implicitly incorporated in the presentation and interpretation of survey results. However, further improvements to formally approach this issue are still required (especially in a landscape scale).

Certainly, density estimations are a unique and irreplaceable tool to represent the overall patterns of the identified set of artifacts, in terms of degree of clustering and distribution. However, it lacks the capability to take into consideration variations in the assemblage other than abundance. Addressing this concern should involve the inclusion of qualitative variables of the surface compound's constituents (fabrics, chronology, function…). Assemblage composition varies not only quantitatively but qualitatively across the study area. Thus, considering both elements simultaneously requires a multifaceted method, in particular if we intend to recognize spatial changes.

In addition, aside from the recording phase, the analytical and interpretative potentialities of medium and low density areas are rarely fully developed (de Haas 2012, 56; Mayoral *et al.* 2006; Mayoral and Sevillano 2013). On the contrary, artifact concentrations have been traditionally targeted as the natural way of occurrence of archaeological phenomena on the surface. Thus, they are still considered the central focus of attention in the majority of research designs. As Van de Velde noted (2001), it is far from uncommon to find general classifications

1 Instituto de Arqueología de Mérida, Consejo Superior de Investigaciones Científicas. Plaza de España, 15. 06800 Mérida (Spain) luissevper@gmail.com.

of datasets which are defined in negative terms: what is the main objective of the work (sites, places of special interest, foci of activity...); and, on the other side, what does not conform the previous category and may or may not be object of further attention (background noise, off-site, isolated finds/scatters...).

Our approach lies in an inclusive notion of the archaeological record. Surface findings are considered the observable trace of a particular set of actions, executed either in a concentrated or a dispersed way through the whole landscape. If the objective of the research is to understand the historical trajectory of the latter, dispersed material must be accounted for, in addition to obstructive concentrations.

This proposal is not intended to set aside the remarkable shortcomings of low density artifact distributions (*i.e.* small size of the samples, high erosion rate). Alternatively, it tries to make the most of its constrained possibilities. For that reason, the main purpose of this paper is to test the whole dataset against a group of analytical procedures to examine its compositional complexity. Hopefully, this method will provide us with stronger means to differentiate, compare and rank patterns among not so clear-cut artifact assemblages.

Materials and methods

The surface survey strategy

If, as already stated, the archaeological record can be described by its attributes of abundance, composition and distribution, a proper surface sampling design must provide guidance for tracking methodically these features across the study area. That premise remained as a central concern in the design and organization of the ZUVAS sampling strategy (Figure 1). The main lines of the fieldwork system are summarized in the plan outlined below:

1. Maintenance of a homogeneous coverage by the adoption of a classical systematic design (organized using parallel tracks within field boundaries).
2. Fixed distance of 10 meters between tracks. These paths were ideally the center of a two meter wide transect. Therefore, the inspected space accounts for roughly 20% of the accessible land.
3. Surveyors were asked to report every archaeological item identified, whose individual approximate location was recorded with the assistance of handheld GPS. This strategy, established for previous survey experiences developed in the Guadiana Basin (Mayoral *et al.* 2009), has also been successfully implemented in other agricultural landscapes of the Iberian Peninsula (García Sánchez and Cisneros 2013; García Sánchez, in this volume; Grau Mira 2013).
4. Classification of artifacts using a simple set of pre-established categories, matching the most common kinds of finds in the region.
5. Those findings which could be ascribed to specific categories within typological grouping systems were collected and studied in a more detailed way in the laboratory.

6. Concerning south-eastern plots surveyed (delineated with a black dashed line in the figures), one final clarification must be added. There, the material compound can be interpreted as the surface evidence of a highly prominent

residential nucleus (datable finds range from the first century B.C. to the V century A.D.). At this place, the recording strategy was adapted to the exceptionally high densities encountered; and in such a way that the ceramic building material (ancient brick, imbrex and tegula) was not geo-referenced. Accordingly, neither the density nor the other quantitative indices explored in the subsequent paragraphs should be directly compared between those two distinct datasets.

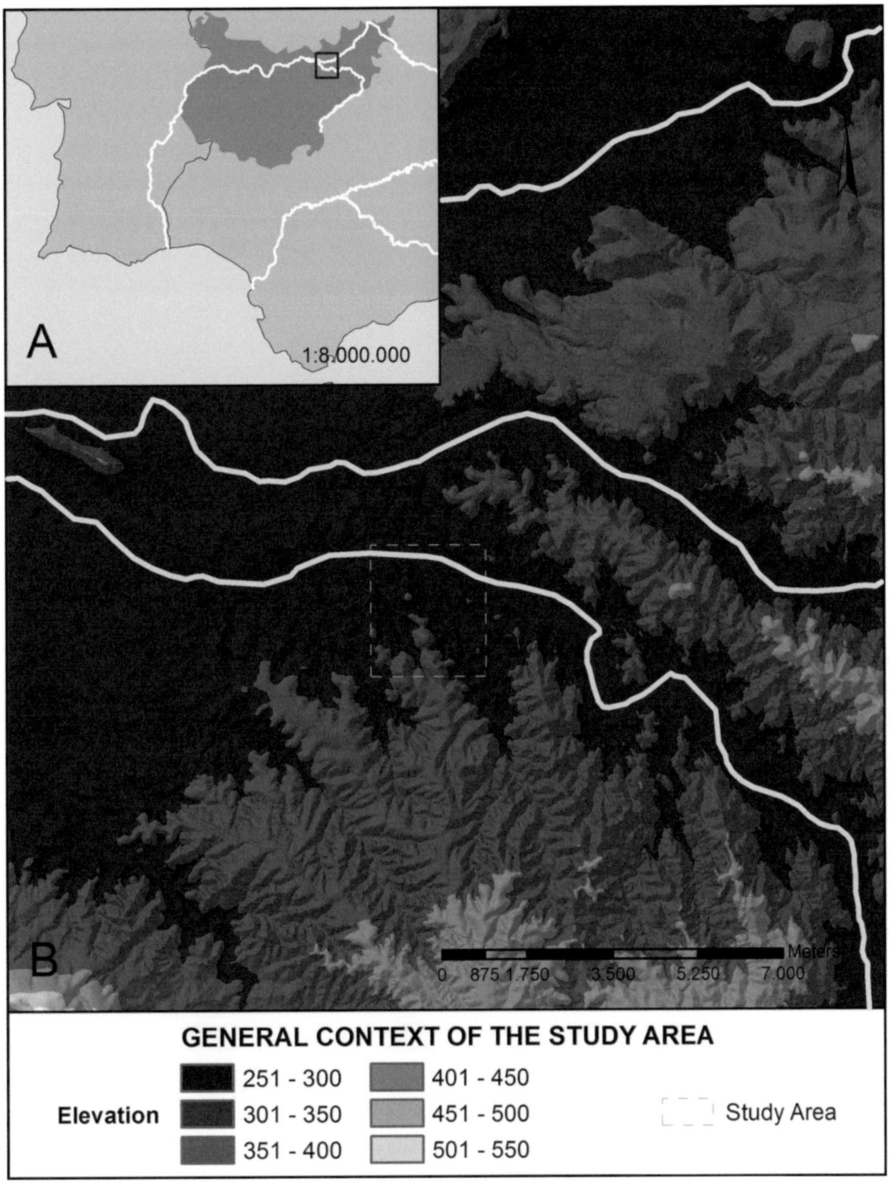

Figure 1: Location of the research area: A. Geographical location (Badajoz, SW Iberian Peninsula); B: DEM and main river network of the Middle Guadiana Basin highlighted in A.

The result of this procedure was the generation of an exhaustive corpus of information (circa 30,000 geo-referenced surface items). It adopts the form of an artifact-based spatial database which enables us to explore geo-quantitative as well as a geo-qualitative approaches to archaeological assemblages. Thus, if surface archaeological record is regarded as the realization of a set of inhomogeneous spatial processes, our dataset can be considered an incomplete point pattern derived from it. Besides, multiple analyses of the variables linked to georeferenced observations are derived from or can be displayed in terms of continuous surfaces. As a consequence, the so-called modifiable areal unit problem (Openshaw 1984) is theoretically avoided because any pattern detected is not a consequence of arbitrarily defined survey units.

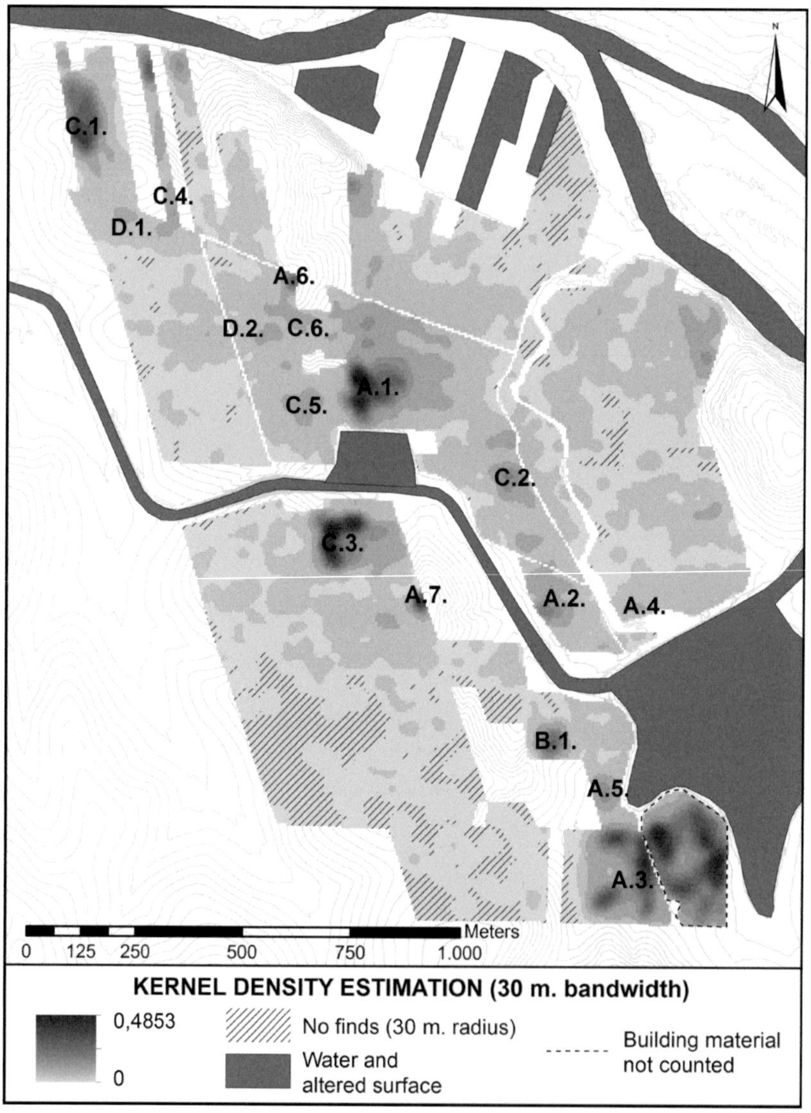

Figure 2: Surface findings Kernel Density Estimation (30 meters bandwidth-Gaussian kernel function). Codes refer to locations mentioned in the text.

An approach to the analysis of the results

As already stated, it is assumed that the most straightforward way to explore systematic survey results is the visualization of abundance data across the study area (normally coming in the form of density maps). Hence, Kernel Density Estimation has been proven to be a valuable tool for ascertaining spatial variations in the surface archaeological record (Figure 2). In fact, some potentialities of this procedure in relation to similar strategies have been already discussed for different survey experiences (Mayoral *et al.* 2009, 15-16; Mayoral and Sevillano 2013). In this case, as the recording strategy was based on pre-established categories, density estimations could be built for every class of artifact considered.

Nevertheless, unlike physical quantities, qualitative variables are not self-evident and need an ad hoc definition. Different variables approach us to unique dimensions of the archaeological record. In this sense, the broad fabric categories employed here are informative about some facets of the surface compound. However valuable, it must be admitted that historical conclusions gained from the analysis of these broad categories should be complemented by a more qualitative approach. Functional, chronological or typological classifications would link us to different dimensions of the assemblage. Even so, the analysis of such a grouping system in this paper is a consequence of the kind of items recorded as well as the adopted sampling strategy. Actually, only a percentage slightly under 4.5% of the findings was collected and positively classified within Chrono-typological schemas. Thus, aside from the biggest concentrations, we are dealing with small amounts "diagnostic" items. This situation is the prevailing trend in the surface archaeology of the Mediterranean area (Winther-Jacobsen 2010, 269). In contrast, the broad classes of materials employed here, though less informative about function or chronology, allows using the whole recorded sample.

Unveiling the composition of the dataset

The analysis of the diversity of artifact aggregates poses special interest because it brings together information about composition and abundance. In fact, it has been suggested that methods linked to the concept of diversity summarize for qualitative variables what variance does for quantitative ones (Patil 2002, 555). However, it is not a closed or unambiguous concept and, accordingly, it demands a definition of the notion assumed. In this paper we consider three interconnected indices, which stand out as the preferred systems to make the compositional arrangement of populations explicit: richness, equitability or evenness and diversity.

The starting point: measuring richness

Richness is the oldest and simplest measurement of assemblage variability. It is expressed by an integer which accounts for the different categories a group of observations can be classified into. Then, before it can be analyzed, it is necessary (1) to define a set of closed categories, and (2) to understand the mechanics of assignment of each individual to the appropriate type.

Despite being a simple concept, its use can be problematic because it is highly affected by sample size biases. It has been largely recognized that richness is strongly influenced by the total number of observations, an element which is not independent of the invested sampling effort. We are always dealing with an incomplete set, and hence we do not know the real number of classes which occur in the target population either globally or in particular spots. Then, since this measurement gives just as much weight to the commonest as to the rarest material types, it is extremely sensitive to undetected classes. Therefore, it might favor the underestimation of true population variability. Still, the measurement can be of interest as a preliminary approach to the compositional structure of the assemblage.

Richness of ZUVAS dataset is plotted in the maps of Figure 3. The values express the range of different artifact categories within a 30 meter radius from each pixel. As expected, it displays a strong connection with abundance. In this sense, the comparison of Figures 2 and 3 reveals a clear spatial correlation between the richest assemblages and the most conspicuous artifact concentrations.

In map A of the Figure 3 artifact range is displayed for all types accounted for. The values of map B are obtained when contemporary pottery classes are excluded (white glazed crockery, glazed pottery and other regional cooking and tableware). Despite being non majoritarian (proportionally, the three categories represent less than 1.5 percent of the total), they bear a valuable influence at few locations. In any case, both figures make explicit a coherent trend of richness growth or

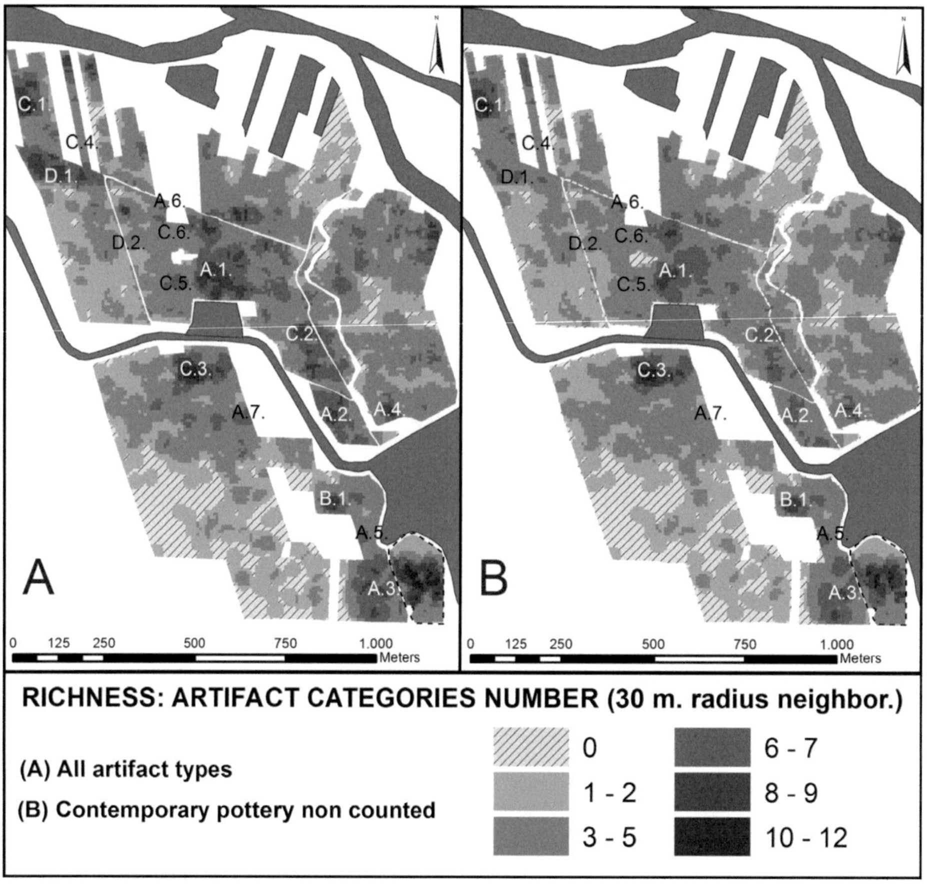

Figure 3: Richness: range of artifact broad categories within a radius of 30 meters.

decay created by the distributions of ancient materials. The majority of artifact concentrations, regardless of their heterogeneous natures, are easily differentiable from the material evidence of dispersed activity areas.

Compositional complexity: diversity and evenness measurement

Diversity indices enable us to gauge the connection between the presence of a number of artifact classes and their quantitative contribution. They are traditionally defined as the conjoint measurement of the number of categories accounted for (richness) and its evenness, which is to say the categories equitably measured by their relative abundance (Tuomisto 2012, 1203). Thus, it is intuitively assumed that a population or a sample is more diverse as (1) they contain more types and/or (2) the relative abundance of detected types is more evenly distributed.

Theoretically, compositional diversity poses no problem when all classes are equally abundant. However, when this is not the case, the weight we give to rarity is crucial. Diversity is a multidimensional attribute of a population. Then, the selection of a procedure to measure it brings the risk of distorting our perception of the object of interest, since a multidimensional property is projected in a one-dimensional ordinal scale (Patil 2002, 555). This fact has been largely recognized and different authors have pointed out its conceptual as well as procedural consequences from different disciplines (Hulbert 1971; Jost 2006; Patil 2002), but also from an archaeologist point of view (Ringrose 1993; Kintigh 1984). A proposed solution to get around this inconvenience is the use of diversity profiles instead of one-way approaches (Hulbert 1971; Chao and Jost 2015, 2). Then it is possible to assess the answers provided by a set of observations when confronted to formulas that give different weight to relative abundance.

Two diversity measures have been applied in a local manner to our dataset: Shannon Entropy and Simpson Index. Both of them are among the oldest and still most used diversity indices (Marrugan 2004, 106-110, 114-116). The first one, formally, quantifies the uncertainty in ascribing an observation to one of a series of categories. It is more sensitive to richness and less influenced by dominant categories than Simpson Index. On the other hand, the latter measures can be described as the probability of any two individuals drawn at random to belong to the same type. It has deserved positive feedbacks (Marrugan 2004, 101; Ringrose 1993), because it is more robust against sampling bias issues than Shannon's formula (it is less affected by the presence/absence of rarer categories) and an unbiased estimator can be derived (Hulbert 1971).

The results of Shannon and Simpson indices are not directly comparable. Thus, to facilitate this task, in this paper they are displayed after a conversion to their "effective number of types" or "true diversities" sensu Jost (2006). This procedure makes possible to observe applied indices under a linear scale (Jost 2010, 209), which depends on the weight given to relative abundances. Thus, diversity values are expressed and interpretable in the same units of the number of categories.

Finally, due to the fact that the implementation of diversity and evenness measurements for our dataset has followed the same plan, we will briefly define the second notion and how it was computed before discussing the results.

The evenness or equitability of an assemblage refers to how equally the relative abundances of different categories are distributed. It can be described according to multiple criteria (Smith and Wilson 1996; Heip *et al.* 1998, 80-83; Tuomisto 2012). One way to derive a measurement of evenness is the normalization of the diversity value, dividing it by richness (Tuomisto 2012, 1207). The maximum possible value, 1, defines a completely even population, where different groups are equally abundant. On the other hand, evenness is closer to 0 when categories show remarkably dissimilar quantitative trends (i.e. those populations formed by one or few dominant categories and a number of less frequent classes). Because it is calculated as a diversity derived product, evenness inherits from it the weight placed on categories relative abundance.

The implementation of both measurements in our dataset is based on a four-step scheme: (1) creation of a neighborhood for each artifact detected (estimation of the number and type of observations within a radius); (2) removal of points whose sample size neighborhood is below a specific threshold; (3) computation of diversity/evenness analysis for each neighborhood independently; (4) creation of a smoothed surface with the results of the previous steps for visual purposes, as observed in Figures 4 and 5. Thus, what we are mapping is an interpolation of

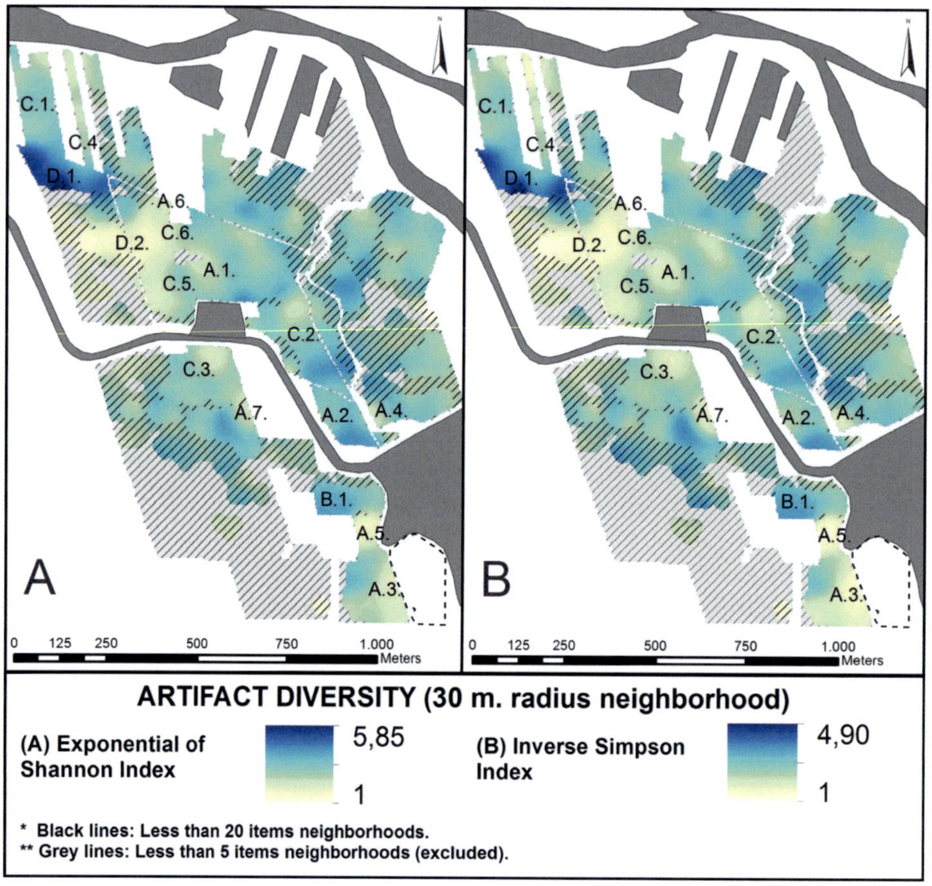

Figure 4: Smoothed surfaces displaying local artifact diversity values (30 meters radius neighborhood): A. Exponential of Shannon Index; B. Inverse of Simpson Index.

the values attached to individual observations neighborhoods. This procedure has been applied because it allows keeping the point based arrangement in which the dataset was constructed. Besides the spatial resolution of the results is less coarse than the one obtained when using other tested approaches, such as the creation of quadrats or any artificial partitioning of the surveyed area.

It is clearly visible in the images that Simpson Index and its derived evenness produce lower values. As expected, the more weight we give to the most abundant species, the lower the diversity value obtained. Nevertheless, the global trend observable in maps A and B remains unchanged. Still, pointing out the differences between the two computations has some interest, because its comparison highlights where the relative abundance of dominant classes has stronger influence. For that reason, the use of ratios created by the division of one value by the other has been proposed (Tuomisto 2012, 1208-1209).

In this case of study, a 30 meter radius was considered appropriate for examining global trends. Radius of smaller size would point out localized patterns of interest otherwise indistinguishable. However, there is a controversy between the spatial scale of the analysis and a minimum number of observations for each neighborhood. Because our data strongly varies in terms of abundance across the

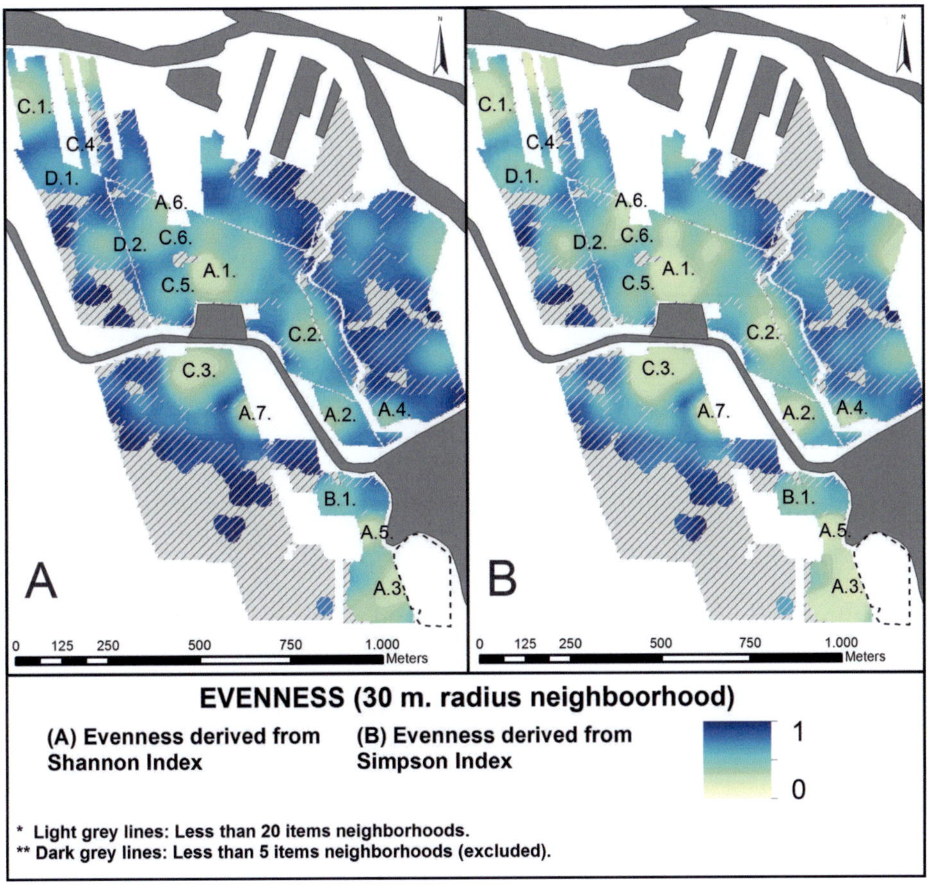

Figure 5: Smoothed surfaces displaying local evenness values (30 meters radius neighborhood): A. Evenness derived from the results of Shannon Index; B. Evenness derived from the results of Simpson Index.

space, a minimum sample size threshold is desirable. As the threshold is widened, more points are excluded and a smaller area is accounted for. For example, in Figures 4 and 5, neighborhoods with less than five individuals are left out and those populated by less than 20 observations are displayed within a line pattern. We are aware of the potential influence of sample size biases, which may be particularly strong at scarcely populated locales. For that reason, the results have been compared with the equivalents for Simpson and Shannon "effective number of types" for rarefied samples (Chao *et al.* 2014).

Discussion of the results

In the following section we will inspect conjointly the analysis of richness, evenness and diversity values obtained for our dataset. They will be interpreted in relation to the abundance and distribution of the artifacts detected, taking into account the main classes of findings and its chronology.

Focusing on prominent clusters of artifacts, as we saw, an escalation of richness is clearly identifiable. This fact is repeatedly complemented by a decrease in evenness (see maps in Figure 5). However, dissimilar spatial variations of both phenomena determine that, when visualizing diversity indices values, three distinct behaviors can be identified. Firstly, there is a set of locales featured by low values of effective number of categories. They correspond with the core and/or immediate surroundings of density peaks which can be interpreted as foci of activity of different nature and chronology (they are marked in Figures as: A.1., A.2., A.3. for conspicuous Roman residential-productive area; A.4., A.5. refer to Roman isolated auxiliary structures; and A.6., A.7. for Contemporary isolated auxiliary structures). The small diversity values are regulated by the acute quantitative disproportion between the abundance of frequent and rare classes of artifact (compare evenness maps of the Figure 5). So we can conclude that these denser neighborhoods contain more categories than its surroundings, however when picking a small sample from them, fewer groups would be identified if compared to their immediate surroundings.

Location B.1. (see Figures 4 and 5), a medium size roman artifacts concentration, makes an exception in our dataset. It follows the opposite trend to that one described in the previous paragraph. In this case, the composition is still more diverse than its neighborhoods because the rise of richness is not accompanied by an extreme decrease of evenness. Yet compositional equitability is slightly lower than that from the surroundings.

Marks C.1., C.2. and C.3. are placed on high density artifact distributions also taken as conspicuous Roman foci of activity. However, unlike the previous compounds, these scatters can hardly be outlined from their surrounding haloes in terms of the employed diversity estimates. Here, the halo-core composition is comparable, except for the northeastern C.2. vicinity, where the recorded data behaves distinctly. Such differences, in contrast, are better understood taking into account evenness.

Additionally, a number of late republican artifact clusters (not very dense though) can be included in this section. Again, their compositional complexity does not outstand from their neighborhoods, neither with regard to diversity values, nor in terms of evenness (they correspond to the codes C.4., C.5., C.6.

in the maps). However, the detected pattern might be due to the fact that the coarse dimension of the archaeological record analyzed in this paper fails to provide proper estimates for featuring the composition of deeply eroded modest concentrations (in particular for some chronologies). In this respect, it should be noted that ceramic building material was introduced in the middle Guadiana Basin in the transition from late republic to early imperial times. From that period onwards it has remained as a main contributor in the material compound.

Logically, higher proportions of building artifacts could be expected in scatters interpreted as the surface remains of architectural structures. In contrast, high pottery to tile ratio in low density scatters might be indicative of assemblages constituted by the repetition of non-spatially concentrated practices, such as manuring (de Haas 2012, 67). Medium or high artifact distributions displaying the same ratio can correspond to refuse dumps, so any kind of relationship (not necessarily in terms of spatial proximity) is expected between them and other foci of activity. In other instances, this pattern had been created by the combination of particular sets of practices carried out in relation to structures which depend only to a limited extend on ceramic building material.

Figure 6 plots those neighborhoods where pottery is the majoritarian component of the assemblages against those areas where ceramic building material (brick, imbrex-curved roof tile and tegula) is dominant. It is a product of the comparison of kernel density estimations built for both sets of categories using the same parameters used elsewhere before in this paper (Gaussian kernel function and 30 meter bandwidth). Excluding loci C.4., C.5. and C.6. previously exposed, the results are quite enlightening about the prevalence of material buildings in the vast majority of artifact concentrations and their surroundings up to different distances (sometimes more than a hundred meters apart). Once this distance has been exceeded, pottery vessels dominate. Still, a more detailed comment would be desirable for every foci of activity as they display different patterns.

The "islands" created at the core or immediate surroundings of loci C.2. and C.3. are particularly interesting. Arguably, they can be linked to processes of different nature articulated independently within the activity foci. Also remarkable is the configuration of C.1. The area to the north of the cluster core is dominated by building material. This fact cannot be interpreted without taking into account the existence of a gently slope which connects the higher southern spaces with the Zújar river. Thus the upper-left corner of the surveyed space presents a rich compound of artifacts relocated after being laterally displaced, mainly downhill. In opposition to this, only a minority set of architectural remains would have been moved upwards by post-depositional forces.

The expressed ideas can also be useful for the exploration of the compositional complexity of the so-called site 'haloes' (Bintliff and Snodgrass 1988). In our study case, such places relate consistently to intermediate to high diversity values according to the estimators used. In general terms, a fairly wide range of artifact categories was identified, but the contribution of different groups' relative abundance displays intermediate evenness values. However, it increases as we move from the core of artifact concentrations. This behavior could be linked to assemblages located around built spaces. An intense and recursive set of actions

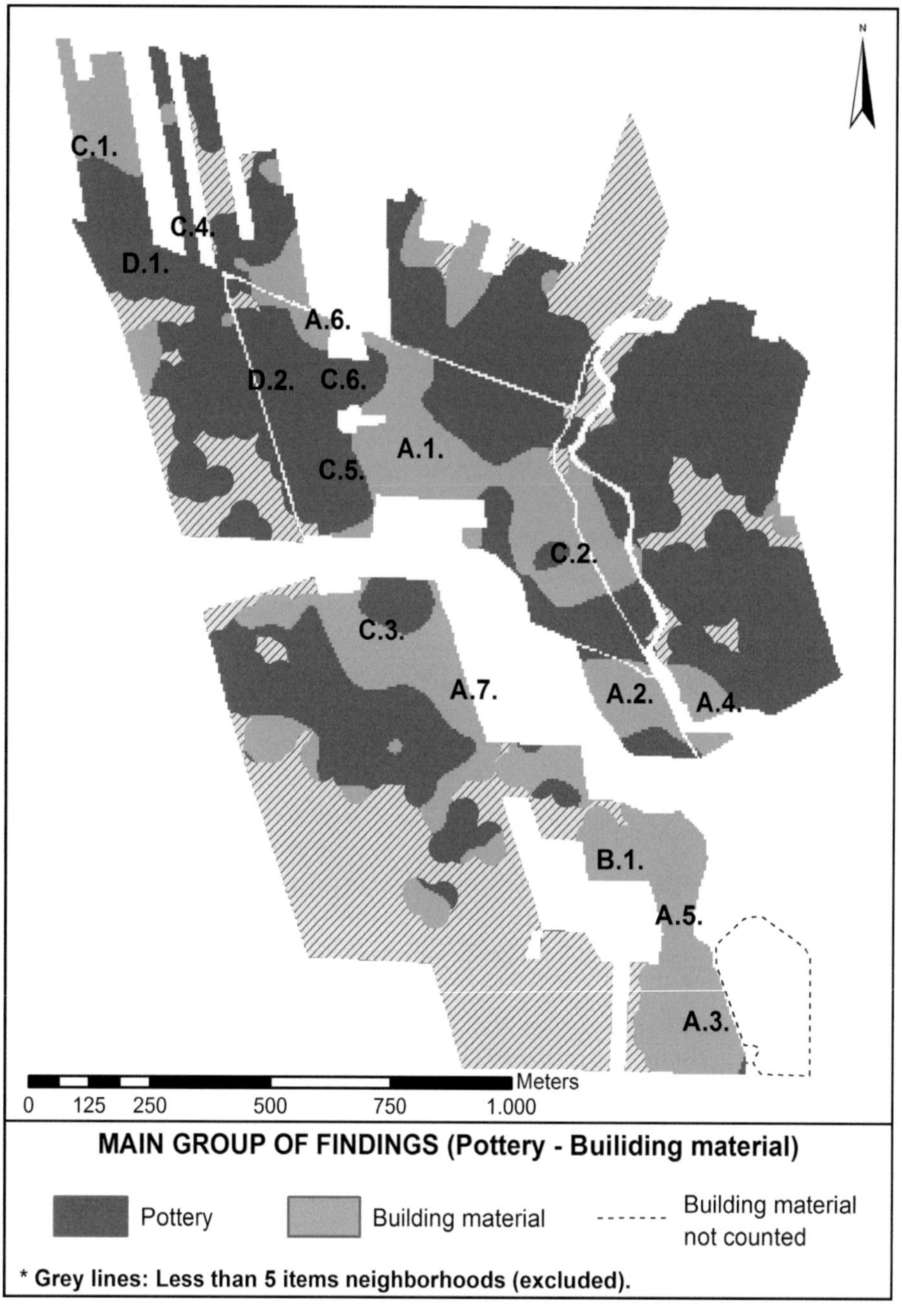

Figure 6: Dominant artifact group: pottery vessels against ceramic building material.

is expected to have taken place there (refuse dumps, infield manuring, post-depositional transportation of elements from denser areas…). Then, the material set they involve would possibly be less intentionally sorted than the one coming from the nuclei of activity (i.e. higher proportions of building material or separated functional areas are expected to occur in the latter case whether we can detect them or not). In any case, it is clear that the phenomenon of haloes has also left a mark in the compositional arrangement of the surface record. Though it is normally identified by progressive decreasing density rates, it is also traceable by the visual inspection of any of the compositional indices explored in this paper.

It has been suggested that large periods of exposure to plowing and other erosion processes could have led to an increase of evenness in the assemblage composition (Mlekuž and Taelman 2012, 129). By the same token, at least for specific contexts and spatial scales, low equitability and wider ranges of categories could be linked to compounds recently incorporated to the plow zone. Both arguments assume that "unaltered" assemblages would be more uneven and that there exists a high spatial variability in the compositional complexity of the surface record. None of these premises is of general application, though they could be plausible for many contexts. For instance, the interpretation of a number of places included in the groups A and C previously described could benefit from this hypothesis. However, it is difficult to validate the applicability of these hypotheses.

Finally, a framework less intelligible is linked to diversity values derived from low density assemblages. At those areas, there is a strong local variability which, in general, can be assigned to (small) changes in richness. Relative abundances of material types are more homogeneous in low density areas, and consequently evenness reaches systematically higher values (compare Figures 2 and 5). Fewer artifacts of fewer types have arrived there, however no dominant class is noticed.

One additional cause can be assigned to heterogeneous diversity patterns detected for less abundant neighborhoods: sample size biases in connection to procedural issues confronted in the development of surface survey. Inconsistencies in assemblage compositional complexity might result from variations in surface visibility. This topic is not covered in this paper, however it would be interesting to test if differences in land use or the presence of vegetation significantly lead to a more diverse or homogeneous dataset.

Yet, notwithstanding what has been said above, it is worth mentioning two clearly divergent situations. The first example concerns to a plot of land showing the highest effective number of types (indicated in maps with the code D.1.). It contains medium to low densities of ancient artifacts, added to which there is also a scarce distribution of contemporary sherds. The material sets from both periods create a rather evener than uneven composite. This framework, in addition to the range of artifacts detected, ends up with the highest "number of equivalent types" of the study area. There are other spots of low density and high diversity (not individualized in the maps, but easily identifiable for instance in the surroundings of nuclei A.2. and C.2.), which are caused mostly by the contribution of ancient materials.

Finally, the pattern recognized in locale D.2. deserves a clarification. It is the only place where low artifact densities are matched with low and constant diversity values. Observing the richness (low to intermediate) and diversity (low), it can be argued that any of the material classes is constantly dominant (mainly coarse pottery) over a small number of categories (among them we can place the ceramic building material). Matching this information with the diagnostic sherds recovered, which mostly date back to late republican period, it emerges a coherent pattern of intense occupation of this space during that time. Other low density scatters also display the same behavior at different points of the surveyed space. The difference now lies in the absence of noticeable later inclusions. Hence, in terms of chronological complexity, "palimpsest" D.2. allows for a much more straightforward assessment of its compositional and distribution patterns. This clearer image, also supported by low artifact heterogeneity, would favor the identification of smaller and isolated foci of activity, hardly noticeable by density figures. According to the line of reasoning previously stated, a richer and less even compound would be detected at those spots in comparison to their immediate neighboring areas, where activities developed in a dispersed way would have been carried out. Thus, if such spatial contrasts have not been completely blurred by post-depositional processes, it might reasonably be expected to find local differences in the composition of the scatters, even if densities remain similar.

Waagen (2014) discussed this topic and proposed suggestive ideas in relation to richness and low density areas. His findings encouraged us to re-inspect a number of scarcely populated areas doubling the sampling intensity (40% coverage) in order to gain some information about five scarce but varied artifact scatters (whose thorough valuation is not included in this paper due to space limitations). The results of this exercise are not yet conclusive, because their interpretation is heavily dependent on local factors. Nevertheless, three of the re-surveyed spaces intersect with D.2. area, and in two of them, it is feasible to suggest that the localized peaks of variability can hardly be ascribed only to diachronic superposition of spatially dispersed activities.

Conclusions

There is not a straightforward method to infer the connections between the heterogeneity of an assemblage and the multi-causal processes connected to its formative history. Then, it is suggested that, to overcome some of the shortcuts of the surface archaeological record, there are relevant techniques to explore patterns in the composition between and within (not so much diagnostic) artifact aggregates.

In this paper, a number of methods have been applied to analyze the compositional complexity of ZUVAS dataset. They stand as supplementary tools to estimate and summarize different facets of the archaeological record. Their implementation have provided us the means (1) to evaluate the variability of surface scatters; (2) to compare the values obtained and understand them in relation to other areas of the surveyed space; and (3) integrate the results within wider approaches to the dataset.

Our tests have deliberately chosen to employ a spatial scale which focuses on medium size patterns because at this stage we are interested in describing the main phenomena occurring in our study area. However, this should not be at odds with parallel approaches, interested in smaller/intra-site variations. In any case, the picture obtained already outlines in considerable detail the main local patterns. Furthermore, we are aware that the kind of employed categories offers a coarse grained model. Nevertheless, surface archaeology lacks the means to efficiently produce more specific catalogues at a landscape scale (i. e. the problem of the scarcity of "diagnostic" artifacts collections in lower density areas). In that sense, as long as sample size is sufficiently large, the compositional complexity of other dimensions of the surface archaeological record could also be analyzed specifically.

Acknowledgements

I wish to express my gratitude to Maria José Ramos, Cristina Mena, David Arjona, Sergio Quintero and Jairo Naranjo for the invaluable support during the development of ZUVAS fieldwork. I also would like to thank Dr. Victorino Mayoral for his helpful comments on this manuscript.

References

Bintliff, J.L., Snodgrass, A. 1988. Off-Site Pottery distributions: A Regional and Interregional Perspective. *Current Antropology* 29/3, 506-513.

Chao, A., Gotelli, N., Hsieh, T.C., Sander, E.L., Ma, K.H., Colwell, R.K., Ellison, A.M. 2014. Rarefaction and extrapolation with Hill numbers: a framework for sampling and estimation in species diversity studies. *Ecological Monographs* 84/1, 45-67.

Chao, A., Jost, L. 2015. Estimating diversity and entropy profiles via discovery rates of new species. *Methods in Ecology and Evolution* 6/8, 873-882.

De Haas, T. 2012. Beyond dots on the map: intensive data survey and the interpretation of small sites and off-site distributions, in: Attema, P.A.J., Schörner, G. (eds.). *Comparative Issues in the archaeology of the Roman rural landscape. Site classification between survey, excavation and historical categories.* Portsmouth, Rhode Island: JRA Supplements 88, 55-80.

García Sánchez, J., Cisneros, M. 2013. An Off-site Approach to Late Iron Age and Roman Landscapes on the Northern Plateau, Spain. *European Journal of Archaeology* 16/2, 289-313.

Grau, I., Carreras, C., Molina, J., De Soto, P., Segura, J.M. 2013. Propuestas metodológicas para el estudio del paisaje rural antiguo en el área central de la Contestania. *Zephirus* LXX, 131-149.

Heip, C.H.R., Herman, P.M.J., Soetaert, K. 1998. Indices of diversity and evenness. *Océanis* 24/4, 61-87.

Hulbert, S.H. 1971. The nonconcept of species diversity: a critique and alternative parameters. *Ecology* 52, 577-586.

Jost, L. 2006. Entropy and diversity. *Oikos* 113, 363-375.

Jost, L. 2010. The relationship between evenness and diversity. *Diversity*, 2.

Kintigh, K.W. 1984. Measuring archaeological diversity by comparison with simulated assemblages. *American Antiquity* 49, 44-54.

Marrugan, A.E. 2004. *Measuring biological diversity*, Oxford, Blackwell Publishing.

Mayoral Herrera, V., Cerrillo Cuenca, E., Celestino Pérez, S. 2009. Métodos de prospección arqueológica intensiva en el marco de un proyecto regional: el caso de la comarca de La Serena (Badajoz). *Trabajos de Prehistoria* 66/1, 7-25.

Mayoral Herrera, V., Chapa, T., Uriarte, A., Nieto, A. 2006. Escuchando el ruido de fondo: estrategias para el estudio de los paisajes agrarios tardoibéricos en la región del Guadiana Menor. *Arqueología Espacial* 26, 87-114.

Mayoral Herrera, V., Sevillano Perea, L.A. 2013. Prospección, Paisaje y el "gran cuadro" de la historia agraria: una revisión crítica y algunas propuestas. *Comechingonia Virtual: Revista electrónica de Arqueología* 17/2, 31-56.

Mlekuž, D., Taelman, D. 2012. Artifact Survey. in: Corsi, C., Vermeulen, F. (eds.). *A Romano-Lusitanian townscape revealed.* Gante: Academia Press, 127-135.

Openshaw 1984. The modifiable areal unit problem, Norwich, GeoBooks.
Patil, G.P. 2002. Diversity profiles, in: El Shaarawi, A.H., Piegorsh, W. W. (eds.) *Encyclopedia of Environmetrics*. Chichester: John Wiley & Sons, 555-561.
Ringrose, T.J. 1993. Diversity Indices and Archaeology, in: Andresen, J., Madsen, T., Scollar, I. (eds.). *Computing the Past. Computer Applications and Quantitative Methods in Archaeology*. Aarhus: Aarhus University Press, 279-286.
Shott, M.J. 1995. Reliability of Archaeological Records on Cultivated Surfaces: A Michigan Case Study. *Journal of Field Archaeology* 22/4, 475-490.
Smith, B., Wilson, J.B. 1996. A consumer´s guide to evenness indices. Oikos, 76, 70-82.
Tuomisto, H. 2012. An updated consumer´s guide to evenness and related indices. *Oikos* 121, 1203-1218.
Van De Velde, P. 2001. An Extensive Alternative To Intensive Survey: Point Sampling in the Riu Mannu Project, Sardinia. *Journal of Mediterranean Archaeology* 14/1, 24-52.
Waagen, J. 2014. Evaluating background noise: Assesing off-site data from field surveys around the Italic Sanctuary of S. Giovanni in Galdo, Molise, Italy. *Journal of Field Archaeology* 39/4, 417-429.
Winther-Jacobsen, K. 2010. The classical farmstead revisited. Activity differentiation based on a ceramic use-typology. *The Annual of the British School at Athens* 105, 269-290.

Notes on contributor

Luis Antonio Sevillano Perea is a PhD student at the Department of Prehistory and Archaeology at the University of Sevilla and former holder of a JAE-Predoc scholarship at the Mérida Institute of Archaeology (CSIC). Integrated within the framework of ancient rural communities' studies, his research focusses on the technical facets of archaeological survey and the theoretical implications associated with the interpretation of the surface archaeological record.

Experiments on Roman surface scatters through digital survey methods

Study cases from Odra-Pisuerga region (Burgos, Spain)

Jesús García Sánchez[1]

Introduction

Literature in intra-site survey archaeology often has the objective of determining whether surface scatters are a trustworthy proxy for inferring past human behavior. This contribution aims to defend the potential of non-destructive tools for understanding the nature of the surface archaeological record. These tools have been used to assess regional historical processes in many fields, including demography (Bintliff 1997), cultural colonization (Vermeulen and Mlekuz 2012), relationships between the sacral world and the settlement pattern (Pelgrom and Stek 2010), landscape exploitation theoretical models (Tol *et al.* 2011), regional consumption habits (Winther-Jacobsen 2008), and inclusively global mapping projects of the entire Roman world (Mattingly and Witcher 2004).

One of the most critical points of archaeological sampling is the common but unsolvable use of modern unit boundaries, which shapes the results in terms of how the data is aggregated artificially, rather than by an underlying past structure which in excavation archaeology helps to interpret pottery assemblages. An interesting example is the aggregation of data by primary or secondary deposition, which is a more feasible option than interpreting scatters through supposed domestic use, like the old-fashioned Pompeii premise. Several factors such as annual plowing, land use, land reclamation (Pérez *et al.* 2013) or terracing (Bevan and Conolly 2011) have to be studied to interpret whether surface scatters replicate primary deposition context or are instead the stochastic result of the long-term destruction

1 Leiden University, Faculteit Archeologie. Van Steenis Building, Einsteinweg 2, 2333 CC Leiden (The Netherlands) jesus.garciasan@gmail.com.

of assemblages (Fentress 2000). Other factors such as sample size (Schiffer *et al.* 1978; Terrenato 2004), find size (Ammerman and Fieldman 1978), multi-period horizontal stratigraphies on isolated sites (Witcher 20012: 26), lack of variable definition and other methodological decisions (sampling designs, collection method) dramatically affect how the the archaeology is recorded and therefore its interpretation. In order to address the importance of intra-site research and the comparability of total collections and sampling survey, Bintliff (2013:196) has criticized the lack of comparisons between different methods of the same data. This paper aims to address this lack, using quantitative techniques to detect the influence of collection and recording technique in the characterization of rural Roman sites.

This paper concentrates on artefactual intra-site surveying. Analysis of survey methodology should be complemented by consideration of the material element of ancient human behaviours (Haselgrove 1985: 11). Household assemblages studies are not only a method for studying pottery typology, but also offer different research perspectives with which to link variation in the surface material to different domestic contexts. Nevett (2010) has stressed the importance of household assemblages for critical research on Roman space and its diverse archaeological problem set, from consumption habits to trade, from women's roles in the rural socio-economic environment to the Late Roman rural elite.

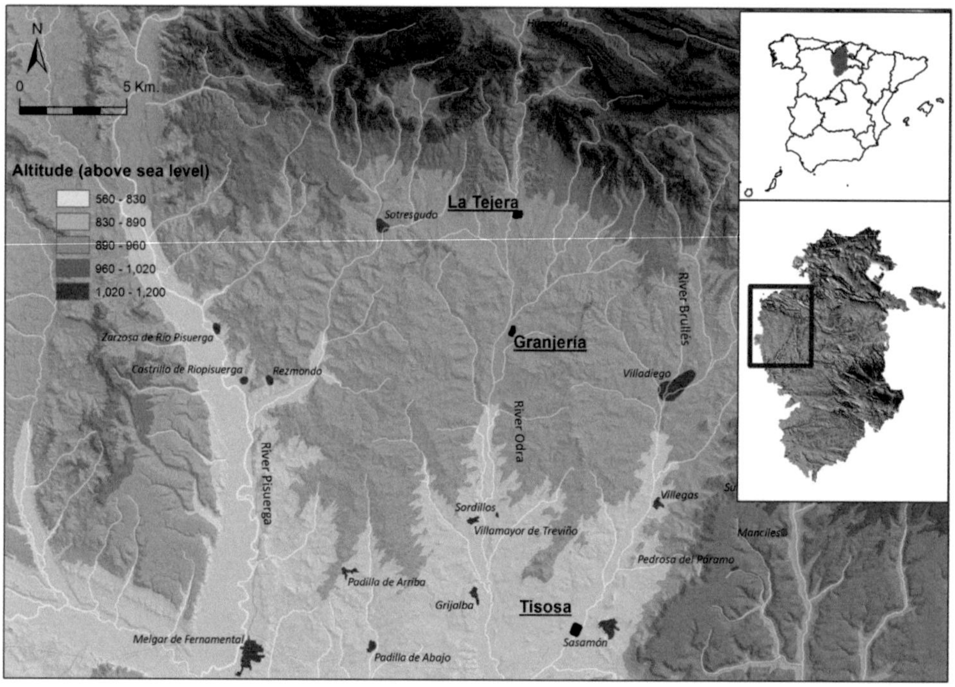

Figure 1: Map of the study area. Sites mentioned in text are bold and underlined.

Late Roman sites from the Odra-Pisuerga region (FIG 1) had a primary role in the organization of the landscape from the third century AD onwards; they were involved in same social and political process of land-ownership expansion and monumentalization of countryside residences followed by other Iberian regions (Chavarría 2004). Large sites such as our case study probably had a residential area, *pars urbana*, surrounded by industrial facilities. This approach to Late Roman domestic and industrial context does not aim to detect classical spaces such as those referred to by Vitrubius, but to offer a systematic understanding of the assemblage collected in the field by different quantitative methods, and to explore the spatial capabilities of such datasets. The methodologies considered in the present paper are a GPS approach to artefacts density (CPM) and a more detailed intra-site survey organized in grids. Both methods were put into practice in 2011 and 2012 in three Roman sites: Tisosa (Sasamón), La Tejera (Villavedón) and Granjería (Sandoval de la Reina). From those three we have selected the two first to analyze datasets by a Wilcoxon test and only the former (Tisosa) to experiment with the HJ Biplot method to compare statistically the outcome of the different applied survey methods.

CPM (Code Per Material) Survey Method

The first method proposed is a high detailed hand-held GPS survey of intra-site surface scatters to explore the relationship between the obtained archaeological sample and the original morphology of sites. Its main aim is to study the spatiality of artefact distributions on intra-site oriented surveys. That methodology has been named Code Per Material or CPM, due to the creation of a code list within the GPS device (Garmin GPS 60cx map) to represent different kinds of artefacts observed on the surface, from different classes of building material to table wares (sigillata) and coarse wares (cooking ware and dolia). The complete research on this site and other statistical approaches to the surface scatters recorded by the CPM methodology has been published previously (García-Sánchez 2013, García-Sánchez and Ezquerro 2014) and the aim of the present chapter is not to repeat discussion about the practicalities of the method. The CPM survey stresses the capabilities of the digital record in the definition of site character, and also the processes that affect the surface record such as horizontal displacement and scatter fragmentation. In the Spanish research agenda, the spatiality of single artefacts has been studied in the survey of micro-regions in Extremadura using GPS techniques (Mayoral *et al.* 2009). That experiment was implemented in regional contexts, while specific places with representative density peaks over the background noise were studied by sampling techniques.

The high degree of standardization of Roman pottery makes this procedure easy to achieve (Winther-Jacobsen 2008: 22). The code list is open and can be enlarged during the survey process, only common-sense communication between surveyors is required to inform the rest of team members about the appearance of new categories of artefacts. Each field walker is asked to record with their own hand-held GPS every sherd (or archaeological artefact) found on the surface and label it with a particular code/ symbol for each artefact category thus we quickly obtain a point cloud with spatial attributes, which can be filtered afterwards in a

GIS environment. It is important to stress that during the recording phase of the survey only a few selected materials (diagnostic forms, decorated sherds, special finds) are collected and brought to the laboratory for further analysis. The vast majority of artefacts are left in situ. This decision also helps to reduce laboratory processing time.

Figure 2: Intensiveness of CPM survey in La Tejera, surveyors walking with 2 m separation.

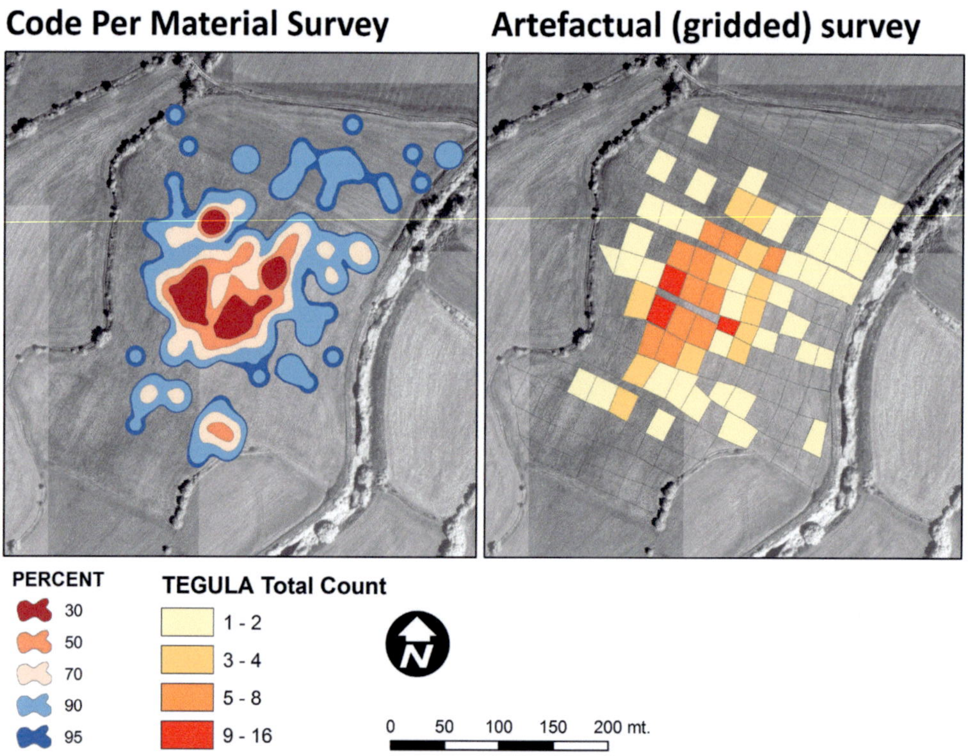

Figure 3: Total count of Tegulae fragments in the CPM survey (left) and in the Artefactual gridded survey (right).

The resulting point cloud has to be split by the attribute field that represents the typology of artefact and afterwards we are able to work separately with each layer of information. It allows comparison of the presence or absence of different categories of artefacts which might indicate variability in the use of both domestic spaces or dwelling activities in the surroundings of sites (FIG 3).

The CPM technique is appropriate for recording smooth density surfaces with high detail (GPS devices use WAAS-EGNOS correction and the average error of about 1 to 2 meters when the point cloud is recorded under the same satellite ephemeris). These artefact densities can be easily integrated within specific GIS software comparison with other data sources such as aerial photography or geophysics surveys (Gillings 1996). After filtering by typology or artefact category, items can be spatially analyzed and the probability of appearance assessed using kernel estimation (Brunsdon 1995). We chose an analysis based on a Percent Volume Contours (PVC) calculation to smooth the output based on probabilistic densities (expressed in percentages).

CPM in *Segisama* hinterland and in the Odra river valley (Burgos)

The CPM procedure was design to acquire a detailed scatter map of a newly discovered Early Roman site during the campaign of an off-site survey (García-Sánchez and Cisneros 2013) in the hinterland of Roman *Segisama* (Sasamón). During the autumn of 2011 the CPM methodology was implemented on several fields around the site-core. CPM produced information about the core area of these buildings and differential patterns of pottery distribution for assessing human behavior and environmental development in the formation of the archaeological record (LaMotta and Schiffer 1999).

The Odra valley (La Tejera and Granjería Late Roman sites) also exhibited great potential for the study of the late rural settlement pattern, due to increased occupation of extensive dry agricultural areas by rural elites that began in the late third century AD. We encountered several Roman sites alongside the terrace above the valley floor, probably as a deliberate strategy for managing river movements and floods. Indeed paleo-river beds can be detected by means of aerial photography (cropmarks) and geological surveys (pebble matrix soils). These sites could have different social functionalities according to size and surface finds; some of them seem to be large farming households with private and productive spaces, while others are much smaller with poor remains, mainly coarse ware pottery, loom weights, and very few Late Hispanic *sigillata*.

The CPM methodology was applied over an area of around four hectares at La Tejera (Villavedón) and 0.96 hectares at Granjería (Sandoval de la Reina); only the results of the former are reported in this paper. This procedure aims to define the site boundaries and check for the presence of different cores of human activity across the area that has been related to the Roman estate. The point cloud serves to delimit the most suitable areas for an intra-site grid survey, which recorded cultural material in very high detail.

From sherd scatter to artefactual survey

After this CPM approach to the site core of Tisosa and La Tejera, we assess the suitability of "gridding" with a detailed sampling and collection strategy that allows us to extract the whole potential from the artefact scatter.

The grid was set out over the entire area where a dense concentration of archaeological material was detected by the first CPM approach to both sites. In order to understand the relationship between site and environment, the survey grid was slightly larger than the core area, so that the edge of the site could also be determined. The grid unit is 10x10 metres in size in La Tejera and Granjería and 20x20 metres in Tisosa, in order to achieve a good resolution without holding down the level of random variation in datasets, as proposed by Ammerman and Fieldman (1978: 735). This grid is geo-located by GPS, thus any information was recorded digitally, which allows further integration with a wide range of spatial data, from CPM survey to geo-referenced aerial photos, or legacy datasets.

Each grid is surveyed for 10 minutes by a team member who collects any kind of archaeological material, regardless of its size or function, including building materials. Thus every unit will be surveyed in a standardized fashion (size, time, collection) and the only possible bias could come from surveyors or changes in ground visibility. At La Tejera and Granjería, building material was counted and weighed in the field, and eventually left in situ, while at Tisosa tegulae (plain tiles) and imbrices (curve tiles) were counted separately with manual clickers. Wares (FIG 4), metal, glass and so on are labeled, packed and studied in the laboratory to assess chronological information and functional variability across the site.

Figure 4: Different types of material collections in La Tejera (Villabedón).

Comparing CPM data with traditional grid survey

One of the procedure objectives is to compare differences between the results of CPM methodology (building materials, tegulae and imbrices) and the artefactual Grid survey of different building materials. A visual comparisons of the overall results obtained from La Tejera (Villavedón) can be inferred from FIG. 5 where an statistical map is plotted and from FIG. 6 where a boxplot summarized the distribution of datasets for each variable. However a statistical approach to extract a quantitative value of such differences between methodologies can also be obtained for each variable (e.g. each type of artefact recovered).

Thanks to the spatial attributes of both CPM and Artefactual survey shape files, we are able to compare the results by aggregating the data spatially, following the survey units from the Artefactual survey. Units are individuals (*n*) of an X_{nxp} matrix, while artefact quantitative data are variables (*p*) influenced by the methodology applied. An appropriate test for studying the influence of methodology on the record is the Wilcoxon signed-rank test used to compare two sets of scores from the same population, also employed for similar comparing purposes by Ammerman and Fieldman (1978: 735). The Wilcoxon test analyses the difference between values sampled on different occasions (different methods in this experiment) from the same population, using a Null Hypothesis as follows: median difference is 0, so there are no differences between the data pairs. In archaeological language, this hypothesis explores whether different collection methods can record scatters comparable in quantitative terms regardless the applied method. The Alternative Hypothesis claims that there is a difference between means, or in archaeological terms: the result of different methods applied to the same context cannot be compared in quantitative terms.

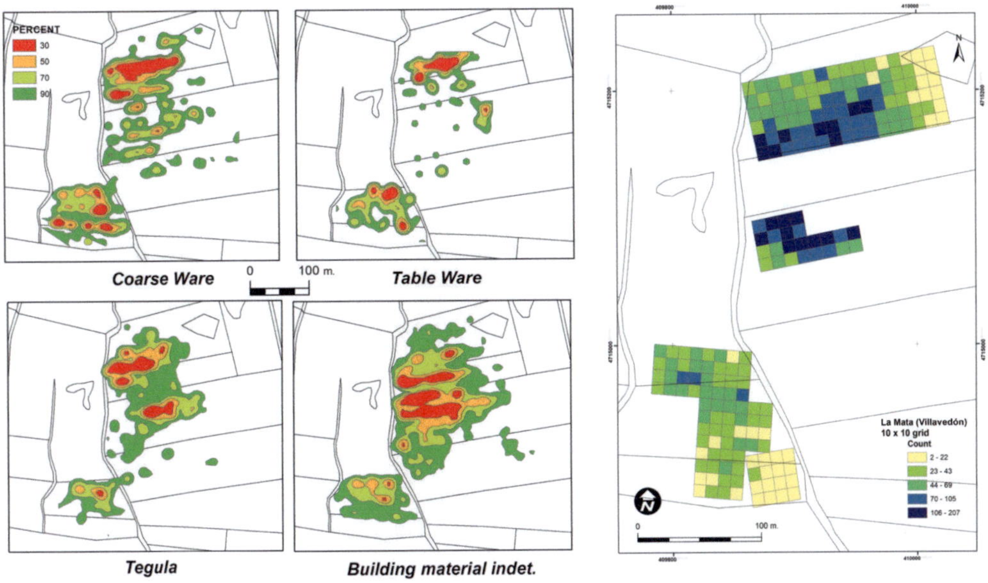

1. Percent Volume Contour (La Mata) after CPM data

2. Total Count of building material from artefactual survey

Figure 5: Mapping of spatial datasets from CPM (left) and Artefactual survey (right) in La Tejera.

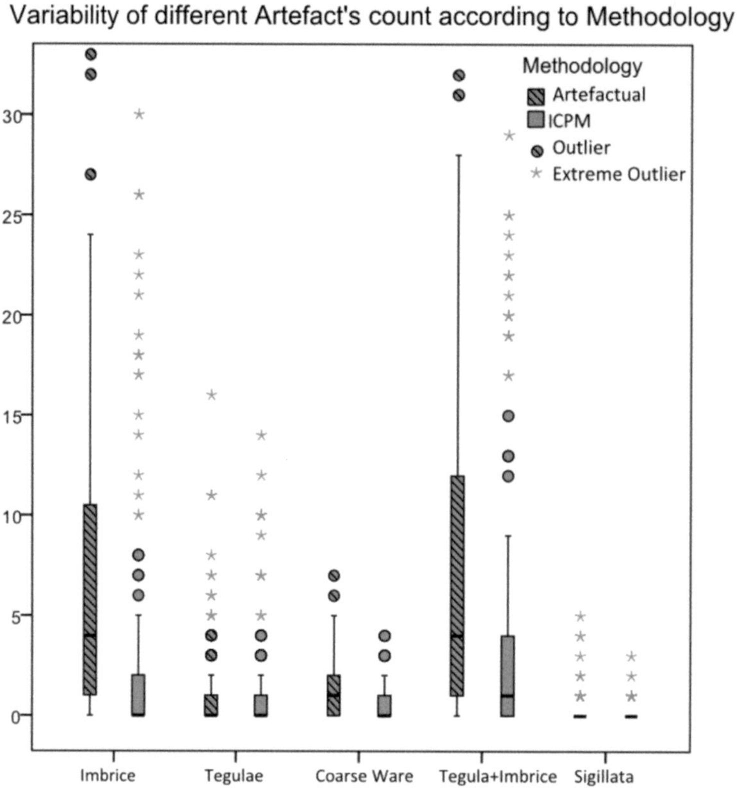

Figure 6: Data distribution of survey unit according to CPM and Artefactual Survey methodologies for the Tisosa site.

Summing up, the main goal of performing a Wilcoxon test is to check statistically how comparable our survey methods are. And if they are not, to examine the biases present in our data and established which kinds of archaeological-historical questions each method can solve, such as chronological definition, site function, and analysis of variability in site assemblages.

The test was first applied to the building material scatters from La Tejera. This site was first surveyed with the CPM method, and afterwards an intensive artefactual survey was conducted over the principal archaeological areas. Here, we have information (while pottery collections are still being studied) about two different kinds of architectural artefacts: *tegulae* and *imbrices* recorded by the CPM survey, while the artefactual survey only records building materials but in a more intensive way, with large systematic collections and weights. The test was performed by contrasting all the CPM against the artefactual survey, and finally the sum of CPM Building Material data against the artefactual survey.

The results from the statistical test could be understood as a clear difference between the datasets collected by both methods, i.e. influenced by the intense collection of surface material in contrast with the continuous recording which characterized the CPM. The p value is < 0.01, thus the Alternative Hypothesis is accepted: there is a significant difference in the median of CPM and artefactual

	Artefactual- Building material	Artefactual-Teg+Imbrex	Artefactual-Tegula
Z	-11.864	-12.505	-12.505
P value	.000	.000	.000

Table 1: Statistical output for the Wilcoxon's non parametric test for the building material scatters in La Tejera (artefactual vs. grid survey).

	Tegulae	Imbrices	Imbrex + Tegulae	Coarse Ware	Sigillata
Z	-1.048	-8.722	-8.353	-5.346	-1.65
P value	.294	.000	.000	.000	.099

Table 2:. Statistical output for the Wilcoxon's non parametric test for the building material, coarse ware and sigillata scatters in Tisosa (artefactual vs. grid survey).

survey datasets (TABLE 1). Thus, the results obtained from our two different approaches to surface scatters are not comparable in quantitative terms. The explanation for this is the very detailed collection and counting in each cell of the grid, which results in a large number of materials, that cannot be compared with strategies based in a continuous walk along the site, as CPM survey does.

The same experiment replicated for the Tisosa (Sasamón) provides us with a completely different view (TABLE 2). The Wilcoxon test says that the mean of both samples, CPM and Artefactual count of *tegulae*, is similar, so the Null Hypothesis of no difference between means is accepted (p value > 0.05). We can interpret that the surface record data has not been influenced by the employed methodology, and both surveys revealed similar quantitative data. This result suggests that the documentation method of CPM and the count of building material (without collection) works in the same way and the results are comparable, at least for the easy recognizable shapes of tegula.

Other variables have also been tested for the Tisosa site (TABLE 2 and FIG. 6) (p value< 0.01) between the applied survey methods; only the table ware (Hispanic sigillata) exhibits a different behavior (p value >0.05), though the amount of such material is quite low, and the distribution is quite similar in both datasets, sigillata scatters fit well with the site-core interpreted after the aerial photography (FIG. 3). According to the imbrices scatter, there are strong differences in distribution, however in both methods the imbrices values are grouped in the lower part of the box plot, with extreme outliers corresponding to survey units just overlapping the "interpreted" site core.

Visualizing results using HJ-BIPLOT

Biplot is a rather neglected technique in archaeology despite is natural connections to the Principal Components Analysis (PCA) and its early description by Baxter (1994). Biplot can be defined as a multivariate analytical technique, as proposed by Gabriel (1971) to analyze a data matrix where n individuals and p variables are considered. Its aim is to plot a multivariate matrix in a 2-dimensional space. This particular method can be used in many different ways. One of the most interesting for us, and perhaps for other archaeological survey projects, is the exploratory potential of pottery assemblages (variables) collected from grids (rows) as archaeologist normally do with scatter plots but in a more comprehensive way, mapping all the variables, individuals and inter-relationships within the same graph.

The Biplot ouput allows readers to interpret easily relations between either rows/ individuals (distance is interpreted as dissimilarity), or between columns/ variables plotted as vectors. For vectors we can interpret variability of the datasets through the length of the vector and correlation through the angle formed by two vectors. FIG 7 offers an comprehensive explanation of the visual display and its geometrical interpretation. This kind of data analysis and visualization is extremely useful to explore data matrices such as those that survey archaeology often creates using using pottery classifications. One step further would be the plotting of the HJ-Biplot results on a map in order to preserve the spatial attributes of each recording unit, this goal can be achieved thanks to the creation of clusters with specific software (Multibiplot) and the exportation of clusters to mapping software as GIS. Below is a technical description of the bases of the HJ Biplot, following the seminal paper by Gabriel (1971) and the more recent work of Galindo (1986) that allows Biplot users to obtain a representation where row and columns were represented with same quality.

The original method described by Gabriel differences two Biplot procedures GH Biplot and JH Biplot; the former enhanced the display of rows while the later does the same with columns.

Given markers $a_1,\ldots, a_n$ (rows) and $b_1,\ldots b_n$ (columns), thereafter the inner product resembles the x_{ij} element from matrix X.

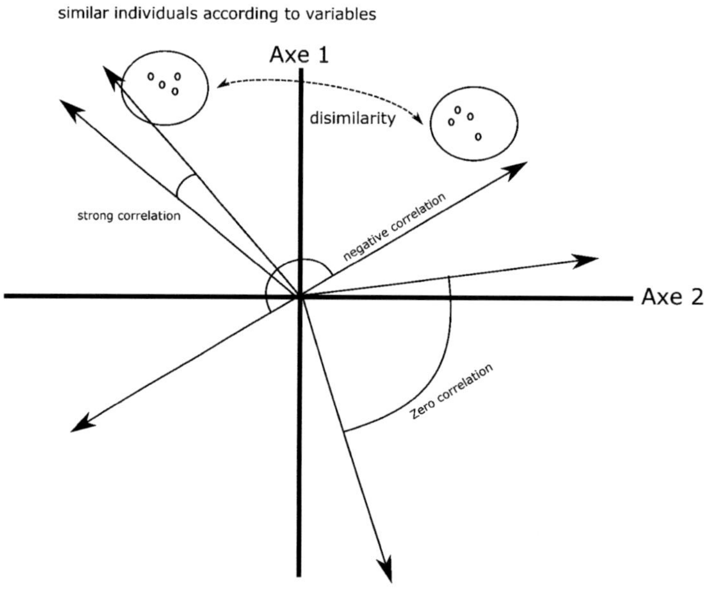

Figure 7: Guide to interpret columns and rows relationships in a HJ Biplot.

If we consider the decomposition of X in single values:

- U is an eigenvector matrix from XXt
- V is an eigenevector matrix from XtX
- D is a diagonal matrix build up with single values from X

We can say that X in the form of X=ABt, nothing else to choose, A=ABα and B=VD1-α. If α=0 we have a GH Biplot (column's display enhanced) and if α=1 we have a JK Biplot (row's display enhanced). Otherwise, if we use A=UD and B=VD as markers we obtain a HJ Biplot with both rows and columns enhanced (Galindo 1986).

Apart from the technical parameters underlying the matrix algebra and the extraction of coordinates for the Biplot representation is important to note that the most important thing about this technique is its applicability to any kind of data.

HJ Biplot Interpretation

For the HJ Biplot of the Artefactual survey (FIG 8) we obtain two axes that explained 82% of the cumulative variance. The graphical output shows a very intense relationship between the building material categories (tegulae/TEG and imbrex/IMB) and the table ware which includes mainly sherds of sigillata (TW). In opposition CW appears in a different direction in an angle of 90º which can be interpreted as an absence of any relationship with the rest of categories. The odd position of CW category along the Axis 2 (TABLE 3) could also represent a bias in the interpretation of very abraded materials that could be confused with sub-modern wares, also present in the area due its vicinity to the modern village and a strong process of artificial modification of the river bed has dramatically changed the landscape in the last 50 years. Thus we found a relatively high quantity of red-fabric sherds with eroded surfaces that makes difficult to ascribe to a chronological period. Nonetheless the most sub-modern pottery like glazed or Chinese crockery can be easily recognized. Apart from specific coarse ware like kitchen ware or few pieces of amphorae, the survey couldn't recover significant or diagnostic wares other than Table Ware (Hispanic sigillata). It should be stressed that this failure is probably due to the nature of the surface record, maybe related to the original pottery assemblage rather than survey methodology. In support of our approach and of the success of the survey methodology we can mention that during the grid survey we were able to recover several *tesserae* of different colors that weren't found in the first CPM survey of the site (FIG 9), that introduces new insights into the site chronological duration and functionality (García-Sánchez and Cisneros 2014).

Method	Column	Axis 1	Axis 2	Axis 3
Artefactual survey	CW	68	930	2
Artefactual survey	TW	633	2	354
Artefactual survey	IMB	755	19	161
Artefactual survey	TEG	867	11	15
CPM survey	CW	586	128	267
CPM survey	TW	552	230	206
CPM survey	IMB	614	169	29
CPM survey	TEG	626	152	

Table 3: HJ Biplot contributions for Artefactual and CPM survey in Tisosa.

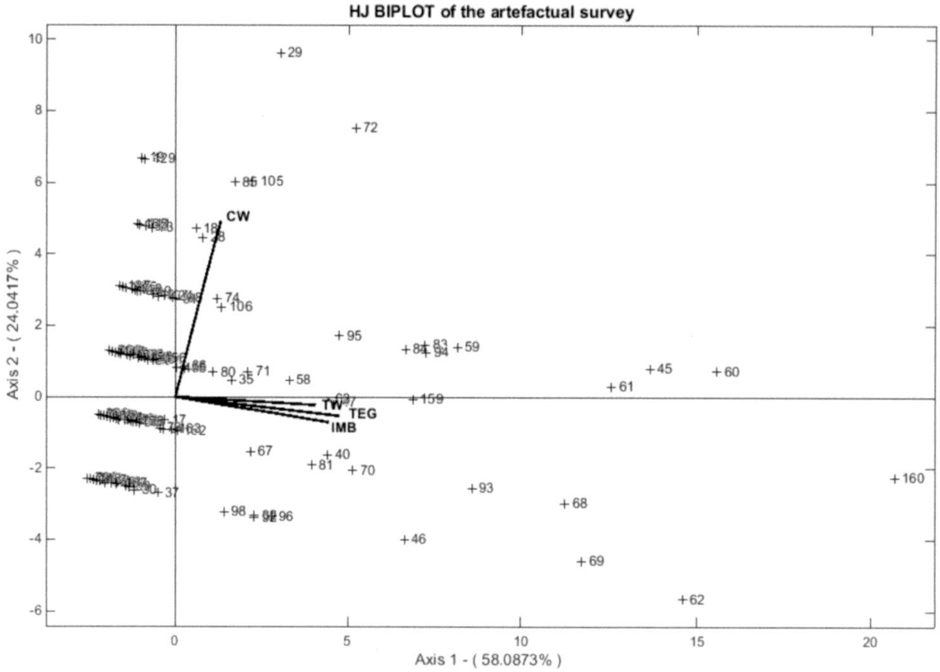

Figure 8: HJ Biplot for Artefactual survey of Tisosa.

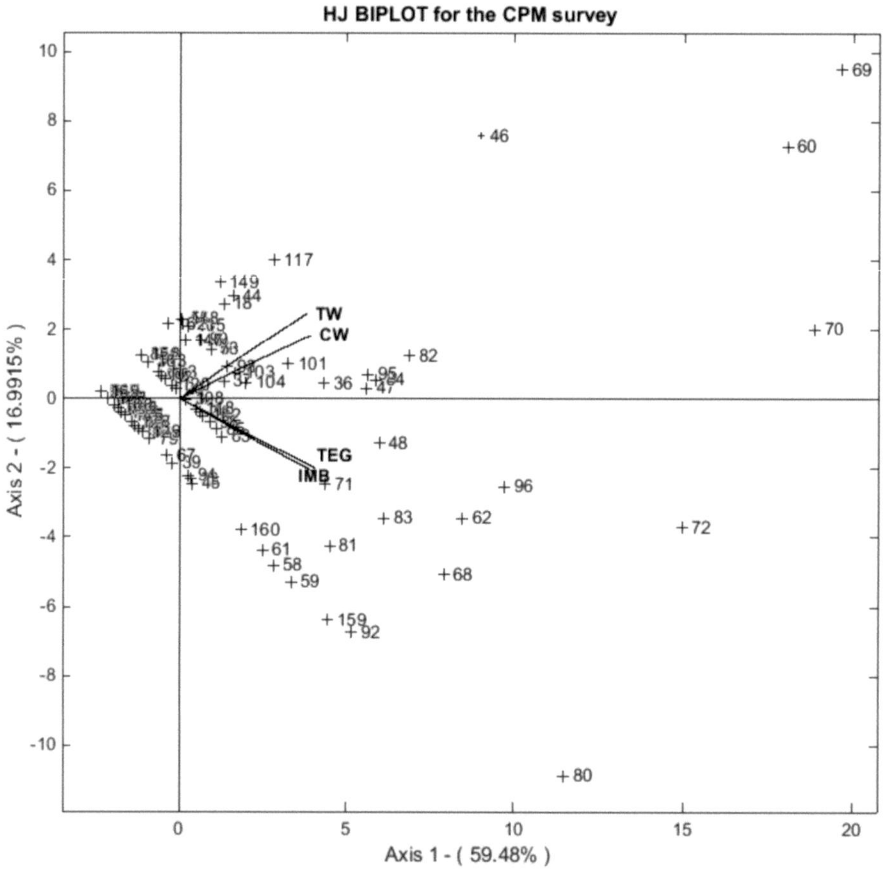

Figure 9: HJ Biplot for CPM survey of Tisosa.

Table 4 and FIG 10 plot the HJ Biplot data and visualization of the CPM survey. In that case we can observe a distinct behavior of two very well-defined classes of pottery. First the Table Ware and Common Wares have similar representations in both direction and length of vectors, expressing a high direct correlation. In opposition, the building material (IMB and TEG) also appears in the same position with a high correlation between them and with a weak correlation with the Table and Common Wares. This confirms clearly the impression of FIG 3, where we can observe how the building materials spread across a large area meanwhile different kind of wares appear within the interpreted site boundaries. Nevertheless the fragmentation of any kind of pottery also creates difficulties in the correct interpretation of any given tile's form. As such, a general category of building materials may be more appropriate in the future.

Conclusions

We use the same experimental approach to field survey that has started several decades ago in relevant works (Ammerman and Fieldman 1978; Terrenato and Ammerman 1996; Van Leusen 2002) from the developments of new methods based on digital tools that makes the treatment of datasets and its quantitative analysis more flexible. Our research is aimed at the development of methodologies to study surface scatters and at combining these with the output of other non-destructive techniques, namely oblique aerial photography. A digital methodology with the advantages of GPS devices allows us to compare statistically the quantitative results of two different methodologies, one based on a quick definition of scatters by means of a GPS code list, and another one conducted in a traditional fashion with collection and counting strategies to achieve a large dataset of household assemblages. Although statistically the results of both surveys are not possible to compare, both are reliable for different stages of research. CPM, for a quick but detailed first characterization of a site with smooth and understandable density maps, and the artefactual survey for research focused on the study of household assemblages, the formation of surface archaeological record and Roman use of domestic space.

Another feasible scenario for intra-site survey is the development of sampling strategies. A probabilistic sample over large sites, their halo and periphery, can result in a qualitative image of Roman sites, and also may enhance the possibilities of working in more sites than a tradition total collection strategy. The last survey campaigns in the Roman Republican landscape of Aesernia (Molise, Italia) conducted by the LERC team from Leiden University can serve as good example of such sampling strategy accompanied of aerial and geophysics survey to gain knowledge about a large selection of republican farms.

Even though Bintliff (2013) has strongly criticized sampling due to the possibility of losing chronological information and then survey reliability. We can consider this fact true for large sites like urban surveys, but small sites have an opposite trend. Most of them are mono-phasic or may be only the site which survives through a couple of centuries before acquiring a larger entity. Such cases have been noted amongst both Late Roman villas in Spain and the large Imperial villas with republican origins in Italy.

A final conclusion is the importance of defining the nature of experimentation and re-survey and its relation with the Baconian and Boylean conception of science. A pillar of the modern tradition of science is the possibility/ necessity of replication of experiments. This is definitely a premise that traditional excavation archaeology as a destruction technique cannot support. Even survey, traditionally considered as non-destructive technique, also modifies the surface record each time that a team of conscious surveyors steps into the plowed field, without considering other factors that influence presence of sherds in the surface such as plowing or soil displacement. Some parts of the surface record will survive in their "original" form but the removal of diagnositic pieces changes the remaining assemblage. A good example is the interest in certain diagnostic pottery, like *terra sigillata*, black-gloss, painted ware, etc. Its systematic total collection could eventually represent the partial destruction of the diagnosticity of the horizontal stratigraphy. It's important to consider that sites are not a continuous spring source of materials, which emerge after each plowing action. Surveys fulfill the requirements of controlling the techniques and controlling variables, but we can fail in assessing the replication of experiments and obtaining similar in our desire of re-survey as an experimental approach to intra-site survey.

Acknowledgements

The municipalities of Villadiego and Sasamón, as well as the local action group for rural development Adeco-Camino, have contributed with depth interest to our work in the area providing funding, human help, and a welcoming environment for scientific research.This paper was finished thanks to a postdoctoral position in the NWO funded project Landscapes of Early Roman Colonization in the Leiden University and in the Royal Dutch Institute in Rome (KNIR).

References

Allison, P. 1999. (ed.). *The Archaeology of Household Activities.* New York: Routledge.

Ammerman, A.J., and Fieldman, M.W. 1978. Replicated Collection of Site Survey. *American Antiquity* 43(4), 734-740.

Baxter, M.J. 1992. Archaeological Uses of the Biplot – a Neglected Technique?, in: Lock, G. and Moffet, J. (eds.). *CAA91. Computer Applications and Quantitative Methods in Archaeology 1991,* 141-48. BAR International Series. Oxford: B.A.R., 577.

Bevan, A. and Conolly, J. 2011. Terraced fields and Mediterranean landscape structure: An analytical case study from Antikythera, Greece. *Ecological Modelling* 222, 1303-1314.

Bintliff, J. 1997. Regional survey, demography and the rise of complex societies in the ancient Aegean. Core-periphery, neo-Malthusian and other interpretive models. *Journal of Field Archaeology* 24, 16-39.

Bintliff, J. 2013. Intra-Site Artefact Surveys, in: Corsi, C., Slapsak, B. and Vermeulen, F. (eds.). *Good Practice in Archaeological Diagnostics.* London: Springer International Publishing, 193-207.

Brunsdon, C. 1995. Estimating probability surfaces for geographical point data: An adaptive kernel algorithm. *Computers & Geosciences* 21, 877-894.

Fentress, E. 2000. What are we counting for?, in: Francovich, R., Patterson, H. and Barker, G., (eds.). *Extracting Meaning from Ploughsoil Assemblages.* Oxford: Oxbow Books, 44-52.

Gabriel, K.R. 1971. The Biplot graphic display of matrices with application to Principal Components Analysis. *Biometrika* 58: 453-67.

Galindo, M.P. 1986. Una alternativa de representación simultánea : HJ-Biplot. *Questió* 10 (1): 13-23

García Sánchez, J. 2013. Metodologías de prospección a escala regional y artefactual en la comarca. La prospección del Ager Segisamonensis; comarca Odra-Pisuerga (Burgos). *Complutum* 24, 9-28.

García Sánchez, J. and Ezquerro, A. 2014. New techniques for artefactual surveying: GIS-GPS methodology for the study of Roman habitational contexts, in: García Moreno, A., García Sánchez, J., Maximiano A. and Rios Garaizar, J. (eds.). *Debating Spatial Archaeology. Landscape and Spatial Analysis in Archaeology.* Santander: Instituto de Investigaciones Prehistóricas, 225-30.

García Sánchez, J. and Cisneros, M. 2013. An Off-Site Approach to Late Iron Age and Roman Landscapes on the Northern Plateau, Spain. *European Journal of Archaeology* 16 (2), 289-313.

García Sánchez, J. and Cisneros, M. 2014. Tisosa: Un establecimiento suburbano de Segisamo (Sasamón, Burgos). *Archivo Español de Arqueología* 87: 123-140.

Gillings, M. 1996. The Utility of the GIS Approach in the Collection, Management, Storage and Analysis of Surface Survey Data, in: Bintliff, J., Kuna, M. and Venclová, N. (eds). *The Future of Surface Artefact Survey in Europe.* Sheffield: Sheffield Academmic Press, 105-120.

Haselgrove, C. 1985. Inference from Ploughsoil Artefact Samples, in: Haselgrove, C., Millet, M., and Smith, I. (eds.). *Archaeology from the Ploughsoil Studies in the Collection and Interpretation of Field Survey Data*. Sheffield: University of Sheffield, 7-29.

LaMotta, V.M. and Schiffer, M.B. 1999. Formation process of household assemblages., in: Allison, P. (ed.). *The Archaeology of Household Activities*. London: Routledge.

Mattingly, D. and Witcher, R. 2004. Mapping the Roman World: The Contribution of Field Survey Data, in: Alcock, S.E. and Cherry,J.F. (eds). *Side-by-Side Survey. Comparative Regional Studies in the Mediterranean World*. Oxford: Oxbow Books, 173-186.

Mayoral, V., Cerrillo, E. and Celestino, S. 2009. Métodos de prospección arqueológica intensiva en el marco de un proyecto regional: el caso de la comarca de La Serena (Badajoz). *Trabajos de Prehistoria* 66, 7-25.

Nevett, L. 2010. Domestic Space in Classical Antiquity, Key Themes in Ancient History. Cambridge: Cambridge University Press.

Orejas, A. and Ruiz del Árbol, M. 2008. Territorio y dominio en las villas romanas: el fundus de Veranes, in: Fernández Ochoa, C., García-Entero, V., and Gil Sendino, F. (eds.). *Las Villae Tardorromanas En El Occidente Del Imperio: Arquitectura Y Función. IV Coloquio Internacional de Arqueología En Gijón*. Gijón: Trea, 167-192.

Pelgrom, J. and Stek, T. 2010. A landscape archaeological perspective on the functioning of a rural cult place in Samnium: field surveys around the sanctuary of S. Giovanni in Galdo (Molise). *Journal of Ancient Topography* 20, 41-102.

Pérez Álvarez, J.A., Mayoral, V., Martínez del Pozo, J.A. and de Tena, M.T. 2013. Multi-temporal archaeological analyses of alluvial landscapes using the photogrammetric restitution of historical flights: a case study of Medellin (Badajoz, Spain). *Journal of Archaeological Science* 40, 349-364.

Schiffer, M.B., Sullivan, A.P. and Klinger, T.C. 1978. The Design of Archaeological Surveys. *World Archaeology* 10, 1-28.

Terrenato, N. 2004. Sample Size Matters! The Paradox of Global Trends and Local Surveys, in: Alcock, S.E. and Cherry, J.F. (eds.). *Side-by-Side Survey: Comparative Regional Studies in the Mediterranean World*. Oxford: Oxbow, 36-48.

Terrenato, N. and Ammerman, A.J. 1996. Visibility and Site Recovery in the Cecina Valley Survey, Italy. *Journal of Field Archaeology* 23, 91-109.

Tol, G. 2012. A Fragmented History. A methodological and artefactual approach to the study of ancient settlement in the territories of Satricum and Antium. Groningen: Barkhuis.

Tol, G., De Hass, T. and Attema, P. 2011. Investing in the Colonia and Ager of Antium. *Facta* 5, 111-144.

Van Leusen, P.M. 2002. Pattern To Process: Methodological Investigatios into the formation and interpretation of Spatial Patterns in Archeological Landscapes. Groningen: Rijksuniversiteit.

Vermeulen, F. and Mlekuz, D. 2012. Roman towns and the space between them: a view from northern Picenum, in: Vermeulen, F., Burgers, G.J., Corsi, C. and Keay, S. (eds.). *Urban Landscape Survey in Italy and the Mediterranean*. Oxford: Oxbow Books, 207-222.

Winther-Jacobsen, K. 2008. The Role of Pottery in the Interpretation and Classification of Rural Sites in Cyprus, in: Vanhaverbeke, H., Poblome, J., Vermeulen, F., Waelkens, M. and Brulet, M. (eds.). *Thinking about Space: The Potential of Surface Survey and Contextual Analysis in the Definition of Space in Roman Times, Studies in Eastern Mediterranean Archaeology*. Turnhout: Brepols, 21-30.

Witcher, R.E. 2012. That from a long way off look like farms': the classification of Roman rural sites, in: Attema, P. and Schorner, G. (eds.). *Comparative Issues in the Archaeology of the Roman Rural Landscape. Journal of Roman Archaeology Supplementary Series 88*, Portsmouth: Thomson-Shore, 11-30.

Notes on contributor

Jesús García Sánchez is a Postdoc in Leiden University for the NWO-funded Landscapes of Early Roman Colonization. He works on survey methodologies and visual and statistical interpretation of intra-site and off-site survey data, mostly in Hellenistic, pre-Roman and Roman period sites. Moreover he is interested in other non-destructive techniques like aerial photography and ALS/LiDAR for the study of archaeological landscapes. Currently he conducts research in Central-South Italy, in the Roman colonies of Aesernia (Molise), Venusia (Basilicata) and the Tappino Valley. Previously he has directed other survey and excavation projects in Burgos, La Rioja, Murcia (Spain) and collaborated with others in Italy, Greece and Israel.

Scope and limitations of airborne LiDAR technology for the detection and analysis of Roman military sites in Northwest Iberia

José Manuel Costa-García[1] *and João Fonte*[2]

Remote sensing: towards a new cost-effective methodology

The Roman military sites are frequently characterised by the perishable nature of their structures and material culture. This situation can easily become extreme in the particular case of the marching camps (*castra aestiva*) (Peralta Labrador 2002; Jones 2012). Since these sites are almost invisible in the modern landscape, the use of remote sensing techniques for their detection and study has proved fundamental. Although the integration of some of these tools -i.e. aerial photography- into the archaeological studies of the Iberian Peninsula started several decades ago, the development of the Roman military archaeology as an autonomous discipline demanded a significant renovation and adaptation of the common methodological approaches.

More recently, we have contributed through the *Romanarmy.eu*[3] research project to the development of a cost-effective methodology for the detection and analysis of Roman military sites in Northwest Iberia. This method combines remote sensing techniques (historical and modern aerial photography, satellite imagery, airborne LiDAR), and GIS with more conventional archaeological field survey techniques, alongside place-names distribution and ethnography (Menéndez Blanco *et al.* 2013; Menéndez Blanco *et al.* 2016b). All these tools provide a new and qualitatively

1 Departamento de Historia, Universidade de Santiago de Compostela. Praza da Universidade, s/n 15703 Santiago de Compostela (Spain) josemanuel.costa@usc.es.

2 Instituto de Ciencias del Patrimonio (Incipit), Consejo Superior de Investigaciones Cientificas (CSIC). Avenida de Vigo, s/n. 15705 Santiago de Compostela (Spain) joao.fonte@incipit.csic.es.

3 Romanarmy.eu is an independent and self-financed project formed by several researchers particularly interested on the archaeological study of the Roman conquest of the northwest of the Iberian Peninsula: http://romanarmy.eu/en/.

differential approach to the sites, allowing us to study their spatial, locational and morphological characteristics. They lead thus to a more integrated approach, since each technique could potentially reveal different features (Crutchley 2009).

Within this context, airborne LiDAR allows us to see and understand landscapes from a different perspective. We understand the landscape as a result of the co-evolution of natural and cultural processes. In this way, each period of activity generates its own concrete landscape forms, partly based on the footprints left by the different physical remains that once existed in a particular space. However, a landscape is not just a messy amount of archaeological features, but rather a palimpsest of elements with different temporality and spatiality we need to read and understand (Mlekuž 2013a, 2013b). By picking those little pieces of information we can try to individualise and reconstruct ancient landscapes. For that reason, we need to develop different strategies to increase the visibility of more faint topographic evidence, or what we might call "archaeological topographies" (Opitz and Cowley 2013). Airborne LiDAR is indeed a very powerful tool for mapping the different traces that are engraved on the terrain.

Finally, we must stress that the north-western area of the Iberian Peninsula is not topographically uniform. The vast plains of the Northern Spanish Plateau are dominated by large states and cereal farming. The mountainous areas of Asturias, western León, eastern Galicia and north-eastern Portugal are frequently forested if not dedicated to pastures. The central and western areas of Galicia as well as northern Portugal are occupied by smallholdings with diversified crops, meadows and forests covering valleys, hills and low mountains. This requires the use of a diversified methodology. By analysing the response of every single technique to different landscapes, terrain types and vegetation canopies, we can outline its potential and limitations.

Historical aerial photography and Roman military archaeology

The use of remote sensing techniques is not new in Roman military archaeology, including in Iberia. Indeed, several *ad hoc* flights over Roman military sites were made before World War II, also in Spain (e.g. Almagro Basch 1943; González Reguero 2007). Besides that, a large amount of aerial photographs commisioned for different purposes have also been extensively used by archaeologists: the aerial photos from historical flights -mainly from the 1930s onwards- and satellite imagery archives are good examples of that (Hanson and Oltean 2013). Any user can access these materials easily today. Consequently, it is possible not only to adopt an important diachronic perspective of the sites and landscapes, but also to digitally reconstruct nowadays lost archaeological landscapes by using photogrammetric software (Cerrillo-Cuenca and Sanjosé 2013, Charro Lobato 2015; Risbøl *et al.* 2015).

In the Iberian Peninsula there have been *ad hoc* photographic flights over Roman republican military sites such as Numantia (González Reguero 2007: 239) and Cáceres el Viejo (Almagro Basch 1943). The archaeological use of the "*Vuelo General de España Serie B*" (USAF AST6 54-AM-78), commissioned by the Spanish government to the United States Air Force (USAF), also led to the discovery of new camps as those of Castrocalbón (Loewinsohn 1965), Valdemeda (Sánchez Palencia 1986), O Cornado (Gago Mariño and Fernández Malde 2015),

Monte dos Trollos (Costa-García *et al.* 2015b) or A Chá da de Santa Marta (Costa-García *et al.* 2015b; *Orejas et al.* 2015).

More recently we have used Structure from Motion (SfM) photogrammetry (Verhoeven *et al.* 2012) to orthorectify and georeference historical aerial photos from several historical flights (SPLAL, RAF and USAF[4]) from the 1940s and the 1950s in Portugal (Redweik *et al.* 2010). This allowed us to obtain new cartographic data, namely Digital Surface Models (DSM) and orthophotos. We have also used historical aerial photos taken by the USAF during the 1940s and the 1950s in Spain: series A (1945-46) and B (1956-57) (Pérez Álvarez *et al.* 2013; Pérez Álvarez *et al.* 2014), recently made available by the Spanish Geographic Institute (IGN) through web-mapping services. Roman military sites such as Campos (Blanco-Rotea *et al.* 2016) and Alto da Cerca (Fonte and Costa-García 2016) were analysed doing an intensive use of these tools.

Modern aerial photography and satellite imagery

Aerial photography also played an important -though frequently underestimated-role in the awakening of the Roman military archaeology as a discipline in Spain. The planning of flights with an archaeological aim during the decades of 1990 and 2000 allowed not only the discovery of new camps but also the detailed study of other sites previously known (Del Olmo Martín 1995; García Merino 1996; Peralta Labrador 2011). In the last years, the popularity of multi-purpose aerial photography and satellite imagery has increased at an exponential rate. The possibility of locating traces of hidden structures determined their extensive use in archaeology (Ceraudo 2013).

The open access to data granted by the Spanish National Plan of Aerial Orthophotography (PNOA) or Google Inc. -among others- has allowed the development of a systematic survey method (Menéndez Blanco *et al.* 2013). This is especially effective in mountainous areas without dense vegetation canopy or in plain regions covered with crops. In some cases the ancient earth ramparts are still surviving; in others the trenches were tracked due to the differential accumulation of moisture in the topsoil. The planning of field surveys focused on those potential sites led to the discovery of many *castra aestiua* in the northern areas of Iberia (González Álvarez *et al.* 2008, 2011; Menéndez Blanco *et al.* 2011a, 2011b; Celís Sánchez *et al.* 2015; Hierro Gárate *et al.* 2015).

Airborne LiDAR

In forested areas with dense vegetation cover, the identification of archaeological features is still very problematic (Doneus *et al.* 2008). The introduction of airborne LiDAR has helped to overcome this problem because of its unique capability to penetrate vegetation canopies, making it possible to document the underlying topographic surface and identify any cultural remains on it (Opitz and Cowley 2013).

---

4 *Sociedade Portuguesa de Levantamentos Aéreos Limitada* (SPLAL) from between 1937 and 1952; British Royal Air Force (RAF) from 1947; United States Air Force (USAF) from 1958. The photos have to be purchased to the Portuguese Army Geospatial Information Center (CIGeoE).

The identification of archaeological features on LiDAR-derived Digital Terrain Models (DTM) is very dependent on visualisation techniques that can enhance our perception and interpretation of anthropogenic features in the high-resolution digital topographic models. Most archaeological applications of airborne LiDAR were undertaken mainly using one type of visualisation technique -shaded relief modelling- where the visibility of potential archaeological features depends to a large extent on the chosen illumination angles that may hinder their detection (Devereux *et al.* 2008).

Different methods have been proposed to highlight microtopographies, from simple hillshading techniques to more complex calculations like those based on a simple combination of slope and hillshade (Doneus *et al.* 2011), Principal Components Analysis (Devereux *et al.* 2008), Sky View Factor (Kokalj *et al.* 2011; Zakšek *et al.* 2011), Local Relief Models (Hesse 2010) or Openness (Doneus 2013). These visualisation techniques have been investigated and analysed (Štular *et al.* 2012; Chalis *et al.* 2012; Bennett *et al.* 2012; Kokalj *et al.* 2013) and the results highlight that no single method outperforms the rest in all types of terrain. Therefore, a combination of these techniques is the only way to obtain a maximum volume of information on potential archaeological traces in relief.

Trend removal procedures (Štular *et al.* 2012) are among the more effective for archaeology. They are based in the theoretical assumption that when a smoothed surface is compared to its corresponding real topography, local small-scale topographic features are contrasted from large-scale landscape forms. One advantage of this technique lies in the richness of the generated data (based on relative heights) because it allows to discriminate between positive and negative microtopographies at a local scale, representing real changes in elevation rather than calculations based on steepness and direction of slope or exposure to light.

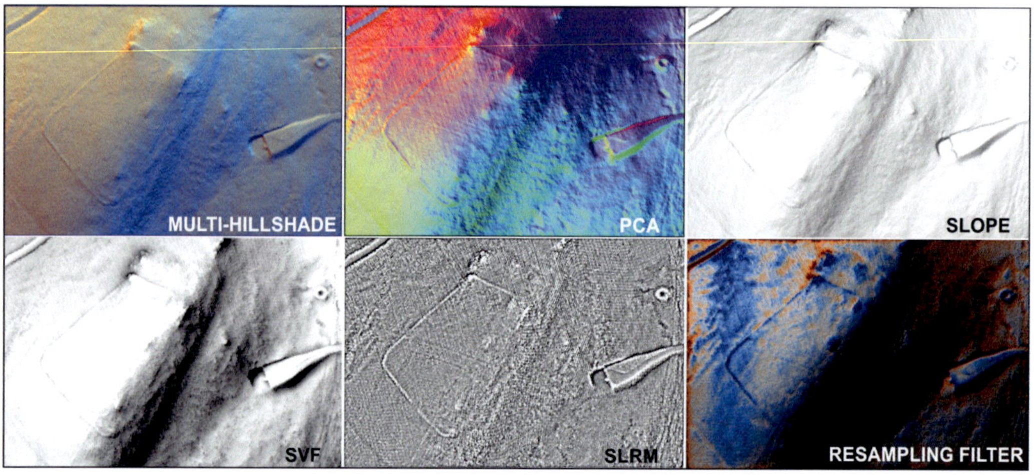

Figure 1: The Roman camp of Moyapán. Different LiDAR visualization techniques.

In Spain an open-access LiDAR coverage of almost the entire territory is available.[5] LiDAR data are already classified. This means that we can directly obtain a DTM from the ground points, although in certain cases it is advisable to do some additional classification (in our case, using the software suite LAStools).[6] In most cases we have used hillshade and slope as standard visualisation techniques, although in other we have applied more complex solutions such as the Resampling Filter available in SAGA GIS software (Conrad *et al.* 2015), Sky View Factor (Štular 2012; Zakšek 2011) or Local Relief Models (Hesse 2010) (fig. 1). The Resampling Filter is a trend removal technique that allows representation of local small-scale elevation differences, in a similar way to Local Relief Models (Hesse 2010).

Geographic Information Systems

GIS allow us not only to integrate different geographic data and to study them together, but also to compensate the main limitations of each technique and to strengthen the global archaeological interpretation. The purpose of this effort is dual: on the one hand, to better characterise the sites (the morphological and locational studies are very important in this case); on the other hand, to understand the landscape in which they are integrated (spatial analysis).

Spatiality is intrinsic to most archaeological data and GIS and geospatial technologies provide an ideal tool for the combination and quantification of different spatial variables, allowing us to create meaning and knowledge from the spatial dimension of archaeological data (McCoy and Ladefoged 2009).

Discussion

Since 2007, within the *Romanarmy.eu* project several Roman military sites have been detected in northwest Iberia by following this non-invasive and integrated methodology (Costa-García *et al.* 2015a; Costa-García *et al.* 2015b; Blanco-Rotea *et al.* 2016; Fonte and Costa-García 2016; González Álvarez and Menéndez Blanco, 2007; Gago Mariño and Fernández Maldeand 2015; González Álvarez *et al.* 2008; Menéndez Blanco *et al.* 2015a; Menéndez Blanco *et al.* 2014; Menéndez Blanco *et al.* 2016; Menéndez Blanco *et al.* 2011b; 2011a) (fig. 2). This effort has not only quantitatively and qualitatively enriched our knowledge about the Roman military presence in the area but also allowed the opening of new research perspectives regarding the military occupation of these territories (see Costa-García in this volume).

Airborne LiDAR has been for us one of the most proficient remote sensing techniques for the identification and analysis of Roman military sites (fig. 3). We are dealing with severely eroded and poorly surviving, almost invisible structures. In addition, they show varying morphologies, although the classical "playing-card" layout is predominant. For this reason it is important to consider the correlation of different factors when dealing with LiDAR data: time of data collection, point

5 *Plan Nacional de Ortofotografía Aérea* (PNOA), *Instituto Geográfico Nacional* (IGN): http://pnoa.ign.es/coberturalidar. The average point density is 0.5 points per square meter, although this varies depending on the geographic area. In most of the cases, we were able to obtain a 1-meter DTM.

6 LAStools, "Efficient LiDAR Processing Software", obtained from http://rapidlasso.com/LAStools.

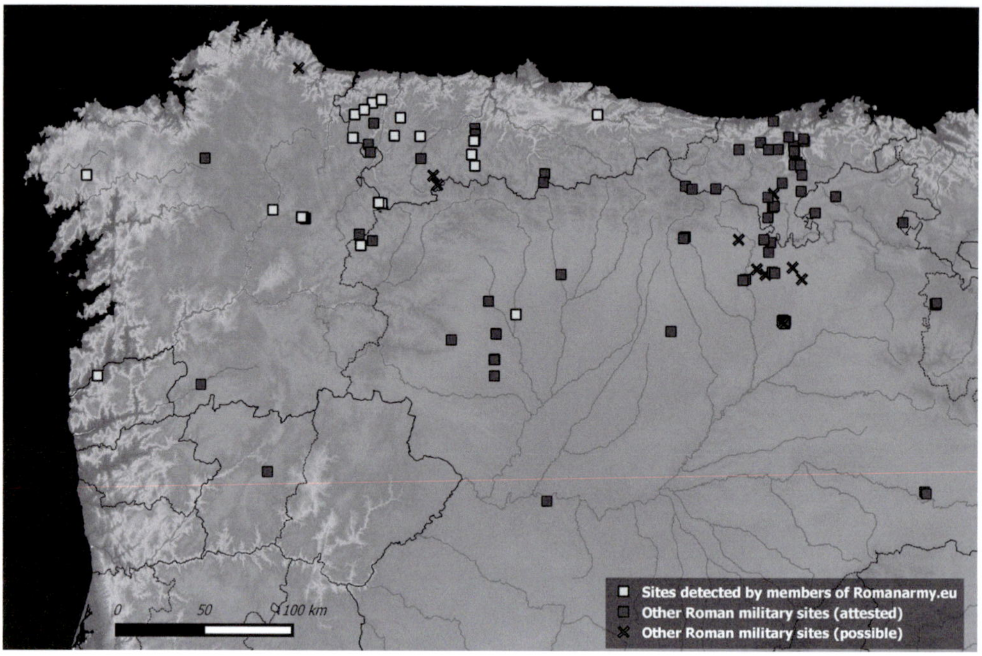

Figure 2: Roman military sites in Northwest Iberia. In yellow, sites located by the Romanarmy.eu research group.

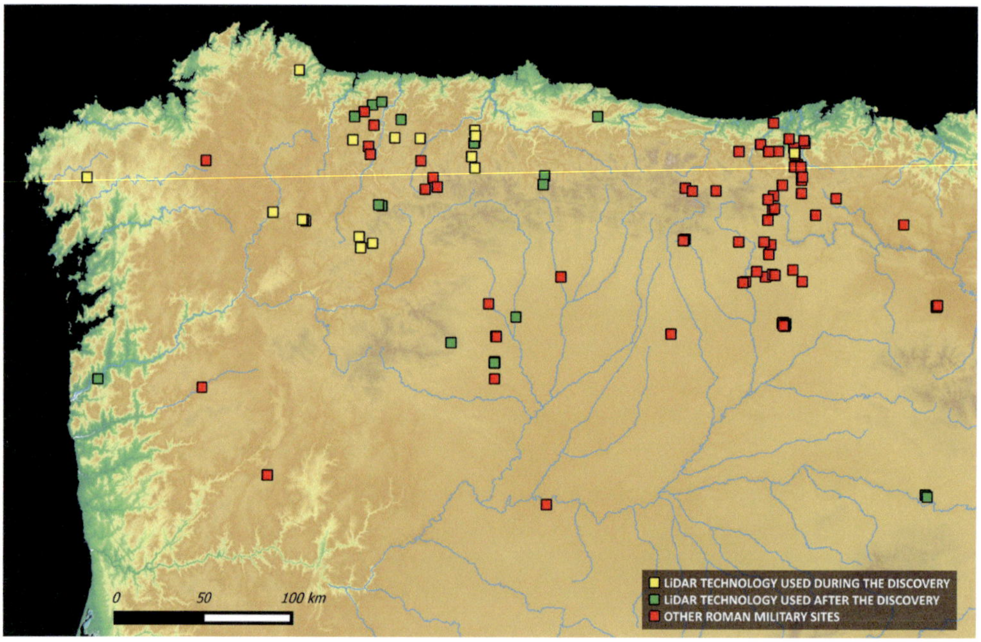

Figure 3: The use of LiDAR technology for the detection and analysis of Roman military sites in Northwest Iberia.

density (Bollandsås *et al.* 2012), classification of the point cloud, interpolation quality and DTMs spatial resolution, terrain type and vegetation cover, as well as the physical properties (size and shape) of the cultural remains (Risbøl *et al.* 2013).

Due to space restrictions, we have selected only a limited number of sites to be analysed here. They have been classified in three big groups that show some similar characteristics in order to stress the differential contribution of airborne LIDAR in contrast with other remote sensing techniques.

LiDAR in lowland areas

The three Roman military enclosures of A Chá de Santa Marta (Lugo, Spain) have suffered the severe impact of the mechanical ploughing and the intense reforestation carried out in this area from the 1960s onwards. The central camp, named Cabianca, was perfectly recognizable in the USAF photos of 1957, but its traces can be hardly detected in modern orthophotos or satellite imagery (fig. 4). The ancient rampart was completely flattened before we could survey the site in late 2014. Only the sunset light allowed us to detect some elevations of about 10-15 cm that resembled lineal structures. Fortunately, the visibility of the enclosure improved after applying the Resampling Filter of SAGA GIS to the LiDAR-based DTM, which was able to enhance those micro-topographic details barely detectable *in situ*.

LiDAR also showed its potential when reviewing another set of camps discovered fifty years ago in Castrocalbón (León, Spain) (Loewinsohn 1965, Costa-García 2015, Costa-García and Casal García 2015, Costa-García 2016). The ploughing activities in this area did not become completely mechanised, so we were able to clearly detect a set of negative and positive structures (fig. 4). They are close to an old path which follows the outline of a Roman road recorded in the Antonine Itinerary (Wess. 422.2-423.5). Some of the ditches are quite clogged and it seems that they were the places from where the materials for the construction of the ancient road were extracted (Moreno Gallo 2011). Other rectilinear trenches draw an enclosure which follows the classical playing-card shape, revealing the presence of a Roman camp previously documented thanks to historic aerial photography. South to this position, several positive structures were detected as well. The presence of a Roman camp and a probable watchtower was already attested here, but the existence of soft terrain elevations configuring another possible enclosure is completely new. Besides, the detailed analysis of the former camp revealed the presence of four *clauiculae*-type entrances.

Unfortunately, a LiDAR-based DTM can also lead us to an erroneous record of the archaeological features. In this way, it is important not to fall in self-confidence or comfort when using these tools: these visualisations must be checked against the results obtained by applying other remote sensing techniques as well as the data recorded on the field.

A good example of how these risks can occur is the Roman camp of Los Llanos (Soria, Spain). This big enclosure was detected through aerial photography in the late 2000's (Didierjean 2008), thanks to the existence of two parallel dark lines which were identified as a *fossa duplex* or double trench. The oldest graphic records of the place already showed these archaeological features. Although the LiDAR visualisations allowed us to document the existence of a severely damaged, inner rampart, only one depression could be observed in the outside by using these resources (fig. 4). There is an easy explanation for this presumed abnormality: after the initial filling of the ancient trenches, the secular ploughing activities flattened out the ground, the remains of the earth bank which originally divided the ditches were destroyed and consequently a single soft depression was formed (Costa-García and Casal García 2015). In this way, while aerial photography is reflecting the effect of the differential accumulation of moisture in the area occupied by the trenches, the LiDAR-based DTM strictly represents the modern ground relief.

The camp of Huerga de Frailes (León, Spain) (Menéndez Blanco *et al.* 2011b) also reflects the limitations of LiDAR technology in plain areas covered with cereal crops. Given that the archaeological survey on the field attested the complete erosion of the ramparts and the filling of the single trench detected by aerial photography, no success was expected when applying LiDAR visualisations. However, the differential growing of the crops over the ditches was enough to detect some traces in the visualisations derived from both the DTM and the DSM (Costa-García and Casal García 2015).

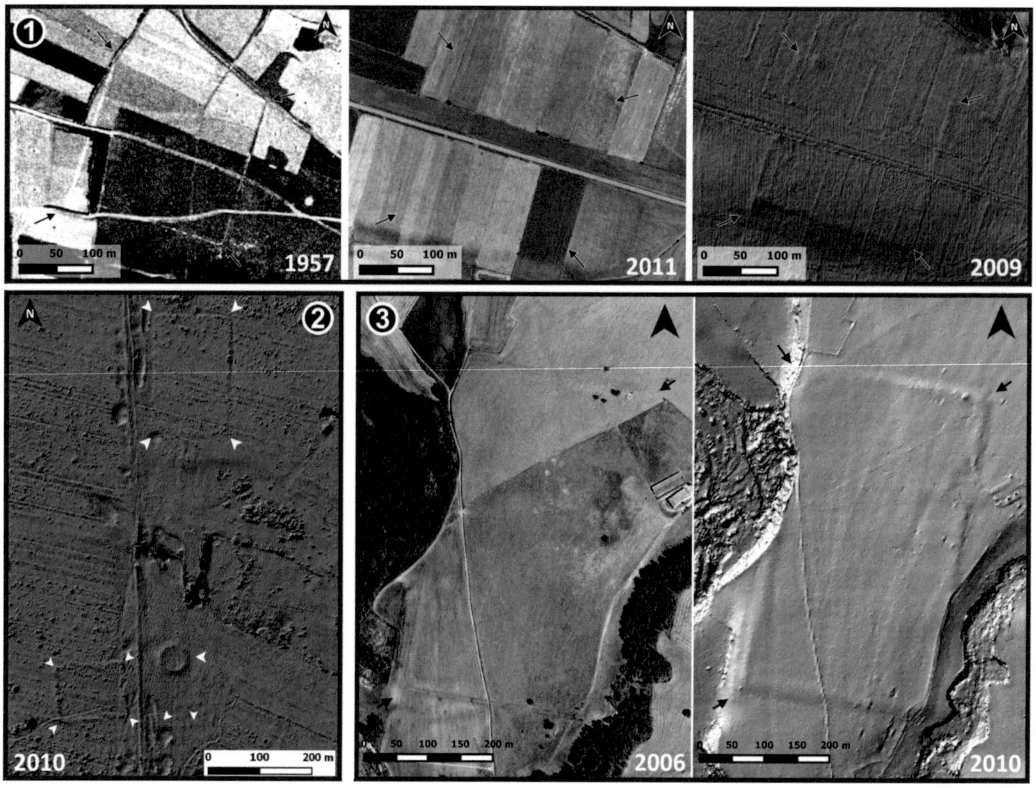

Figure 4: The enclosures of Cabianca (USAF aerial photography, PNOA aerial orthophoto and Resampling Filter visualisation) (1), Castrocalbón (Resampling Filter visualisation) (2) and Los Llanos (PNOA aerial orthophoto and Multi-hillshade visualisation) (3).

LiDAR in forested areas

From the 1960s onwards several areas of Northwest Iberia have been affected by different projects of reforestation. The impact of these policies on the archaeological heritage cannot be generalised, since it depends on various factors such as the type of species introduced, the reforestation techniques or the actual degree of preservation of the remains before the process. The site of Valdemeda, discovered long time ago after reviewing the USAF aerial photos of 1957 (Sánchez Palencia 1986), was considered lost for archaeological research due to the aggressive reforestation. Although not very well preserved, the use of LiDAR technology has revealed that the Roman military camp is still there (Costa-García 2015, Costa-García and Casal García 2015, Sánchez Palencia and Currás Refojos 2015) (fig. 5). Moreover, it is possible to appreciate the effects of two types of reforestation plans: the zone occupied by grown trees allow the penetration of a greater number of laser pulses and this implies the production of a more precise DTM, while the area covered by lower, younger trees produces a 'noisy' picture. This situation can also reflect the introduction of modern machinery in the second sector which could have severely damaged the archaeological features.

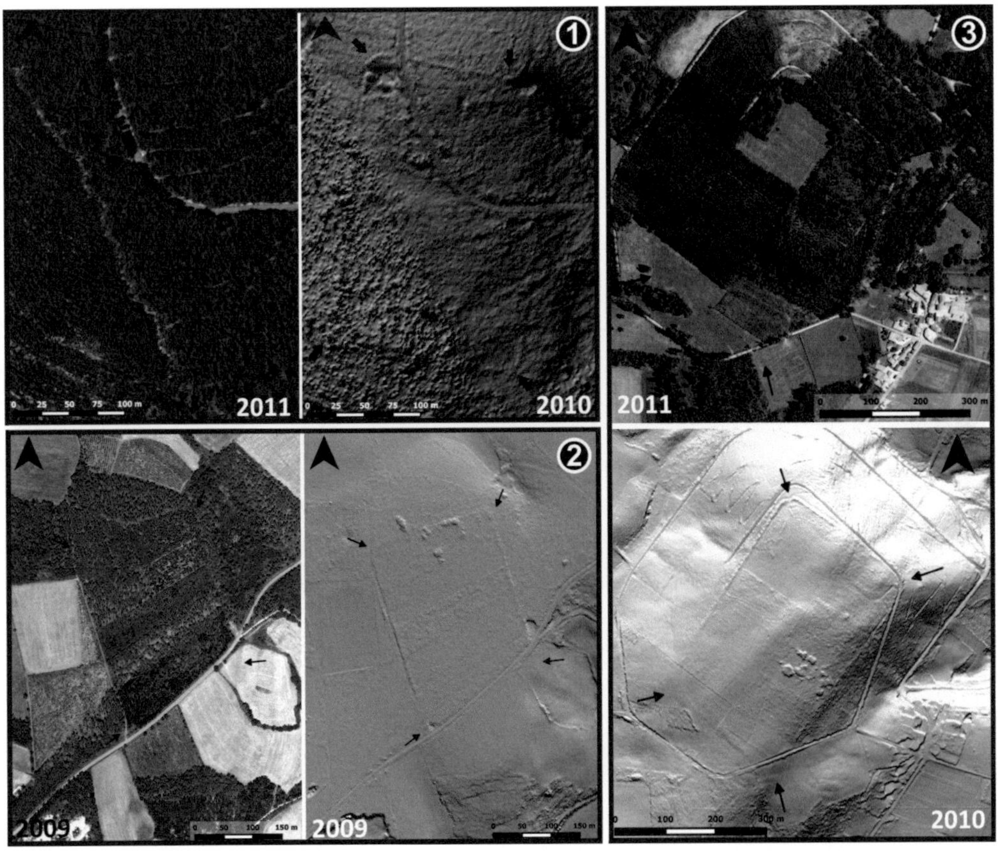

Figure 5: The sites of Valdemeda (PNOA aerial orthophoto and Resampling Filter Visualisation) (1), Monte dos Trollos (PNOA aerial orthophoto and Slope visualisation) (2) and O Cornado (PNOA aerial orthophoto and Multi-hillshade visualisation) (3).

The enclosures of Monte da Chá and Monte dos Trollos (Lugo, Spain) show a different degree of deterioration in the same geographical area (Costa-García *et al.* 2015b). Both Roman camps are perfectly distinguishable in the historical aerial photos of the 1950s. The former has been almost obliterated by reforestation and re-parcelling activities. Still, LiDAR visualisations show at least some additional traces which we could not observe on the field until 2014. Regarding the latter, its defences are surprisingly well preserved in the areas covered with trees, while the modern paths have practically wiped out the ancient trenches in the western sector (fig. 5). In this way, the enclosure detected in LiDAR visualisations shows almost the entire shape and limits documented in the aerial photography six decades ago. In this case, we can clearly appreciate the dissimilarities between a traditional reforestation plan and another one which introduces heavy machinery.

The situation of O Cornado (A Coruña, Spain) (Gago Mariño and Fernández Malde 2015) is quite different due to the smallholding: some parcels have been reforested, while others are ploughed (fig. 5). The LiDAR-based DTM outperform modern aerial photography and satellite imagery, and reveal how the constant ploughing have soften the relief of the ancient ramparts and trenches (Costa-García *et al.* 2015, 2015b). On the contrary, they still preserve a good size in several areas below the tree canopy.

LiDAR in upland areas

The upland regions have been generally less affected by anthropogenic activities than the plains or the valleys, so several of the best preserved Roman military camps are located in such areas. That is the case of enclosures like El Chao de Carrubeiro, Moyapán or El Mouru, among others (González Álvarez and Menéndez Blanco 2007; González Álvarez *et al.* 2011, 2012; Martín Hernández 2015; Menéndez Blanco *et al.* 2014). These sites were detected by using modern aerial photography, but the application of LiDAR technology has helped not only to better define their defences but also to detect new archaeological features previously unknown, such as *clauiculae*-type entrances or new ramparts (Costa-García 2015, Costa-García *et al.* 2015b) (fig. 6). Moreover, the systematic use of LiDAR-based DEMs in archaeological survey has allowed the discovery of new sites in areas where the review of aerial photography and satellite imagery was unsuccessful.[7]

But the archaeological structures of these regions are not free from human aggressions. The reforestation, the building of roads, the opening of firebreak or the installation of TV and radio masts are concentrated in the same mountain ranges where the Romans established their camps about two thousand years ago. The site of El Pico El Outeiro (Asturias, Spain) (Menéndez Blanco *et al.* 2014) perfectly exemplifies this situation and the traces of the ancient ramparts are only partially distinguishable in LiDAR visualisations (fig. 6). Fortunately, it was possible to

7 The discovery of sites such as A Penaparda (Lugo, Spain), El Xuegu la Bola (fig. 6), El Cueiru, El Chao or Resieḷḷa (Asturias, Spain) has been reported to the Regional Cultural Heritage authorities of those territories between the years 2015 and 2016. They are also included in some forthcoming papers.

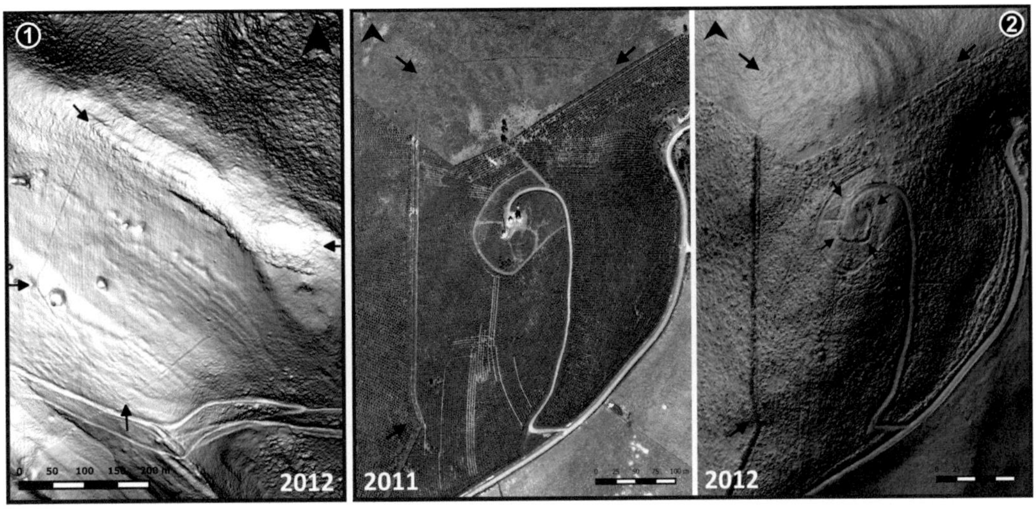

Figure 6: The enclosures of El Xuegu La Bola (Slope visualisation) (1) and El Pico El Outeiro (PNOA aerial orthophoto and Resampling Filter visualisation) (2).

detect a small *castellum* here thanks to the combined use of this technology and the USAF aerial photos of 1946 (Costa-García *et al.* 2015b).

Grazing activities can also make LiDAR-based remote sensing technology difficult in these areas. If in the lowlands the presence of "false friends" usually derives from the regular parcelling of modern times, the existence of livestock enclosures can usually lead to the incorrect identification of Roman military sites in mountainous landscapes. The grazing and opening of pastures continuously erodes the soil, destroying the remains of earth banks and trenches. Paradoxically, these activities allowed us to partially observe the defences of the Roman camp of Serra da Casiña (León, Spain) by using aerial photography (Menéndez Blanco *et al.* 2016). However, it was thanks to the LiDAR-based DTM that we could reconstruct the whole perimeter of the site and also attest its poor preservation. Other enclosures such as A Pena Dereta (Asturias, Spain) are in a very similar situation (Menéndez Blanco *et al.* 2014).

Conclusions

The reason why airborne laser scanning is being increasingly used in archaeological research is no longer a secret: it allows the generation of high-resolution topographic maps, even under a dense vegetation cover. Nevertheless, we must remember that airborne LiDAR has a specific temporal nature, since it documents the surface conditions at the time of data collection. In this way, its potential dramatically increases when combined with other geospatial datasets that can better contextualise ancient practices and post-depositional processes (Randall 2014). In addition, field validation of the digital data interpretation is also very important in order to reduce the risk of "false friends". Nonetheless, we also tried to demonstrate here that there are different variables (time of data collection, point density, classification of the point cloud, interpolation quality and DTMs spatial resolution, terrain type and vegetation cover and size and shape of the archaeological features) that could influence the quality of LiDAR data and the derived products. Therefore, this technique is not the definitive answer to every research question that deal with a Landscape Archaeology perspective. It is simply a new way to see and understand archaeological landscapes from topography: it enables the detailed recording and new interpretation of the archaeological topography (Opitz and Cowley 2013).

Regarding the detection and study of Roman military sites in Northwest Iberia, airborne LiDAR has proved to be a powerful archaeological tool. Is has clearly contributed to the exponential increasing of discoveries in the last years and its potential is far from being run out. Moreover, the announcement of new LiDAR flights in Spain to be commissioned by the IGN through the PNOA project looks very promising. Yet, several studies have demonstrated that this tool cannot be used as a standalone technique and let alone replace the field survey. This is a key principle that should be extended to every remote sensing technique in order to avoid future misinterpretations.

Acknowledgements

We would like to thank César Parcero Oubiña, Victorino Mayoral Herrera and Pastor Fábrega Álvarez for their kind invitation to participate in this publication and for all their contribution to the education of a new generation of archaeologists in the field of geographic information technologies. Also, to the other members of the RomanArmy.eu project for their efforts in the discovery of new Roman military sites in Northwest Iberia. To Ioana Oltean for the help with the English language revision and insightful comments. Finally, to the Incipit-CSIC and the research group Síncrisis of the University of Santiago de Compostela for their support during the completion of this work.

References

Almagro Basch, M. 1943. La colaboración de la aviación española en el campo de la arqueología. *Revista Ampurias* 5, 247-249.

Bennett, R., Welham, K., Hill, R. and Ford, A. 2012. A Comparison of Visualization Techniques for Models Created from Airborne Laser Scanned Data. *Archaeological Prospection* 19, 41-48.

Challis, K., Forlin, P. and Kincey, M. 2011. A Generic Toolkit for the Visualization of Archaeological Features on Airborne LiDAR Elevation Data. *Archaeological Prospection* 18, 279-289.

Bollandsås, O.M., Risbøl, O, Ene, L.T., Nesbakken, A., Gobakken, T. and Næsset, E. 2012. Using airborne small-footprint laser scanner data for detection of cultural remains in forests: an experimental study of the effects of pulse density and DTM smoothing. *Journal of Archaeological Science* 39, 2733-2743.

Celís Sánchez, J., Valderas, A. and Muñoz Villarejo, F.A. 2015. Localización de un nuevo conjunto de campamentos romanos (Castra Aestiva) en la Vía XVII, in: *V Jornadas de Jóvenes Investigadores del valle del Duero*. Abstract available at: http://arqueologiavalledelduero.jimdo.com/resumenes/romanizaci%C3%B3n/

Ceraudo, G. 2013. Aerial Photography in Archaeology, in: Corsi, C., Slapšak, B. and Vermeulen, F. (eds.), *Good Practice in Archaeological Diagnostics: Non-invasive Survey of Complex Archaeological Sites*. New York: Springer, 11-30.

Cerrillo-Cuenca, E. and Sanjosé, J. 2013. Mapping and interpreting vanished archaeological features using historical aerial photogrammes and digital photogrammetry, in: Contreras, F., Farjas, M. and Melero, F.J. (eds.), *Proceedings of the 38 the Annual Conference on Computer Applications and Quantitative Methods in Archaeology, CAA2010*. BAR International Series 2494, 43-46.

Charro Lobato, C. 2015. Historical aerial photographs to recover a lost landscape using digital photogrammetry: a case study of the Iron Age site of Cerro de la Mesa (Alcolea de Tajo, Toledo, central Spain), in: Ivanišević, V., Veljanovski, T., Cowley, D., Kiarszys, G. and Bugarski, I. (eds.), *Recovering lost landscapes*. Belgrade: Institute of Archaeology, 129-139.

Conrad, O., Bechtel, B., Bock, M., Dietrich, H., Fischer, E., Gerlitz, L., Wehberg, J., Wichmann, V., and Böhner, J. 2015. System for Automated Geoscientific Analyses (SAGA) v. 2.1.4. *Geoscientific Model Development* 8, 1991-2007.

Costa García, J.M. 2015. Asentamientos militares romanos en el norte peninsular: aportes de la fotografía aérea histórica, la fotografía satelital y el LiDAR aéreo. *Férvedes* 8, 35-44.

Costa García, J.M. 2016. Presencia militar romana en La Chana (Castrocalbón, León). *Nailos* 3, 47-85.

Costa García, J.M., Blanco Rotea, R., Gago Mariño, M. and Fonte, J. 2015a. Novedades sobre la presencia del ejército romano en el occidente galaico, in: Camino Mayor, J., Peralta Labrador, E. and Torres Martínez, J.F. (eds.), *Las Guerras Astur-Cántabras*. Gijón: KRK ediciones, 285-289.

Costa García, J.M. and Casal García, R. 2015. Fotografía aérea histórica, satelital moderna y LiDAR aéreo en algunos recintos militares romanos de Castilla y León. *Portugalia* 36, 143-145.

Costa García, J.M., Menéndez Blanco, A., González Álvarez, D., Gago Mariño, M., Fonte, J. and Blanco Rotea, R. 2015b. The presence of the Roman army in north-western Hispania. New archaeological data from the ancient Asturian and Galician territories, in: *XXIII Limes Congress 2015 (12/09/2015-23/09/2015)*. Inglostadt: German Limescommission (DLK)-Bavarian State Conservation Office (BLfD). Permanent link: https://youtu.be/e0FdUGQD0BM.

Crutchley, S. 2009. Ancient and modern: Combining different remote sensing techniques to interpret historic landscapes. *Journal of Cultural Heritage* 10, 65-71.

Del Olmo Martín, J. 1995. Arqueologi⊠a aérea en tres núcleos campamentales romanos de Zamora y León. *Brigecio* 4-5, 109-118.

Didierjean, F. 2008. Camps militaires romains et archéologie aérienne: méthodologie et données nouvelles. *Saldvie* 8, 95-115.

Doneus, M. 2013. Openness as Visualization Technique for Interpretative Mapping of Airborne Lidar Derived Digital Terrain Models. *Remote Sensing* 5, 6427-6442.

Doneus, M., Briese, C., Fera, M. and Janner, M. 2008. Archaeological prospection of forested areas using full-waveform airborne laser scanning. *Journal of Archaeological Science* 35, 882-893.

Doneus, M. and Briese, C. 2011. Airborne Laser Scanning in Forested Areas – Potential and Limitations of an Archaeological Prospection Technique, in: Cowley, D. (ed.), *Remote Sensing for Archaeological Heritage Management: Proceedings of the 11th EAC Heritage Management Symposium, Reykjavik, Iceland, 25-27 March 2010*. Budapest: Archaeolingua, 53-76.

Devereux, B., Amable, G. and Crow, P. 2008. Visualisation of LiDAR terrain models for archaeological feature detection. *Antiquity* 82, 470-479.

Fonte, J. and Costa García, J.M. 2016. Alto da Cerca (Valpaços, Portugal): um assentamento militar romano na Serra da Padrela e sua relação com o distrito mineiro de Tresminas. *Estudos do Quaternário* 15, 39-58.

Gago Mariño, M. and Fernández Malde, A. 2015. Un posible recinto campamental romano en O Cornado (Negreira, Galicia). *Nailos* 2, 229-251.

García Merino, C. 1996. Un nuevo campamento romano en la cuenca del Duero: El recinto campamental de Uxama (Soria). *Archivo Español de Arqueología* 69, 269-273.

González Álvarez, D. and Menéndez Blanco, A. 2007. Un nuevu emplazamientu militar romanu n'Asturias: el campamentu de Moyapán (Ayande). *Asturies: memoria encesa d'un país* 24: 16-21.

González Álvarez, D., Menéndez Blanco, A. and Álvarez Martínez, V. 2008. El campamento de Moyapán (Ayande, Asturias). *Férvedes* 5: 363-371.

González Álvarez, D., Menéndez Blanco, A., Álvarez Martínez, V. and Jiménez Chaparro, J. 2011. El Mouru y la presencia del ejército romano en La Mesa, in: Mañana, G. (ed.), *El Camín Real de La Mesa*. Oviedo: Cajastur, 160-168.

González Álvarez, D., Menéndez Blanco, A., Álvarez Martínez, V. and Jiménez Chaparro, J. 2012. Los campamentos romanos de El Mouru (Grau-Miranda, Asturias) en la vía de La Mesa. *Boletín del Seminario de Arte y Arqueología. Sección Arqueología*, 77-78, 245-267.

González Reguero, S. 2007. *La fotografía aérea en la arqueología española (1860-1960): 100 años de discurso arqueológico a través de la imagen*. Madrid: RAH-UAM.

Hanson, W. and Oltean, I. (eds.) 2013. *Archaeology from Historical Aerial and Satellite Archives*. New York: Springer.

Hesse, R. 2010. LiDAR-derived Local Relief Models – a new tool for archaeological prospection. *Archaeological Prospection* 17, 67-72.

Hierro Gárate, J. Á., Gutiérrez Cuenca, E. and Bolado del Castillo, R. 2015. Avances en la identificación de nuevos escenarios del Bellum Cantabricum, in: Camino Mayor, J., Peralta Labrador, E. and Torres Martínez, J. F. (eds.), *Las Guerras Astur-Cántabras*. Gijón: KRK Ediciones, 197-205.

Kokalj, Ž., Zakšek, K. and Oštir, K. 2011. Application of Sky-View Factor for the Visualization of Historic Landscape Features in Lidar-Derived Relief Models. *Antiquity* 85, 263-273.

Kokalj, Ž., K. Zakšek and K. Oštir. 2013. Visualizations of Lidar Derived Relief Models, in: Opitz, R. and Cowley, D. (eds.), *Interpreting archaeological topography: lasers, 3D data, observation, visualisation and applications*. Oxford: Oxbow, 100-114.

Loewinsohn, E. 1965. Una calzada y dos campamentos romanos del conuentus asturum. *Archivo Español de Arqueología* 38, 26-43.
Martín Hernández, E. 2015. El Mouro. Castrametación en la vía de la Mesa (Belmonte de Miranda/Grao, Asturias), in: Camino Mayor, J., Peralta Labrador, E. and Torres Martínez, J. F. (eds.), *Las Guerras Astur-Cántabras*. Gijón: KRK Ediciones, 239-247.
McCoy, M. and Ladefoged, T. 2009. New Developments in the use of Spatial Technology in Archaeology. *Journal of Archaeological Research* 17, 263-295.
Menéndez Blanco, A., González Álvarez, D., Jiménez Chaparro, J. and Álvarez Martínez, V. 2011a. Nuevas evidencias de la presencia militar romana en el extremo occidental de la Cordillera Cantábrica. *Gallaecia* 30, 145-165.
Menéndez Blanco, A., González Álvarez, D., Jiménez Chaparro, J. and Álvarez Martínez, V. 2011b. Un nuevo campamento militar romano en El Páramo leonés: Huerga de Frailes. *Argutorio* 26, 32-35.
Menéndez Blanco, A., González Álvarez, D., Álvarez Martínez, V. and Jiménez Chaparro, J. 2013. Propuestas de prospección de bajo coste para la detección de campamentos romanos de campaña. El área occidental de la Cordillera Cantábrica como caso de estudio. *Munibe (Antropologia-Arkeologia)* 64, 175-197.
Menéndez Blanco, A., González Álvarez, D., Álvarez Martínez, V. and Jiménez Chaparro, J. 2014. Campamentos romanos de campaña en el occidente de Asturias, in: *Excavaciones arqueológicas en Asturias 2007-2012. En el centenario del descubrimiento de la caverna de La Peña de Candamo*. Oviedo: Principado de Asturias, 245-251.
Menéndez Blanco, A., González Álvarez, D., Álvarez Martínez, V. and Jiménez Chaparro, J. 2015. La sierra de Penouta y El Cordal d'Ouroso: una línea de avance del ejército romano en el occidente cantábrico, in: Camino Mayor, J., Peralta Labrador, E. and Torres Martínez, J.F. (eds.), *Las Guerras Astur-Cántabras*. Gijón: KRK ediciones, 261-267.
Menéndez Blanco, A., González Álvarez, D. and Costa García, J. M. 2016. A Serra da Casiña (Valboa, León): un campamento romano en las montañas bercianas. *Revista Arkeogazte* 5, 239-251.
Menéndez Blanco, A., González Álvarez, D., Costa García, J. M., Fonte, J., Gago Mariño, M., Álvarez Martínez, V. 2016b. Tras las huellas del ejército romano: una propuesta metodológica para la detección de asentamientos militares romanos en el noroeste peninsular, in: Rosas, L., Sousa, A.C., Barreira, H. (eds.), *Genius loci. Lugares e significados. Breves reflexões*. Porto: Universidade do Porto, 79-81.
Mlekuž, D. 2013a. Messy landscapes: lidar and practices of landscaping, in: Opitz, R. and Cowley, D. (eds.), *Interpreting archaeological topography: lasers, 3D data, observation, visualisation and applications*. Oxford: Oxbow, 102-116.
Mlekuž, D. 2013b. Skin Deep: LiDAR and Good Practice of Landscape Archaeology, in: Corsi, C., Slapšak, B. and Vermeulen, F. (eds.), *Good Practice in Archaeological Diagnostics: Non-invasive Survey of Complex Archaeological Sites*. New York: Springer, 113-129.

Moreno Gallo, I. 2011. Identificación y descripción de la vía de Astorga a Braga por Chaves. De Asturica a Veniata, in: *Vías romanas en Castilla y León*. Valladolid: Junta de Castilla y León, 2-38.

Opitz, R. and Cowley, D. 2013 (eds.). *Interpreting Archaeological Topography: 3D Data, Visualisation and Observation*. Oxford: Oxbow Books.

Orejas, A.; Sánchez Palencia, F. J., Beltrán, A.; Ron, J. A.; López, L. F., Currás Refojos, B.X., Romero, D., Zubiaurre, E., Pecharromán, J. L. and Arboledas, L. 2015. Conquista, articulación del territorio y explotación de recursos en el límite entre el convento lucense y el de los ástures (Proyecto IVGA), in: Camino Mayor, J., Peralta Labrador, E. and Torres Martínez, J F. (eds.), *Las Guerras Astur-Cántabras*. Gijón: KRK Ediciones, 247-260.

Peralta Labrador, E. 2002. Los campamentos romanos de campaña (castra aestiva): evidencias científicas y carencias académicas. *Nivel Cero* 10, 49-87.

Peralta Labrador, E. 2011. Campamentos romanos en Cantabria. *Castillos de España* 161-3, 23-26.

Pérez Álvarez, J.A., Bascón Arroyo, F.M. and Charro Lobato, M.C. 2014. Photogrammetric Usage of 1956-57 Usaf Aerial Photography of Spain. *The Photogrammetric Record* 29, 108-124.

Pérez Álvarez, J.A., Bascón Arroyo, F.M., Crespo Pérez, F.J. and Charro Lobato, M.C. 2013. Project Casey Jones, 1945-46: el vuelo histórico «fotogramétrico» de la serie A en España y sus aplicaciones cartográficas. *Revista Mapping* 22, 14-24.

Randall, A. 2014. LiDAR-aided reconnaissance and reconstruction of lost landscapes: An example of freshwater shell mounds (ca. 7500-500 cal b.p.) in northeastern Florida. *Journal of Field Archaeology* 39, 162-179.

Redweik, P., Roque, D., Marques, A., Matildes, R. and Marques, F. 2010. Triangulating the Past: Recovering Portugal's Aerial Images Repository. *Photogrammetric Engineering & Remote Sensing* 76, 1007-1018.

Risbøl, O., Bollandsås, O.M., Nesbakken, A., Ørka, H.O., Næsset. E. and Gobakken, T. 2013. Interpreting cultural remains in airborne laser scanning generated digital terrain models: effects of size and shape on detection success rates. *Journal of Archaeological Science* 40, 4688-4700.

Risbøl, O., Briese, C, Doneus, M. and Nesbakken, A. 2015. Monitoring cultural heritage by comparing DEMs derived from historical aerial photographs and airborne laser scanning. *Journal of Cultural Heritage*, 16, 202-209.

Sánchez Palencia, F.J. 1986. El campamento romano de Valdemeda, Manzaneda (León). *Numantia* 2, 227-234.

Sánchez Palencia, F.J. and Currás Refojos, B.X. 2015. Campamentos romanos en zonas mineras del cuadrante noroeste de la Península Ibérica, in: Camino Mayor, J., Peralta Labrador, E. and Torres Martínez, J.F. (eds.), *Las Guerras Astur-Cántabras*. Gijón: KRK Ediciones, 273-284.

Štular, B., Kokalj, Ž., Oštir, K. and Nuninger, L. 2012. Visualization of lidar-derived relief models for detection of archaeological features. *Journal of Archaeological Science* 39, 3354-3360.

Verhoeven, G., Doneus, M., Briese, C. and Vermeulen, F. 2012. Mapping by matching: a computer vision-based approach to fast and accurate georeferencing of archaeological aerial photographs. *Journal of Archaeological Science* 39, 2060-2070.

Zakšek, K., Oštir, K., Kokalj, Ž. 2011. Sky-View Factor as a Relief Visualization Technique. Remote Sensing 3: 398-415.

Notes on contributors

José Manuel Costa-García has a PhD in Archaeology (2013) and BA in History (2006) with an honours degree from the University of Santiago de Compostela. He obtained an Advanced Studies Diploma in Archaeology from the same university (2008). He also got a FPU predoctoral grant (2008-12) and a postdoctoral grant of the Galician Regional Government (2016-2019). Active member of the archaeological research team in the Roman fort of A Cidadela (Sobrado dos Monxes, A Coruña) (2007-2010 and 2016). His specialization area is the study the Roman military presence by combining specific methodologies coming from archaeology, ancient history, epigraphy and new technologies.

João Fonte has a PhD in Archaeology from the University of Santiago de Compostela (2015), having enjoyed a doctoral fellowship from the Portuguese Foundation for Science and Technology (FCT). Advanced Studies Diploma in Archaeology from the University of Santiago de Compostela (2009), with recognition to the master degree by the Faculty of Arts, University of Coimbra (2010). Master in Geographic Information Systems from the Faculty of Arts, University Porto (2009). BA in History, specialization Archaeology from the University of Minho (2006). Now he holds a postdoctoral grant from the Galician Regional Government (2016-2019). Areas of expertise: Landscape Archaeology, geospatial technologies and Late Iron Age-Roman transition in Northwest Iberia.

Making Visible the Invisible: Low Cost Methodologies for the Study of Ancient Carvings

Miguel Carrero-Pazos,[1] Benito Vilas-Estévez[2]
and Alia Vázquez-Martínez[1]

Introduction

3D modelling is now considered to be one of the main trends in current Archaeological Science. Indeed, it is currently one of the most prolific research fields in the discipline. The use of 3D models in archaeology has proliferated not only for recording purposes in excavations but also for use in the development of numerous techniques to implement a better visualisation of 3D archaeological models via many different approaches (Chandler *et al.* 2005; Chandler *et al.* 2007; El-Hakim *et al.* 2004; Duffy 2010; Díaz Guardamino and Wheatley 2013; among others).

In Galicia (located in the northwest of the Iberian Peninsula), the first use of dense photogrammetry and digital image processing took place at the University of Santiago de Compostela in the 1990s, with the study of Bronze Age rock art (Rodríguez Casal *et al.* 1995; García Calviño *et al.* 1995). Unfortunately, these early efforts were not continued and the results of the research remained unpublished. It was not until the first half of this century that the use of laser scanners and photogrammetric techniques started to become widespread in the research community (Mañana Borrazás *et al.* 2009; Ortiz Sanz *et al.* 2010; Gil Docampo *et al.* 2011; Pereira Rodríguez 2012, Vázquez Martínez *et al.* 2015; Vilas Estévez *et al.* 2015, 2016). It was at this time that enthusiasts of these techniques began

1 GEPN-AAT-University of Santiago de Compostela, Faculty of Geography and History. Praza da Universidade, s/n 15703 Santiago de Compostela (Spain) miguel.carrero.pazos@gmail.com; alia.vazquez.mtnez@gmail.com.

2 Department of Cultural Astronomy and Astrology, University of Wales Trinity Saint David vieito4@hotmail.com.

to emerge, as occurred in the case of Bronze Age rock art.[3] In this particular case, both researchers and social organisations clearly pointed out that the archaeology of Galicia was in desperate need of the design of a 3D model-based approach. On the one hand, this would improve recording and study processes in multiple areas of knowledge, such as the study of inscriptions or rock art, while on the other hand, it would contribute towards the dissemination of results among the general public.[4] In this context, the work presented here aims to provide a clear answer to this current trend. We aim to study particular inscriptions in more detail and to obtain a tracing of the study cases.

Previous work

Depicting shape and detail in 3D models

One of the most relevant branches of 3D modelling has been the use of shading to represent shape and detail in 3D models. The first approaches were developed in the late 1970s (Docarmo 1976), starting from technical illustration and continuing with digital cartography (Horn 1981; Imhof 1982), which led to a new research line in computer graphics known as Non-Photorealistic Rendering (NPR), an area which would expand greatly in the 1990s (Gooch 1998; Gooch and Gooch *et al.* 1998; Gooch *et al.* 1999; Gooch and Gooch 1999).

Playing with artificial light in order to generate shadows was one of the first methodologies developed. Miller's work (1994) may be considered as a starting point, with the creation of algorithms to create local and global shadings, a technique which is known as accessibility shading. Ten years later, Pharr and Green (2004) presented the ambient occlusion technique (based on Landis 2002), which produces images in which the concavities are represented by toning down the most irregular parts of the surface. After that, many rendering techniques relied on normal highlighting (Cigoni *et al.* 2005), in which shading can be managed in a single layer in order to exaggerate or lessen the surface details. Algorithms of line reflectance and line drawings have also been developed (Decarlo *et al.* 2003), along with contours (Decarlo *et al.* 2004; Rusinkiewicz 2004) or "suggested contours", which highlight the shape of an object more precisely and consistently (Burns *et al.* 2005; Decarlo and Rusinkiewicz 2007).

Furthermore, attention must be paid to the development of certain approaches involving image-based relighting, such as Polynomial Texture Mapping (PTM) (Malzbender *et al.* 2000, 2001), Reflection Transformation Imaging (RTI)[5] (Malzbender *et al.* 2006), or its virtual counterpart RTI, which combines the

---

3 See for example the work of the "Colectivo A Rula": https://colectivoarula.wordpress.com/.

4 See for example the commendable 3D repository of the Museo de Pontevedra, created with Sketchfab (https://sketchfab.com/museodepontevedra, 3D models made by E. Martínez Soto). These kinds of efforts clearly advance the dissemination of cultural heritage. However, we must point out that they are closed platforms in the sense that the 3D model is not available for download or study. On the other hand, it must be noted that poor control is sometimes exerted over the photogrammetric process, which on some occasions makes them unusable for the purpose of studying, as we have pointed out before (Vázquez Martínez *et al.* 2016).

5 As a guide, tutorials and software for the acquisition and processing of RTI images can be consulted at http://culturalheritageimaging.org/.

processing techniques of reflection with photogrammetry and non-intrusive digitizing. This has generated an advanced level of interaction with 3D models, thereby enhancing the topographical surface (Earl *et al.* 2010).

Mention should also be made of the Morphological Residual Model (MRM), a methodology recently created by H. Pires which allows for a better visualisation of the micro-surface details of a 3D model, thereby enhancing it (Caninas *et al.* 2011; Pires *et al.* 2014; Correia Santos *et al.* 2015; Pires *et al.* 2015). This technique is based on a methodology defined in MeshLab, in which a decimation algorithm is used to produce a coarse mesh similar to the original model but free from small morphological details, thus improving the effects of the smoothing filters which are normally used to calculate the trend surface. The algorithm tends to imitate the effects of the cartographic representation scale, in which morphological details are progressively excluded from the scene as the observer moves away from it (or the opposite) (Pires *et al.* 2015). This kind of methodological approach is also present in other works, such as Ortiz *et al.* (2010) and in the creation of *deviation maps*, which were recently presented by L. Lescop and S. Cassen (2013; Cassen *et al.* 2014, 2015).

One final approach which should be mentioned, originating from a different perspective, is the application of LiDAR visualisation techniques developed in landscape studies. These are applied to the study of small-sized archaeological remains, such as inscriptions or Bronze Age rock art panels. It is in this context that we have developed a novel methodology known as AsTrend (Carrero-Pazos *et al.* 2016). This methodology is based on the combined application of the accessibility shading technique and a trend removal technique over the studied image (Miller 1994; Štular *et al.* 2012).

3D modelling of ancient inscriptions

This paper aims to present a 3D approach to depicting shape and detail in 3D archaeological models of various inscriptions. The two case studies proposed here have been chosen from the collections of the Museo do Pobo Galego and the church of Santiago de San Domingos de Bonaval, both located in Santiago de Compostela (Spain). Permission was granted for the study of these inscriptions by the Museo do Pobo Galego and they were chosen in order to show the possibilities of the post-processing techniques explained below.

The first case is a Roman votive inscription found in Negreira (Santa Eulalia de Logrosa, A Coruña, Spain), which has the following transcription (CIL II 5641= CIRG I, 24):

Iovi / Op(timo) Ma(ximo) / Ma(- – -)

The inscription, carved in granite, is divided into three parts via three horizontal lines. The upper part, which is badly preserved, has a circular focus above and three arcs on the front. Its back is not carved, which is surely a result of it being placed on a wall (Pereira Menaut 1991, 77).

The second example is a floor plaque found in the church of San Domingos de Bonaval, with no transcription available. It was also made in granite (Figure 1).

Figure 1: Orthoimages of both inscriptions.

The methodology used for the creation of the 3D models is dense photogrammetry produced by Structure from Motion. This is a technique based on a passive sensor, which searches the acquisition, treatment and processing of at least two pictures taken in 2D of the same scene in order to produce a three dimensional model (Remondino 2014). The camera we used was a Canon EOS 70D, with a focal length of 18 mm., which allowed us to obtain images at a resolution of 5472 x 3648 (20.0 MP, 3:2) in JPEG format.

Dense photogrammetry allows for the derivation of accurate metric and semantic information from photographs taking into account the calibration of the camera and the orientation of the image. The objective is to create a 3D point cloud which can be converted into a mesh, upon which it is possible to add a photorealistic texture. It is also possible to reconstruct a Digital Elevation Model of the object (Figure 2).

The workflow begins by deriving 3D metric information from a series of images, taking into account the camera calibration and orientation of the images in order to create a 3D point cloud, which is structured and shaped. In the final step, a photorealistic texture can be added to the 3D model for a better view. The main

Figure 2: Digital Elevation Models of the reconstructed inscriptions.

difference between photogrammetry and active range sensors is the 3D point cloud derivation: while range sensors deliver the points directly, in photogrammetry, mathematical processing of the image data is needed in order to derive a sparse or dense cloud (Remondino 2014, 65).

Study Case Information	Votive inscription	Floor plaque
Number of Images	20	20
Resolution	0.00193132 m/pix	0.00113373 m/pix
Tie-points	113854	165881
Projections	308158	447935
Error	0.362572 pix	0.320183 pix
Image overlap	> 9	> 9
Point density	67024.5 points per sq m.	777998 points per sq m.

Table 1: Technical information involved in the photogrammetric reconstruction of both studied cases.

The basic data concerning the acquisition process are summarised in Table 1.[6]

The application of three dimensional models in archaeology has allowed for a widening of the scope of the geometric survey process, providing high resolution models which can be combined with different resources and historical information. From this point of view, the use of 3D models sharply increases the objectivity and archaeological knowledge of the elements involved, compared to the regular two-dimensional pictures or hand drawn models (Guidi *et al.* 2014, 55; Green *et al.* 2014). In this study, we have used Agisoft Photoscan, [7] specialised software for this purpose.[8] Once the images have been uploaded, the software performs a first step to alignment them, using a Scale Invariant Feature Transform (SIFT) algorithm. This work is carried out automatically by the program and includes the calibration and correction of all the images based on their EXIF data (Exchangeable Image File Format, the metadata associated with any digital photography). The main point of the procedure is that it treats the pixels of the images as points, searching for common points among all the photographs and thereby calculating the relative positions of each camera (Pereira Rodríguez 2012, 104). In this process, two kinds of point clouds (basic and dense) can be obtained. Our aim was to carefully document a stone surface which potentially had certain reliefs and engravings, for this purpose the use of the dense point cloud is more relevant. The next step involved the creation of a polygonal model from the point cloud, commonly known as a mesh. In Agisoft Photoscan there is no explicit information about the reconstruction algorithm used, although some authors have suggested that it could be some kind of Poisson reconstruction. The mesh creation is the penultimate step prior to the texturing of the model, which has the aim of providing the 3D model with a hyper-realistic effect (Remondino 2014, 69).

Beyond the bare 3D model: highlighting the details of the carved surfaces

Radiance Scaling

Thanks to the development of Open Source 3D software such as MeshLab (Cigoni *et al.* 2008), the creation of different rendering plugins for enhancing the morphological features of 3D models, such as shaders, has proliferated. One of the most interesting is Radiance Scaling (Vergne *et al.* 2010), which depicts the shape of a surface through shading. The application of these kinds of techniques is nothing new, but it was not until the last decade that they became popular for multiple purposes. In Galicia, for example, the use of a Radiance Scaling shader was recently introduced, and has become popular among researchers, for the study of Bronze Age rock art (Vilas Estévez *et al.* 2015).

---

6 Data extracted from the Agisoft Photoscan report.

7 http://www.agisoft.com/

8 Other choices could include: visual SFM (http://ccwu.me/vsfm/); PhotoSynth Toolkit (http://www.visual-experiments.com/demos/photosynthtoolkit/); Python Photogrammetric Toolbox (http://www.arc-team.com/); PMVS2 (http://www.di.ens.fr/pmvs/).

Radiance Scaling adjusts reflected light intensities in a way which is dependent on both surface curvature and material characteristics. As a result, diffuse shading or highlighting variations are correlated with surface feature variations, thus enhancing surface concavities and convexities instantaneously (Vergne *et al.* 2010, 1). This technique was used for the first time in 2012 on epigraphs in order to facilitate the reading of 3D models (Granier *et al.* 2012). Thanks to its implementation in MeshLab (Cigoni *et al.* 2005) as a shader, a plugin to create an isosurface perturbed by a noisy isosurface, its use is now widespread. This is due to the broad range of possibilities for registering and documenting rock art carvings, as has repeatedly been demonstrated in Galicia (Vilas Estévez *et al.* 2015; 2016) and in Italy (Medici and Rossi, 2015), among other places. As has been proposed in another paper (Vilas Estévez *et al.* 2016), Radiance Scaling has many different options, although, in our opinion, the most useful one is the use of the "Lit Sphere Radiance Scaling" option, wherein a lit sphere encodes the lighting environment and reflective properties into an image of a sphere, while a different sphere can be used for convex and for concave regions (Granier *et al.* 2012) (Figure 3).

Figure 3: Double-sphered Radiance Scaling applied to both study cases in grey tones.

One of the most remarkable aspects of this technique is that it enables an instant and objective visual approach to the 3D model, in which the concavities and convexities are clearly highlighted. However, one of the problems is that, on certain occasions, some carvings are still diffuse and noise is almost always present.

The lower part of the left inscription has a contraction of M – A. This line had previously been detected by Pereira Menaut (1991, 77) (CIRG I, 24), although it had, surprisingly, not been picked up on in previous studies, such as those of Chamoso Lamas (1955, 220) or Fita (1911, 404). The reason for this could be the fact that the carvings on line 3 are more diffuse than the others.

Implementing Radiance Scaling: the use of Ambient Occlusion

Radiance Scaling is clearly one of the most interesting methods available for enhancing the details in a 3D model. However, as has been mentioned previously, in some cases the surface carved is diffuse and does not produce a proper depiction of the details. On the other hand, by definition, a shader is computed by the graphics card of the computer, which means that it cannot be stored as a filter and can only

Figure 4: Ambient occlusion with Radiance Scaling shader, in B/W.

be used for visualisation purposes. Even so, more can still be implemented in the workflow of Radiance Scaling. In this context, we propose the initial application of the ambient occlusion technique to the 3D model in order to help visualise the geometric features of the inscriptions.[9] Ambient occlusion gives a measure of the occlusion of each part of the object, showing the parts which are more self-occluded in a darker tone and those which are more exposed in a lighter tone. In the second step of the process, Lambertian Radiance Scaling can be applied in order to re-enhance the carvings (figure 4).

Ambient occlusion is calculated in MeshLab as colour information, which means it can be mapped and preserved on the 3D model, not only with black and white values. This is one of the main improvements of the approach described as all the changes can be stored in the model. In this way, the 3D model will preserve the information of the colour per vertex, and different palettes can be played with, as is shown in figure 5.

Figure 5: Ambient occlusion with Radiance Scaling shader, in colour.

9 A tutorial of the technical procedure is described in https://www.youtube.com/watch?v=wiNzvlSGLAI

Generating a virtual tracing

Even though some of these figures could already be used as tracings due to their clear ability to show the details of the surface of the 3D model, it is of interest here to describe the methodology which we used to create a virtual tracing. This technique has previously been used elsewhere (Carrero-Pazos *et al.* 2016) in order to simulate a traditional calque.

This method is a response to one of the main challenges in digital rock art techniques today. Although a lot has been achieved in terms of the development of enhanced visualisations of depressions in 3D model surfaces, the real goal is to create methodologies which contribute towards identifying and distinguishing man-made artworks from natural features on stone surfaces. Some efforts have been made in this area to propose automatic vectorisations of 3D enhanced images, such as the Canny edge detector (Canny 1986). However, most authors claim that the final result must always be achieved with multiple methods (a comparison between ranking light-based traditional methods and 3D visualisations) in order to obtain a final single recorded product (Cassen *et al.* 2014).

We propose the use of Adobe Photoshop software and the dodge and burn tool in order to lighten or darken particular areas of the image. The methodological procedure consists of decreasing the light to lighten an area of the image (dodging) or increasing the exposure to darken areas of the image (burning) (figure 6).

These tools are based on a traditional darkroom technique for regulating the exposure in specific areas of an image. The final results can be seen in figure 7.

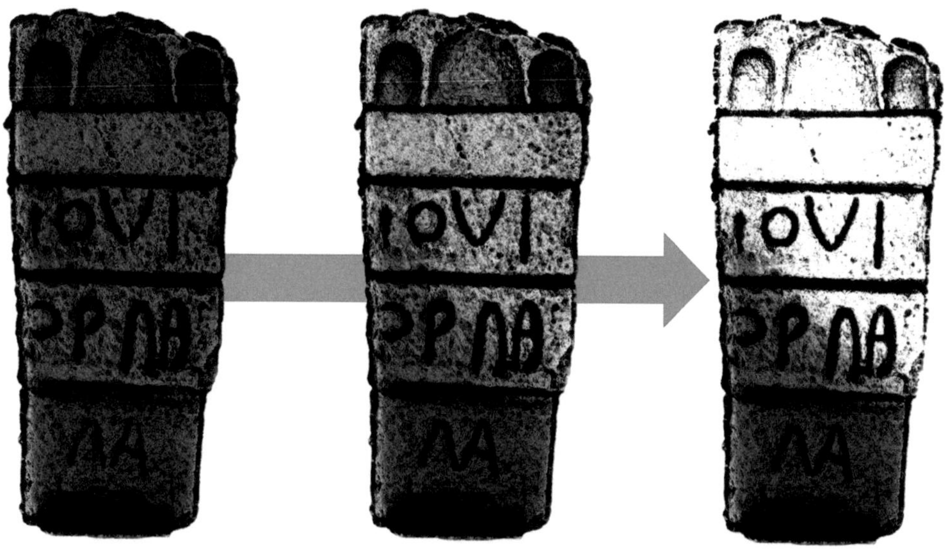

Figure 6: Steps of the methodological procedure applied to the inscription of Santa Eulalia de Logrosa.

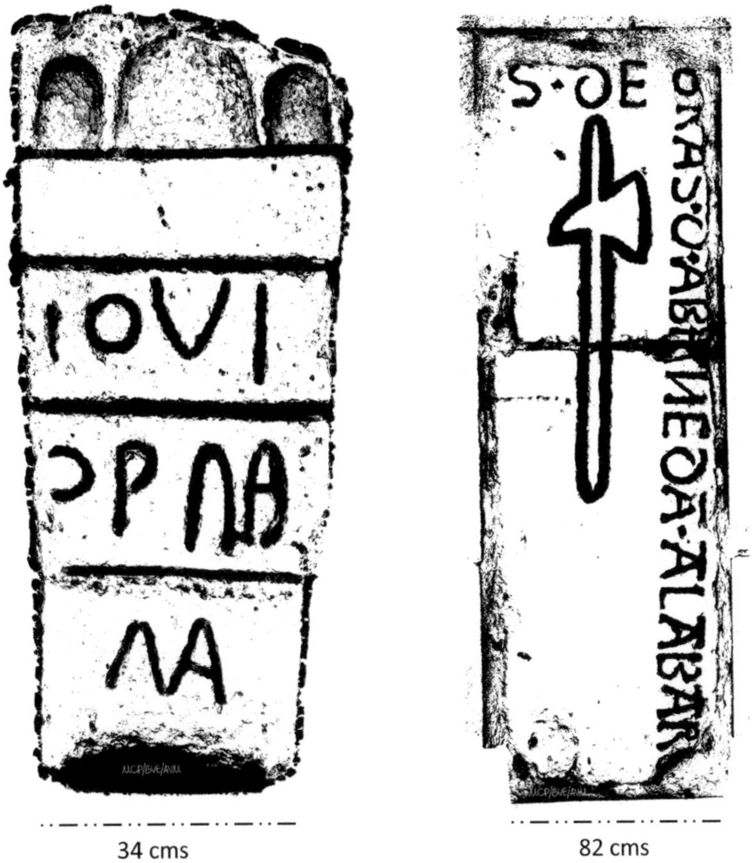

Figure 7: Virtual tracings of both inscriptions.

Conclusions

To a certain extent, our work has revealed that there is no single technique which allows for the creation of a definitive virtual tracing of a 3D carved surface. Rather, the results of many techniques must be compared, as has been pointed out by Cassen *et al.* (2014). Different techniques may be applied, all of them leading to slightly different results. Furthermore, the use of Radiance Scaling contributes towards a reduction in the subjectivity present in traditional methods, which were mainly centred on hand-drawn tracing through an oblique light. We have already seen how other methods can substantially improve the results of this shader and, working together with image analysis techniques, we are able to offer a definitive tracing of the studied inscriptions.

The use of Radiance Scaling in the study of inscriptions is less time-consuming than other methods, such as those explained in the introduction. Finally, it clearly reduces the cost of research due to the fact that MeshLab is an open source software.

Acknowledgements

We are grateful to Victorino Mayoral, César Parcero and Pastor Fábrega for allowing us to participate in this event and to publish the results of our research. The authors would also like to thank the Museo do Pobo Galego for allowing them to study the inscriptions mentioned here.

The research leading to these results has received funding from the European Commission Seventh Framework Programme (under grant agreement FP7-INFRASTRUCTURES-2012-1-313193 (ARIADNE)) through the participation of M. Carrero-Pazos in the TNA 2D/3D Documentation for Archaeology (ISTI-CNR, Visual Computing Lab, Pisa, 2016). We would also like to extend our appreciation to Mateo Dellepiane, Marco Callieri and Roberto Scopigno for their advice, especially concerning the MeshLab processing stages. Some of the material used here comes from the TNA summer school. Of course, any mistakes are our own.

References

Burns, M., Klawe, J., Rusinkiewicz, S., Finkelstein, A., Decarlo, D. 2005. Line drawings from volume data. *ACM Transactions on Graphics (Proceedings of ACM SIGGRAPH 2005)* 24(3), 512-518.

Caninas, J. C., Henriques, F., Batista, Á., Pires, H. 2011. Pedra das Cruzinhas: notícia de um monólito gravado na fronteira entre os concelhos do Sabugal e da Guarda. *Sabucale,* 3, 35-56.

Canny, J. 1986. A computational approach to edge detection. *IEEE Transactions on pattern analysis and machine intelligence,* PAMI 8, NO. 6, 679-698.

Carrero-Pazos, M., Vázquez-Martínez, A., Vilas-Estévez, B. 2016. AsTrend: Towards a new method for the study of ancient carvings. *Journal of Archaeological Science Reports,* Journal of Archaeological Science: Reports, 9, 105-119.

Cassen, S., Lescop, L., Grimaud, V., Robin, G. 2014. Complementarity of acquisition techniques for the documentation of Neolithic engravings: lasergrammetric and photographic recording in Gavrinis passage tomb (Brittany, France). *Journal of Archaeological Science* 45, 126-140.

Chamoso Lamas, M. 1955. *NAH,* 2 (1953), 220-221.

Chandler, J. H., Fryer, J. G., Kniest, H. T. 2005. Non-invasive three-dimensional recording of Aboriginal rock art using cost-effective digital photogrammetry. *Rock Art Research* 22, 119-130.

Chandler, J. H., Bryan, P., Fryer, J. G. 2007. The Development And Application Of A Simple Methodology For Recording Rock Art Using Consumer-Grade Digital Cameras. *The Photogrammetric Record,* 22 (117), 10-21.

Cigoni, P., Scopigno, R., Tarini, M. 2005. A simple normal enhancement technique for interactive non-photorealistic renderings. *Computer & Graphics* 29, 125-133.

Cigoni, P., Callieri, M., Corsini, M., Dellepiane, M., Ganovelli, F., Ranzuglia, G. 2008. MeshLab: an Open-Source Mesh Processing Tool. *Sixth Eurographics Italian Chapter Conference,* 129-136.

Correia Santos, M. J., Sousa, O., Pires, H., Fonte, J., Gonçalves-Seco, L. 2015. Travelling back in Time to Recapture Old Texts. The use of Morphological Residual Model (M.R.M.) for epigraphic reading: four case studies (CIL 02, 02395a, CIL 02, 02395c, CIL 02, 02476, CIL 02, 05607). *Acts of Information Technologies for Epigraphy and Cultural Heritage. Proceedings of the first EAGLE International Conference, Europeana Eagle project. Studi umanistici- Antichistica, Sapienza Universitá Editrice*, 437-450.

Decarlo, D., Finkelstein, A., Rusinkiewicz, S., Santella, A. 2003. Suggestive Contours for Conveying Shape. *ACM Transactions on Graphics (Proc. SIGGRAPH 2003)*, 22 (3), 848-855.

Decarlo, D., Finkelstein, A., Rusinkiewicz, S. 2004. Interactive Rendering of Suggestive Contours with Temporal Coherence. *NPAR '04 Proceedings of the third international symposium on Non-photorealistic animation and rendering*, 15-145.

Decarlo, D., Rusinkiewicz, S. 2007. Highlight Lines for Conveying Shape. *Proceedings of the fifth International Symposium on Non-Photorealistic Animation and Rendering (NPAR)*, 63-70.

Díaz Guardamino, M., Wheatley, D. 2013. Arte rupestre y tecnologías digitales: la aplicación de Reflectance Transformation Imaging (RTI) y escaneado láser 3D al estudio de las estelas del Bronce Final de la Península Ibérica. *Menga. Revista de Prehistoria de Andaluicía* 4, 187-203.

Docarmo, M. P. 1976. *Differential geometry of curves and surfaces,* New Yersey: Prentice-Hall.

Duffy, S. 2010. Polynomial texture mapping at Roughting Linn rock art site. *International Archives of Photogrammetry, Remote Sensing and Spatial Information Sciences* 38(5), 213-217.

Earl, G., Beale, G., Martínez, K., Pagi, H. 2010. Polynomial Texture Mapping and Related Imaging Technologies for the Recording, Analysis and Presentation of Archaeological Materials. *International Society for Photogrammetry and Remote Sensing,* XXXVIII (part 5), 218-223.

Fita, F. 1911. Nuevas lápidas romanas de Noya, Cando, Cerezo y Jumilla. *BRAH* 59, 398-417.

García Calviño, F. J., Casas, A., Gómez González, L., Rodríguez Casal, A. A. 1995. Un sistema de fotogrametría digital en coordenadas homogéneas. Aplicación al arte rupestre. *XXV Reunión bienal de la Real Sociedad Española de Física (18-23 de septiembre de 1995. Santiago de Compostela)*, 657-658.

Gil Docampo, M. L., Ortiz Sanz, J., Martínez Rodríguez, S., Vázquez, B., Rego Sanmartín, T. 2011. The detection of petroglyphs through digital image processing. The particular case of the stone inside the chapel of Saint Bartolomé (Lugo-Spain). *Proceedings of the IMProVe 2011, International Conference on Innovative Methods in Product Design (Venice, Italy)*, 363 – 367.

Gooch, A. A. 1998. *Interactive Non-Photorealistic Technical Illustration,* Utah, Master's Thesis. Department of Computer Science. University of Utah.

Gooch, A. A., Gooch, B., Shirley, P., Cohen, E. 1998. A non-photorealistic lighting model for automatic technical illustration. *Proceedings of the 21st Annual Conference on Computer Graphics and Interactive Techniques (SIGGRAPH 1994, Orlando). ACM*, 447-452.

Gooch, A. A. and Gooch, B. 1999. Using Non-Photorealistic rendering to communicate shape. *In:* Green, S. E. (ed.) *SIGGRAPH'99 Course Notes on Non-Photorealistic Rendering.* New York: ACM SIGGRAPH.

Gooch, B., Sloan, P. -P. J., Gooch, A., Shirley, P., Riesenfeld, R. 1999. Interactive technical illustration. *I3D-ACM*, 31-38.

Granier, X., Vergne, R., Pacanowsky, R., Barla, P., Reuter, P. 2012. Enhancing Surface features with the radiance scaling MeshLab plugin, in: Earle, G., Sly, T., Chrysanthi, P., Murrieta-Flores, P., Papadopoulos, C., Romanowska, I., Wheatley, D. (ed.) *Archaeology in the Digital Era: Papers from the 40th Annual Conference of Computer Applications and Quantitative Methods in Archaeology (CAA).* Amsterdam: Amsterdam University Press.

Green, S., Bevan, A., Shapland, M. 2014. A comparative assessment of structure from motion methods for archaeological research. *Journal of Archaeological Science* 46, 173-181.

Guennebaud, G., Gross, M. 2007. Algebraic Point Set Surfaces. *Siggraph 2007 San Diego, USA,* Retrieved 15/06/2016, from http://www.labri.fr/perso/guenneba/docs/APSS_sig07.pdf.

Guennebaud, G., Germann, M., Gross, M. 2008. Dynamic Sampling and Rendering of Algebraic point Set Surfaces, in: Drettakis, G., Scopigno, R. (ed.) *Eurographics 2008.* Blackwell Publishing Ltd: Computer Graphics Forum.

Guidi, G., Ruso, M., Angheleddu, D. 2014. 3D survey and virtual reconstruction of archaeological sites. *Digital Applications in Archaeology and Cultural Heritage* 2014, 55-69.

Horn, B. K. P. 1981. Hill shading and the reflectance map. *Proceedings IEEE,* 69 (1), 14-47.

Imhof, E. 1982. *Cartographic relief representation,* New York: Walter de Gruyter.

Landis, H. 2002. Production-ready global illumination. *Course 16 notes, SIGGRAPH 2002.*

Lescop, L., Cassen, S. 2013. Gavrinis. The raising of digital stones. *Proceedings of the 2013 Digital Heritage International Congress (Marseille, France, 28 Oct-1 Nov 2013),* 2, 561-568.

Malzbender, T., Gelb, D., Wolters, H., Zuckerman, B. 2000. *Enhancement of shape perception by Surface Reflectance Transformation. Tech. Rep. HPL- 2000- 38R1,* Palo Alto, California: Hewlett-Packard Laboratories.

Malzbender, T., Gelb, D., Wolters, H., Zuckerman, B. 2001. Polynomial texture maps. *SIGGRAPH '01: Proceedings of the 28 th Annual Conference on Computer Graphics and Interactive Techniques (New York, USA, 2001),* ACM Press, 519-528.

Malzbender, T., Wilburn, B., Gelb, D., Ambrisco, B. 2006. Surface enhancement using real-time photogrammetric stereo and reflectance transformation. *Eurographics Symposium on Rendering*, 245-250.

Mañana Borrazás, P., Blanco Rotea, R., Rodríguez Paz, A. 2009. La documentación geométrica de elementos patrimoniales con láser escáner terrestre. La experiencia del Lapa en Galicia. *Cuadernos de Estudios Gallegos* 65, 33-65.

Medici, P., Rossi, G. 2015. Valcamonica 3.0: a new dimension in rock art recording. From tracing to structure from motion and post processing, in: Troletti, F. (ed.) *Research perspectives on prehistoric art. Pre-proceedings of the XXVI Valcamonica Symposium 2015 9-12 September. 1-6 CCSP.* Capo di Ponte: Centro camuno di Studi Preistorici.

Miller, G. S. P. 1994. Efficient algorithms for local and global accessibility shading. *Proceedings of the 21st Annual Conference on Computer Graphics and Interactive Techniques, (SIGGRAPH 1994, Orlando), ACM*, 319-326.

Ortiz, J., Gil, M. L., Martínez, S., Rego, T., Meijide, G. 2010. A simple methodology for recording petroglyphs using low-cost digital image correlation photogrammetry and consumer-grade digital cameras. *Journal of Archaeological Science* 37, 3158-3169.

Pereira Menaut, G. D. 1991. *Corpus de inscricións romanas de Galicia.* Santiago de Compostela: Consello da Cultura Galega.

Pereira Rodríguez, J. 2012. Algunas experiencias con Reflectance Transformation Imaging (RTI) en grabados rupestres. *http://www.jpereira.net/apuntes-breves/algunasexp.*

Pharr, M., Green, S. 2004. Ambient occlusion, in: Randima, F. E. (ed.) *GPU Gems.* Addison Wesley.

Pires, H., Fonte, J., Gonçalves-Seco, L., Correia Santos, M. J., Sousa, O. 2014. Morphological Residual Model. A Tool for Enhancing Epigraphic Readings of Highly Erosioned Surfaces. *EAGLE- Information Technologies for Epigraphy and Cultural Heritage in the Ancient World*, 133-144.

Pires, H., Martínez Rubio, J., Elorza Arana, A. 2015. Techniques for revealing 3D hidden archaeological features: Morphological Residual Models as virtual-polynomial texture maps. *The International Archives of the Photogrammetry, Remote Sensing and Spatial Information Sciences (Papers from 3D Virtual Reconstruction and Visualization of Complex Architectures, 25-27 February 2015, Avila, Spain),* XL-5/W4, 415-421.

Remondino, F. 2014. Photogrammetry. Basic Theory, in: Remondino, F. and Campana, S. (ed.). *3D recording and modelling in Archaeology and Cultural Heritage.* Oxford: Archaeopress.

Rodríguez Casal, A. A., García Calviño. F. J., Gómez González, L. 1995. Fotogrametría y procesado digital de imágenes aplicado al arte rupestre. *I Symposio de Manifestaciones rupestres del Archipiélago Canario-Norte de Africa*, Unpublished work.

Rusinkiewicz, S. 2004. Estimating Curvatures and Their Derivatives on Triangle Meshes. *3DPVT '04 Proceedings of the 3D Data Processing, Visualization, and Transmission, 2nd International Symposium*, 486-493.
Štular, B., Kokalj, Ž., Oštir, K., Nuninger, L. 2012. Visualization of LiDAR-derived relief models for detection of archaeological features. *Journal of Archaeological Science* 39(11), 3354-3360.
Vázquez Martínez, A., Vilas Estévez, B., Carrero Pazos, M. 2015. Sobre as técnicas de reprodución dos gravados rupestres ao aire libre en Galicia. *Férvedes. Revista de Investigación* 8, 17-24.
Vázquez Martínez, A., Vilas Estévez, B., Carrero Pazos, M. 2016. Sobre a técnica fotogramétrica aplicada á documentación da arte rupestre galega: novas propostas sobre vellos rexistros. *X Encontro Arqueolóxico Do Barbanza (Boiro, 27-28 febreiro de 2016)*, Conference presentation. Unpublished.
Vergne, R., Pacanowski, R., Barla, P., Granier, X., Shlick, C. 2010. Radiance Scaling for Versatile Surface Enhancement. *I3D '10 Proceedings of the 2010 ACM SIGGRAPH symposium on Interactive 3D Graphics and Games, ACM New York, NY, USA*, 143-150.
Vilas Estévez, B., Vázquez Martínez, A., Carrero Pazos, M. 2015. The Use of Photogrammetric Techniques for Recording the Rock Art Carving at Campo Lameiro (Galicia, Northern Spain). *43rd Computer Applications and Quantitative Methods in Archaeology "KEEP THE REVOLUTION GOING" Conference (CAA 2015 SIENA)*, Poster presentation. http://www.rupestre.net/tracce/?p=9520.
Vilas Estévez, B., Vázquez Martínez, A., Carrero Pazos, M. 2016. Going Further: (Re)discovering Rock Art Carvings With Photogrammetric Techniques in Galicia (North-western of Iberian Peninsula), in: Ippolito, A. and Cigola, M. (ed.). *Handbook of research on emerging technologies for Cultural Heritage.*

Notes on contributors

Miguel Carrero-Pazos: PhD in Prehistory, University of Santiago de Compostela (2017). Specialist in computationally-informed landscape archaeology, his research focuses on the application of GIS and spatial statistics to model monumental landscapes. He is also interested in 3D imaging analysis and the application of computer tools for the study and 3D representation of rock art engravings.

Benito Vilas-Estévez: Graduate in History, University of Santiago de Compostela (USC), Spain, in 2010, with minors in Archaeology, Ancient History and Prehistory. He studied at the Nova University of Lisbon (Portugal) for one year in 2009/2010. He started an MA in 2012 at the University of Wales Trinity Saint David (UWTSD) in Cultural Astronomy and Astrology, and in 2015 he obtained an MA in *Arqueoloxía e Ciencias da Antigüedade* from the University of Santiago de Compostela, Spain. His master's dissertation is about megalithic astronomy.

Alia Vázquez-Martínez: Graduate in History of Art from the University of Santiago de Compostela (USC), Spain, in 2012, with minors in Ancient History and Medieval Art, Artistic Heritage and the History of Modern Art. She obtained an MA in Arqueoloxía e Ciencias da Antigüedade at the University of Santiago de Compostela, Spain, in 2013, minoring in Prehistory. Currently, under the tutorship of Dr. Ramón Fábregas Valcarce and Dr. Carlos Rodríguez Rellán, and with the aim of pursuing her PhD studies, she is focusing her research on Galician Petroglyphs, thus extending the research she carried out during the course of her Master's studies. She has participated in several National and International conferences.

Section 2

Tools and methods

Procedural approaches

Methods for the evaluation of the visualisation of archaeological sites

Pablo Paniego Díaz[1]

Introduction

Throughout many periods in history, one of the most important factors in choosing a place to settle was the fact of being seen from other places (Criado 1999). At some moments it seems that visual exposure was a priority, while at others discretion, or even occultation, seems to have been more desired.

To gain an insight into these concepts we must carry out objective comparisons, with contrastable and measurable data, which will allow us to compare sites and search for tendencies.

We have considered four possible methods for a better understanding of how sites are seen (though the techniques and methods can be applied to any element within the landscape), each of which has a series of advantages and disadvantages. We have then applied the methods to a series of archaeological sites located around the basin of the Ardila River (a tributary of the Guadiana), and the mountain ranges of Huelva (located in the southwest of the Iberian Peninsula), which were settled during the second half of the 1st millennium BC.

Methods

First of all, it must be specified that, according to Berrocal (2007) and Paniego (2015), we have considered visual dominance at two distance ranges. On the one hand, an accurate range for delimiting the territorial space or area of influence of the sites has been applied (the "dominated landscape"); in other words, within a radius of 2500 m from the observation point. Whereas, on the other hand, the horizontal or "visualised landscape", which refers to the area beyond the visually dominated territory, has been used, up to 7500 m from the point.

---

1 Universidad Autónoma de Madrid. Departamento de Prehistoria y Arqueología. Campus de Cantoblanco 28049 Madrid (Spain) pablo.paniego@hotmail.com.

Relative altitude

This method has been used by many researchers (Mayoral 2004; Parcero and Fábrega 2006; Paniego 2015, among others) mainly due to its ease of application. Relative altitude has been used to evaluate accessibility to sites and as a factor for the evaluation of visual exposure or prominence within the landscape.

To calculate the relative altitude of a site the following formula can be used:[2]

$$\text{Relative altitude} = \frac{\text{Site altitude} - \text{Minimum altitude}}{\text{Maximum altitude} - \text{Minimum altitude}}$$

In order to obtain data, a simple analysis has been carried out: the definition of an area (with a 2500 m or 7500 m radius) followed by a calculation of the minimum and maximum heights within the area in addition to the altitude of the site. No GIS program is needed for this analysis and the data is usually general and not focused on the site which has been studied. Nevertheless, computer programs make it easier to obtain better quality data and make the task less time consuming.

The results of this approach are extremely general and only serve as a broad approximation. To a certain extent, via the use of this simple analysis we can generate an approximate response to certain questions such as accessibility to sites or the evaluation of physical predominance, which, to a certain degree, can be associated with visual predominance, although the latter can also occur in sites with low topographical prominence (Llobera 2005).

Relative altitude of visible areas

Unlike the previous analysis, here we only take into account areas within a line-of-sight from the site under study. Therefore, relative altitude is only computed considering those specific areas.

It must be stated that this method has a series of basic flaws, starting with the fact that not all of the areas which are visible from the observation point have inter-visibility with that point. Nevertheless, we consider that the generated error is acceptable since our objective is to obtain a general, overall idea of the prominence of a site.

Oriented relative altitude

This method is more specific than the previous two. It consists of analysing visibility from a point with a predetermined orientation and visual angle. Inter-visibility between observer and site is a prerequisite for this analysis.

This type of analysis is particularly interesting for defining the relationship between a specific element within the surrounding landscape (for example another settlement, a mountain pass, fords, bridges, roads, etc.) and the site under study.

2 This method has been chosen because the results are quite acceptable. It is a simple formula and initial data are easily obtained. Other possible ways to calculate it can be found in De Reu *et al.* 2011.

It has occasionally been considered that visual dominance from one site to another may indicate political dominance (Zamora 2013). However, we do not believe that this is necessarily a determining factor. In this context, it can be mentioned that, when carrying out detailed studies regarding the relationship between different sites located at different altitudes, not only altitude and/or relative altitude should be taken into account, but also the difference in angle between the different points within the territory being analysed, for which a reduced distance and exact height are necessary.

Furthermore, this analysis does not require the use of a computer program, although the use of certain programs does benefit the quality of the data and the time invested in the analysis. The main benefit of this type of analysis is the specificity of the obtained data.

Total visual prominence

This is a relatively simple method of analysis which consists of the creation of a series of observation points (ideally one for each cell of the raster grid in use) followed by an indiscriminate analysis of accumulated visibility (Llobera 2003; 2006), via which a global image of the studied areas can be obtained. The analysis is rather simple, although the computational time can be long. With this technique, the most visible areas or objects within the landscape can be determined, independently of whether they are physically predominant or not (Llobera 2003).

The difference with the previous method is that appropriate software is required and the analysis is more difficult to carry out. Analytical time depends greatly on the hardware and software available.

	Dificulty	Processing time	Specific/General	GIS Required
Relative altitude	Low	Short	General	No
Visible relative altitude	Medium	Medium	Semi-specific	Recommended
Oriented relative altitude	Medium	Medium	Specific	Recommended
Total visual predominance	Medium	Long	Semi-specific	Yes

Table 1: Characteristics of each analytical method.

Application of methods

In order to carry out this analysis we have taken into consideration the basic measurements proposed by Berrocal (2007) of radiuses of 2500 m and 7500 m from each site.

To illustrate these methods we have applied them to different sites from the second half of the 1st millennium BCE (Fig. 1).

- Balcón de Pilatos (Burguillos del Cerro, Badajoz): a possible Iron Age II site with no apparent subsequent Roman occupation. The site is located upon an isolated hill. Although some archaeological remains have been found at the top, it is possible that the site was located on the slope (Baquedano 1996). Altitude: 545 m.a.s.l.

- Castillejos 2 (Fuente de Cantos, Badajoz): located near a secondary water source. There is a phase of Roman occupation from the Republican period on top of an Iron Age II settlement (Fernández Corrales *et al.* 2004). Altitude: 542 m.a.s.l.
- Castillo de la Morería (Jerez de los Caballeros, Badajoz): this site has been identified as Roman, with a small pre-Roman phase seeming to correspond to one of the *oppida* indicated by Pliny in his description of the *Baeturia*. Although it was, at first, thought to have belonged to Iron Age II, it is now believed, based on new archaeological material recovered, that it was not established prior to the Roman conquest (Carrasco 1991; Rodríguez and Ortiz 2003). Altitude: 489 m.a.s.l.
- Castrejón de Capote (Higuera la Real, Badajoz): a hill-top settlement situated at a river fork and occupied during Iron Age II with continuity up to the late-Republican period, with no occupation after the Sertorian war (Berrocal 2007). Altitude: 446 m.a.s.l.
- El Castañuelo (Aracena, Huelva): located at the top of a hill near the base of the highest elevations of the Aracena Mountain Range. Its chronology is considered to belong between the 5th century BC and the 3rd-2nd centuries BC. (Jiménez Ávila 2009). It is possible that it was one of the first settlements to receive a large influx of population with Celtic roots in the south-western Iberian Peninsula. The site was abandoned when these populations finally became established in the area (ca. 400 BC) Altitude: 607 m.a.s.l.

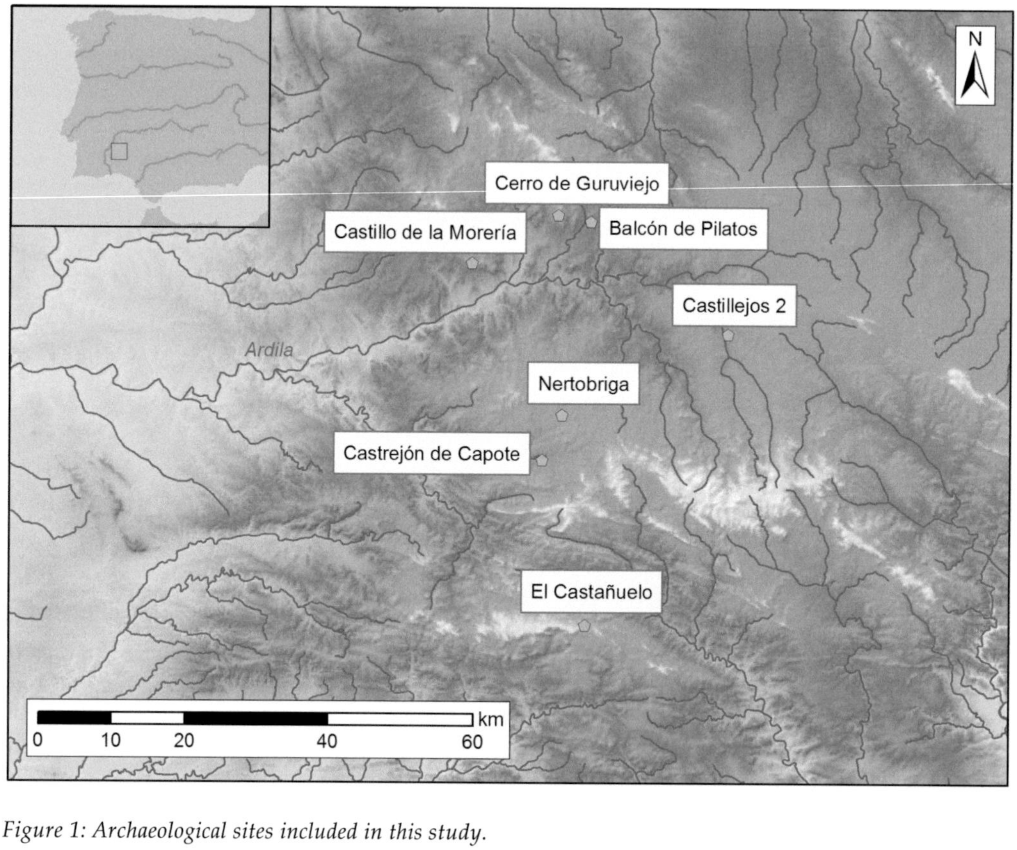

Figure 1: Archaeological sites included in this study.

- *Nertobriga Concordia Iulia* (Fregenal de la Sierra, Badajoz): according to the archaeologists who excavated the site, this is one of the earliest Roman sites in the Ardila basin. It is supposed to have been founded at the end of the 2nd century BC. The settlement is located upon a predominant hill in the landscape (Berrocal *et al.* 2014). This is one of the *oppida* described by Pliny in his description of *Baeturia*. Altitude: 683 m.a.s.l.

The difference in altitude above sea level among these sites is not very high, although this factor does not have much importance as it does not affect the visual prominence of the sites, which is the basic element for the relationship between a site and its surrounding landscape. Furthermore, it is interesting to highlight the fact that these sites are located in a relatively homogeneous terrain, albeit with more open spaces to the North (near the Ardila's basin) than to the South, which has a rougher landscape.

In order to carry out the analysis, we used a DEM with a 5 m resolution produced by the IGN (the Spanish National Geographic Institute), available for free download, and ArcGIS 10.

Relative Altitude (R. A.)

Taking each site, we created two *buffers* at 2500 m and 7500 m and then extracted the maximum and minimum altitude within these buffers. The next step was to calculate the relative altitude using the aforementioned formula. In order to obtain the data rapidly once the *buffers* had been created, we decided to use *extract by mask.*

The results, which oscillate between 0 (the site is at the lowest possible topographical position within the area) to 1 (the site is at the highest possible topographic position within the area), are shown in Table 2.

	R.A. 2500 m	R.A.V.A. 2500 m	R.A. 7500 m	R.A.V.A. 7500 m
Balcón de Pilatos	1	1	0.53	0.52
Castillejos 2	0.33	0.28	0.34	0.17
Castillo de la Morería	0.57	0.45	0.46	0.33
Castrejón de Capote	0.29	0.13	0.31	0.28
Cerro de Guruviejo	0.73	0.72	0.74	0.71
El Castañuelo	0.55	0.32	0.49	0.31
Nertobriga	1	1	0.64	0.55

Table 2: Results of the analysis of the Relative Altitude and the Relative Altitude from visible areas.

Relative altitude of visible areas (R. A. V. A.)

Taking into account that what is really being studied is the visualisation of a site, obtaining the relative altitude considering only the portion of terrain visible from the site may make more sense.[3] The results of this analysis must also be used with caution since a site may be extremely prominent from the point from which it is visible, but it may be visible from just a tiny portion of the landscape around it, making it hard to argue that such a site is prominent.

On the other hand, we cannot ignore the fact that the ideal analysis should not be carried out for places that are visible from a site, but from each of the observation points from which that site is visible. For the sake of simplicity, we have decided to carry out the viewshed analysis only from the site and take it as representative of the potential places from which the site is visible. In our case, this simplified approach falls within a framework in which we are only searching for general trends, and the results we expect to obtain are merely approximations. This is partially because we do not take into account certain factors such as vegetation, the ability of the observer or the weather and environmental conditions (Llobera 2003).

Oriented relative altitude (O. R. A.)

This analysis is an attempt to individualise the relationship between two points in the landscape, one of them being the observer. This method requires more extensive prior preparation, although it is not difficult to carry out and does not consume much time, despite the fact that every point has to be processed and studied individually.

In our example we have chosen four observation points near the limits of the territory under study, located at the cardinal points. This coverage is intentionally low, simply to be able to verify the validity and effectiveness of this analysis. Observation points have been placed at conventional spots in order to avoid biasing the results. In the case of a point not being inter-visible with the site, a different point has not been added.

In our example we have considered that the lowest radius to consider from the observer would be 250 m. The angle of visualisation would be 40º oriented from the observation points to the site. We must also add a favourable compensation (extra elevation) both for the observer and for the observed elements (Conolly and Lake 2009: 376), which cannot always be applied universally and depends on each case.

In order to carry out this analysis in ArcGIS we inputted the following data into the table of attributes of the observation points layer: Radius 1 (250 m), Radius 2 (2500 m or 7500 m, depending on whether we were analysing what was labelled as dominated or visualised landscape), Azimuth 1 and Azimuth 2, which can vary depending on whether the observation point is to the north (160 and 200 respectively), east (250 and 290), south (340 and 20) or west (70 and 110).

3 We are aware that a visibility analysis might be a more efficient way of understanding non-visible areas than visible ones, since areas that can potentially be viewed may not be visible due to multiple factors.

	2500 m	7500 m
Balcón de Pilatos	1; 1	1
Castillejos 2	0	
Castillo de la Morería	0.94; 0.75	0.94; 0.60
Castrejón de Capote		
Cerro de Guruviejo	1; 1; 1	0.84
El Castañuelo	0.75	
Nertobriga		0.75

Table 3: Results of the analysis of the Oriented Relative altitude (in the case of various results within each range these are separated by semi-colon).

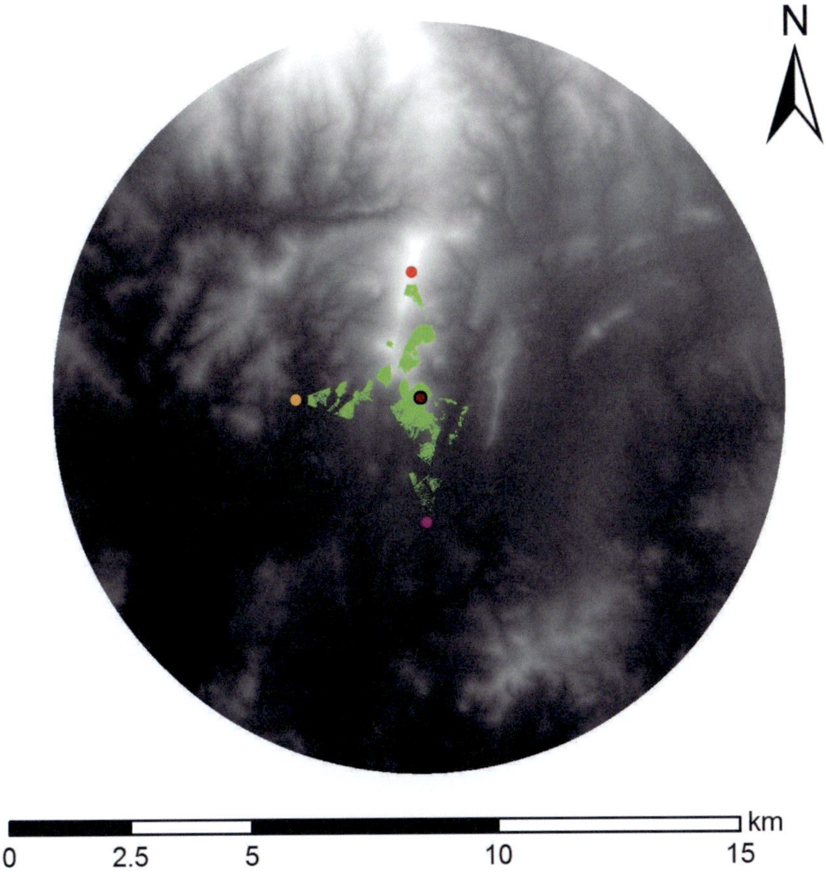

Figure 2: O. R. A. of Cerro de Guruviejo at 2500 m.

Obviously, the azimuth presented in this paper is valid exclusively for the observers located at exactly each point; should the analysis be carried out from a different point, the azimuth would have to be recalculated.

The low number of observation points has affected the results. Therefore, the data presented here must be taken with a certain degree of caution and it should be taken into account that it would be ideal to have more reference points when trying to reach a conclusion. Nevertheless, the results, though scarce, have allowed us to reach a conclusion regarding the efficiency and potential use of these techniques, especially in relation to the other methods used.

Finally, reference must be made to the importance of taking into account the vertical angle generated when one point is observed from another to deduce the actual existence of visual dominance. With this in mind, we have used two sites with inter-visibility in this study, Balcón de Pilatos and Cerro de Guruviejo. We have chosen to measure possible dominance by means of the difference in the angle between the reduced distance and the Euclidian (geometric) angle, in which the higher the degree of subordination, the higher the resulting angle is.

In order to be able to calculate this angle, knowing the reduced distance (r) and the altitude difference (z), it is necessary to obtain the Euclidian distance (e) (Fig. 3). The Z angle is the arctangent.

In the case of Cerro de Guruviejo and Balcón de Pilatos, the altitude difference is 83 m and the reduced distance between both is 3805 m, according to which the resulting difference angle, observed from the lowest point, is barely 1° 14'.

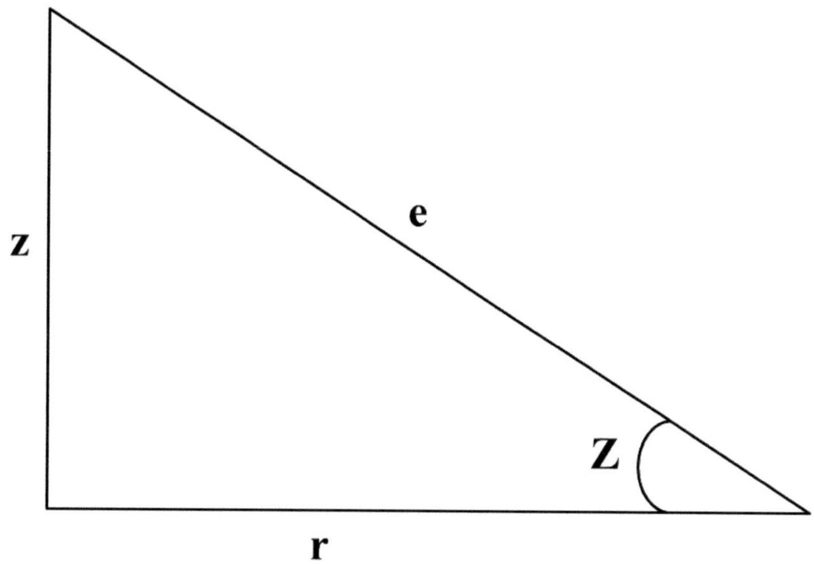

Figure 3: Diagram with the data necessary for obtaining the visualisation angle between two points.

Considering that the ideal analysis, regarding the relationship between two points, should take into consideration both the oriented relative altitude, which tells us how physically prominent one of the elements is regarding the other, and the individual relationship between them both, in which case (as is the case of Burguillos) on some occasions the difference in altitude barely creates a difference in the visual dominance of one point regarding another. We understand this domination exclusively in terms of visual control, without necessarily implying any sort of political or economic relationship.

Visual prominence

As with the previous method, we have chosen to carry out an analysis with two limits defined at 2500 m and 7500 m respectively. Unlike the previous cases, the results are extremely intuitive and can be easily understood with a graphic representation.

In this type of analysis it is advisable to compute only the areas which are to be studied, with the objective of shortening analytical time. The observation points, which can be random, as is our case, can be produced with most standard GIS

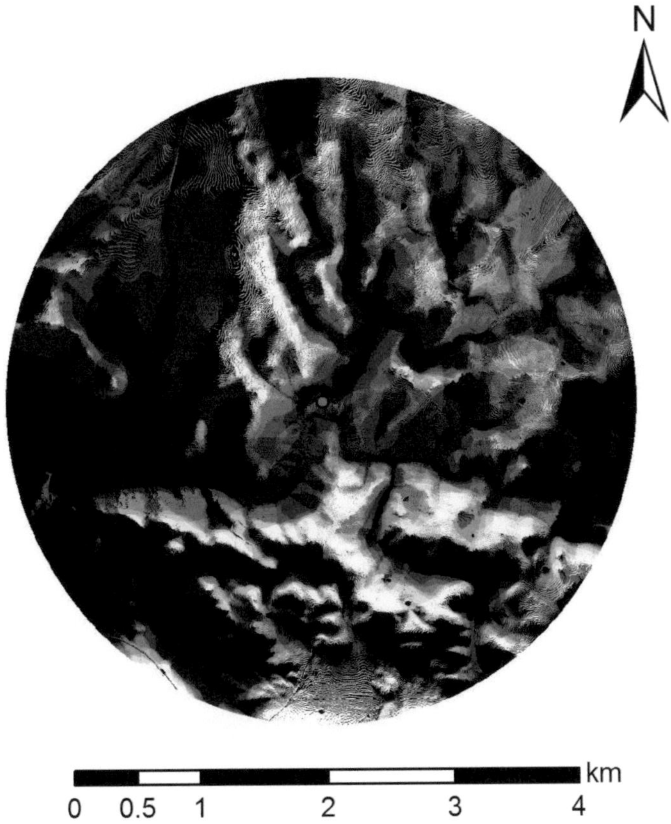

Figure 4: Visual Prominence from Castrejón de Capote.

tools (we chose 225 for the visualised landscape and 25 for the visually dominated landscape). The next step is to execute the visibility analysis (*viewshed*) which allows us to obtain a map showing the most prominent areas. The ideal analysis would be carried out from every raster grid (Llobera 2006), although this would require a long period of time and a powerful computer.

Unlike the previous analyses, this does not take physical prominence into account and allows the identification of sites with a high degree of visibility, even though they may be found in a depressed area within the landscape. In the case of a high index, these sites can be seen from multiple places within the landscape (Llobera 2003). It must be mentioned that within the area studied (the basin of the Ardila River and the mountain range of Huelva), due to its particularities, visual prominence is usually associated to physically prominent places within the landscape. In these examples we have a high index of relative altitude with a high degree of visual prominence.

Comparison

If we compare the values of the two first methods in which relative altitude was used, the importance of relationships of visibility can be seen. Therefore, although relative altitude can give us a general panorama of the visual prominence of a site, we believe that the best way to carry out these analyses is from places from which the site is visible, since in some cases there is a significant variation in the results. We must take into consideration, for instance, that a site may be hidden within the landscape but be prominent from those areas from which it is visible.

There may also be a significant difference in the visual relationship obtained with a general and with a specific analysis. Thus, an element with a low visual prominence may have a larger dominance than initially supposed when using a specific analytical method which relates two sites within a landscape.

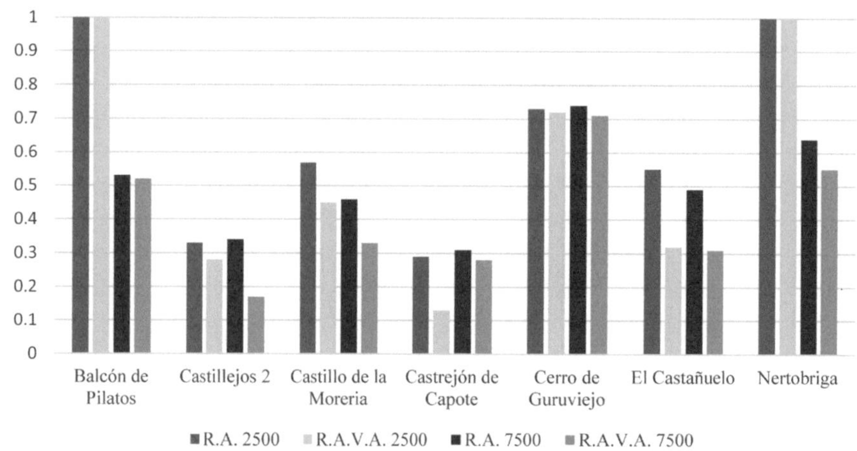

Figure 5: A comparison of relative altitude between the different analysis methods.

Therefore, it is possible to verify the existence of variations depending on the application of one method or another regarding the visualisation of archaeological sites when using relative altitude. We believe that the analyses are more useful when applying them to the study of relationships of visibility rather than to the totality of a predefined area (assuming the error which was previously stated regarding the absence of an exact relationship between what can be seen from two points with inter-visibility). Furthermore, if we are searching for the relationship between two elements, the best option would be to use oriented analysis, taking

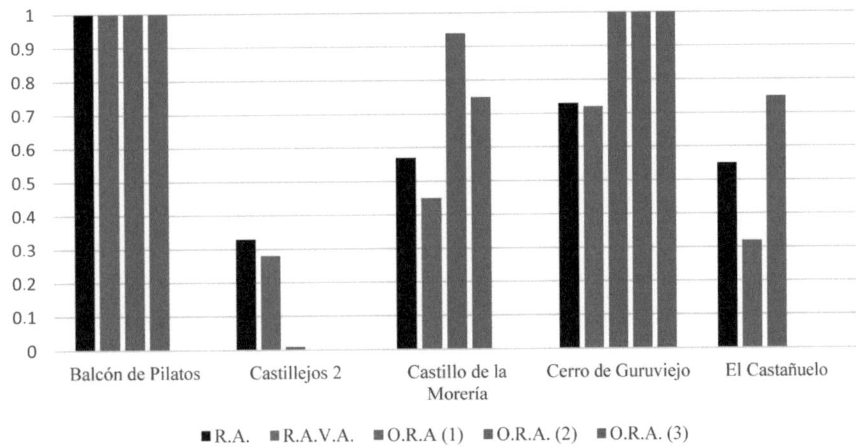

Figure 6: A comparison of the Oriented Relative Altitude at 2500 m.

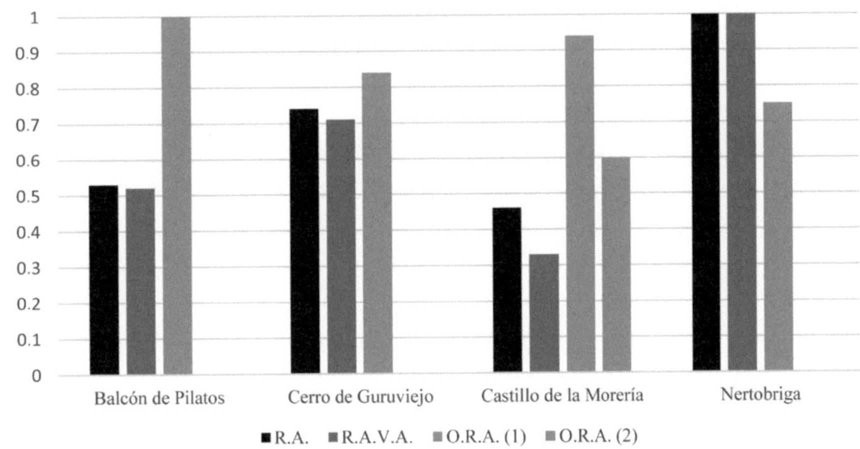

Figure 7: A comparison of the Oriented Relative Altitude at 2500 m (7500 m).

into consideration the different angles generated for assessing the visual dominance of one regarding the other.

As far as total visual prominence is concerned, it is an extremely intuitive analytical method which allows us to obtain a general idea of the possibility of a point being prominent or not from a site within a determined territory. As is the case with general relative altitude, it is mostly a non-specific analysis, which does not mean it is not useful, especially when trying to search for sites based on a series of patterns.

The different results obtained with all these analytical methods do not have an absolute value, with the possible exception of visual prominence, and, therefore, they are mostly useful when carrying out comparisons. Nonetheless, the absolute values may also indicate different tendencies within the territorial organisation when searching for prominence/discretion within a territory.

Conclusions

Taking into account the four methods we have assessed, it can be concluded that total visual prominence seems to be the most efficient method when trying to identify potential settlement locations as far as their visualisation is concerned, as it generates intuitive maps.

Relative altitude generally gives a non-accurate image of reality but does help by giving us an overall idea of the level of concealment of the settlement within the territory. To be able to calculate visualisation from the surroundings, it is more efficient to exclusively use the data of the visible area. On the other hand, relative oriented altitude has a clear advantage when attempting to establish the relationship between two points, with the use of the difference in the degrees between the visualised element and the observer being an open path at this point.

Acknowledgements

I would like to thank the organisers of the seminar for allowing me to participate in this cycle, especially V. Mayoral, C. Parcero and P. Fábrega. I would also like to thank C. Bashore for his help in the translation of this text.

References

Baquedano Beltrán, M. I. 1996. *Informe de la intervención en el castillo de Burguillos del Cerro (Badajoz).* Unpublished.

Berrocal Rangel, L. 2007. El poblado fortificado de El Castrejón de Capote y su paisaje: la fortificación de lo sagrado, in: Berrocal Rangel, L. and Moret, P. (eds.). *Paisajes fortificados de la Edad del Hierro: las murallas protohistóricas de la meseta y de la vertiente atlántica en su contexto europeo.* Madrid: Real Academia de la Historia, 255-280.

Berrocal Rangel, L. *et al.* 2014. Nertobriga Concordia Iulia. El paisaje de un enclave romanizador entre los célticos de la Beturia, in: Salas Tovar, E. (ed.). *La gestación de los paisajes rurales entre la protohistoria y el período romano. Formas de asentamiento y procesos de implantación.* Mérida: CSIC, 143-169.

Carrasco Martín, M. J. 1991. Excavaciones de urgencia en el Casillo de "la Morería" (Jerez de los Caballeros, Badajoz). *Extremadura Arqueológica* 2, 559-576.

Conolly J. and Lake, M. 2009. *Sistemas de información geográfica aplicados a la arqueología.* Barcelona: Bellaterra.

Criado Boado, F. 1999. *Del terreno al espacio: planteamientos y perspectivas para la Arqueología del Paisaje.* Santiago de Compostela: Grupo de Investigación en Arqueología del Paisaje.

Fernández Corrales, J. M. *et al.* 2004. Los Castillejos de Fuente de Cantos: un conjunto fortificado de época protohistórica en el sur de Badajoz. *Revista de Estudios Extremeños* 60/3, 913-924.

Jiménez Ávila, J. J. 2009. El poblado de El Castañuelo (Aracena) y el Post-Orientalizante en la Sierra Norte de Huelva. *IV Encuentro de Arqueología del Suroeste*, 3-33.

Llobera, M. 2003. Extending GIS-based visual analysis: the concept of "visualscapes". *International Journal of Geographical Information Science* 17/1, 25-48.

Llobera, M. 2006. Arqueología del paisaje en el siglo XXI. Reflexiones sobre el uso de los SIG y modelos matemáticos, in: Grau Mira, I. (ed.). *La aplicación de los SIG en la arqueología del paisaje.* Alicante: Universidad de Alicante, 109-124.

Mayoral Herrera, V. 2004. *Paisajes agrarios y cambio social en Andalucía oriental entre los períodos ibérico y romano.* Madrid: CSIC.

Paniego Díaz, P. 2015. *Arqueología y Estudio del Territorio del cerro de Guruviejo (Burguillos del Cerro, Badajoz).* Madrid: La Ergástula.

Parcero Oubiña, C. y Fábrega Álvarez, P. 2006. Diseño metodológico para el análisis locacional de asentamientos a través de un SIG de base "ráster", in: Grau Mira, I. (ed.). *La aplicación de los SIG en la arqueología del paisaje.* Alicante: Universidad de Alicante, 69-90.

Rodríguez Díaz, A. y Ortiz Romero, P. 2003. Defensa y territorio en la Beturia: castros oppida y recintos ciclópeos, in: Morillo Cerdán, A. *et al.* (eds.). *Defensa y territorio en Hispania de los Escipiones a Augusto: (espacios urbanos y rurales, municipales y provinciales).* León: Universidad de León and Casa de Velázquez, 219-251.

Zamora Merchán, M. 2006. Visibilidad y SIG en Arqueología: mucho más que unos y ceros, in: Grau Mira, I. (ed.). *La aplicación de los SIG en la arqueología del paisaje*. Alicante: Universidad de Alicante, 41-54.
Zamora Merchán, M. 2013. Análisis territorial en arqueología: percepción visual y accesibilidad del entorno. *Comechingonia* 17/2.

Notes on contributor

Pablo Paniego Díaz is a graduate in History from the Universidad Autónoma de Madrid and gained an MA in Archaeology and Heritage from the same university. His research mainly focuses on the changes which took place between the 3rd century B.C. and the turn of the Era in the Ardila basin (located in the southwest of the Iberian Peninsula), particularly as far as the analysis of settlement patterns and their diverse relationships with the landscape are concerned. He is currently carrying out his PhD studies under the supervision of Sebastián Celestino Pérez and Luis Berrocal Rangel.

Landscapes on the move. Digitally exploring the relationship between megaliths and mobility in Northern Cáceres (Spain)

Jose M. Señorán Martín[1]

Introduction: Landscape archaeology, barrows and mobility as an interpretive framework for megalithism in Extremadura

The configuration of space reflects how a community is reproduced and, as a consequence, how it represents itself (Criado and Mañana 2003, 103). A landscape is built by human groups through their work and production (Vicent 1991) and their modes of perception arising from their experience (Tilley 1994). The ways of moving through a landscape (Ingold 2011; Llobera 2015) are an essential mechanism for domesticating it and making it suitable for human inhabitation (Bender 1993).

The creation of monumental architecture implies the transformation of space in a landscape through artificial structures (Gil 2003, 26). Therefore, it could be stated that monumental architecture (in our case megaliths) consciously determines and regulates the experience, thus imprinting a particular ideology (Bradley 1993). Its location articulates the existing social relationships and power structures (Tilley 1994).

The processes of construction and acculturation of a space can be analysed from the perspective of Landscape Archaeology, along the lines proposed by F. Criado (1999). In this way, we can analyse the mechanisms used by past societies to define their existence in the world. Therefore, studies on perception, especially those related to mobility, are fundamental to understanding this issue.

Landscape can be regarded as both a material and imaginary social construction (Criado 1999, 5), which is in permanent motion and rooted in culture (Criado 1993). Its construction also depends on relationships of power (Sharp *et al.* 2000).

---

1 C/ Los Celtas, 50. 10810 Montehermoso. Spain.

A landscape is the materialisation of a prevailing ideological discourse, which is controlled, selected and redeployed (Foucault 2008). Therefore, a landscape does not have a universal essence. It is not a pre-existing reality (Gell 1992, 231), but is shaped by personal experiences. Our perception of it allows us to sort and classify the facts of experience (Hernando 1999, 28). These experiences are essential in the socialisation of any group. Mobility, memory and daily routine have a role to play in the creation of identity in any given landscape. This experience and routine has been called "habitus" by Bourdieu (1977). It refers to the "embodied" experience of the landscape, in which there is no difference between mind and body, nature and culture (Thomas *et al.* 2001, 547).

In this context, megaliths can be understood as physical manifestations of the symbolic use of power by a particular community or social group. They legitimise a series of power relationships (Bradley 2000, 158) and perpetuate an ideological discourse (Shanks and Tilley 1987, 68). In short, we can define the emergence of monumental architecture as the "representation of this new way of being-in-the-world" (Criado 1999, 35).

Megalithism in Northern Cáceres: A brief history

The first references to megaliths in Extremadura can be found in 15th century texts. However, it was not until the end of the 19th century that archaeologists such as Jose Ramón Mélida and George and Vera Leisner carried out studies of a more scientific nature.

The last decades of the 20th century bore witness to an unprecedented advance in studies relating to megalithism, beginning with the publication of the doctoral thesis of P. Bueno Ramirez (1984). Her work includes numerous contributions on megalithic art (Bueno and Balbín 1992), settlements (Bueno *et al.* 2000) and the role of menhirs (Bueno 1984; Bueno and Balbín 1994).

In addition, the excavations carried out by Dr. Ruiz Gálvez (2000) in the megalithic nucleus of Montehermoso are worthy of mention. Mention should also be made of the publication of a series of theses in recent years. These studies analyse megaliths from the theoretical perspective of Landscape Archaeology (López-Romero 2005; Fernández Freire 2008). Furthermore, studies relating to the emergence of the Neolithic in the region are also significant (Cerrillo 1999; 2005; 2006).

As far as chronologies are concerned, studies conducted in recent decades enable us to understand in greater detail the megaliths and the Neolithic period in the region. The old hypothesis about the existence of a late megalithism in the Bronze Age (Almagro Basch, 1962a; 1962b) is largely dismissed nowadays, as there are no more than ten radiocarbon dates available for our region. The only studies which show us a chronological series based on C14 dates are those carried out in the necropolis of Montehermoso (Ruiz-Gálvez 2000) (table 1). This series suggests continuity in the use of the megaliths from the beginning of the 5th millennium cal BC to the Late Bronze Age. In addition, the association of megaliths with settlements from the Copper Age, such as the site of El Canchal in Jaraiz de la Vera (Bueno *et al.* 2000) can be highlighted.

Site	Context	Lab number	Material	Date BP	Cal BC (95% prob)	Reference
MH4 (2)	Dolmen	GrA 15938	charcoal	7960±60	7056-6685	Ruíz-Gálvez, 2000
MH4 (1)	Dolmen	GrA 15903	charcoal	5000±60	3948-3662	Ruíz-Gálvez, 2000
MH4 (3)	Dolmen	GrA 15941	charcoal	4860±60	3765-3525	Ruíz-Gálvez, 2000
MH11 (4)	Dolmen	Ua 17766	charcoal	4965±75	3946-3642	Ruíz-Gálvez, unpublished
MH11 (1)	Dolmen	Ua 17763	charcoal	4920±70	3942-3553	Ruíz-Gálvez, unpublished
MH8 (1)	Dolmen	Ua 17768	charcoal	5040±70	3971-3667	Ruíz-Gálvez, unpublished
Joaniña	Dolmen	Sac-1380	charcoal	5400±210	4706-3777	Unpublished
Joaniña	Dolmen	Sac-1381	charcoal chamber floor	3840±170	2867-1881	Oliveira, unpublished
Trincones	Dolmen	Beta-197160	charcoal	3600±60	2136-1773	Bueno *et al.* 2004
HM11 (2)	Dolmen	Ua 17664	charcoal	2200±65	396-95	Ruíz-Gálvez, unpublished

Table 1: summary of C-14 dates.

Geographical context. Mountains, rivers and meadows

In this paper, we shall investigate the likely relationship between monuments and mobility on a macro scale. By scale here we do not mean merely the geographic degree of detail, but also categories such as social scale (individual, family or group) or the scale of the theoretical and methodological models. In other words, we shall approach this issue from a very general point of view, simply attempting to lay the very basic foundations of the possible relationship between monuments and mobility. Our aim is not to explore the individual and detailed connections between people and single monuments, but only to examine to what extent such a widely claimed relationship can be supported in this particular case study.

In this case, a macro-geographic scale will be used, including the majority of the province of Caceres, with the notable exclusion of the extreme southwest. The border of the study area is the Salor River, a natural border. The remainder of the study area is defined by the administrative boundaries of the province, known as Alta Extremadura.

This is an area which extends over about 20,000 square kilometres. It presents a relief moulded from Precambrian materials, arranged in vast east-west swathes, which facilitate mobility in that direction. The north-south mobility is limited by natural boundaries represented by mountain ranges, as well as the Tajo River. Its basin includes tributary rivers flowing from the Central System range, such as the Tiétar and Alagón, which have a nivo-pluvial flow regime. This means that the waters are higher in the spring, after the melting of the snow and the rains of the winter. The second hydrographic maximum occurs during the autumn, with the return of the rains. The summer is characterised by severe drought. During the winter they suffer their second minimum value.

Megaliths, numbers and rasters

The spatial analysis of past or current landscapes requires the construction of a spatial database, the components of which are selected depending on the detail or accuracy of the analysis to be performed (Vicent 1991). In this case, extensive use of GIS has been made, which allowed me to work with different data types depending on their nature and to give them meaning, producing interpretations which allow us to contextualise them in a pattern of rationality (Parcero 2002, 60).

There are two basic components of the spatial database used for the analysis described here:

- A Digital Elevation Model (DEM): a 5 m cell size model produced and distributed by the CNIG (the Spanish National Centre for Geographic Information). The points have an accuracy of 50 cm-100 cm (RMSZ) according to specifications. The DEM was produced in 2010.
- The distribution of megaliths in the study area: the study area includes the whole province of Caceres with the exception of the southwest (where the necropoles of Valencia de Alcántara and Cedillo are located). This area has been excluded because, to the west of the Salor River, it seems to be more connected to the architectural dynamics of the Portuguese Alentejo, in addition to the fact that it has been widely studied in previous archaeological doctoral dissertations (Bueno Ramirez 1988; López-Romero 2005). The Regional Archaeological Map catalogues a total of 275

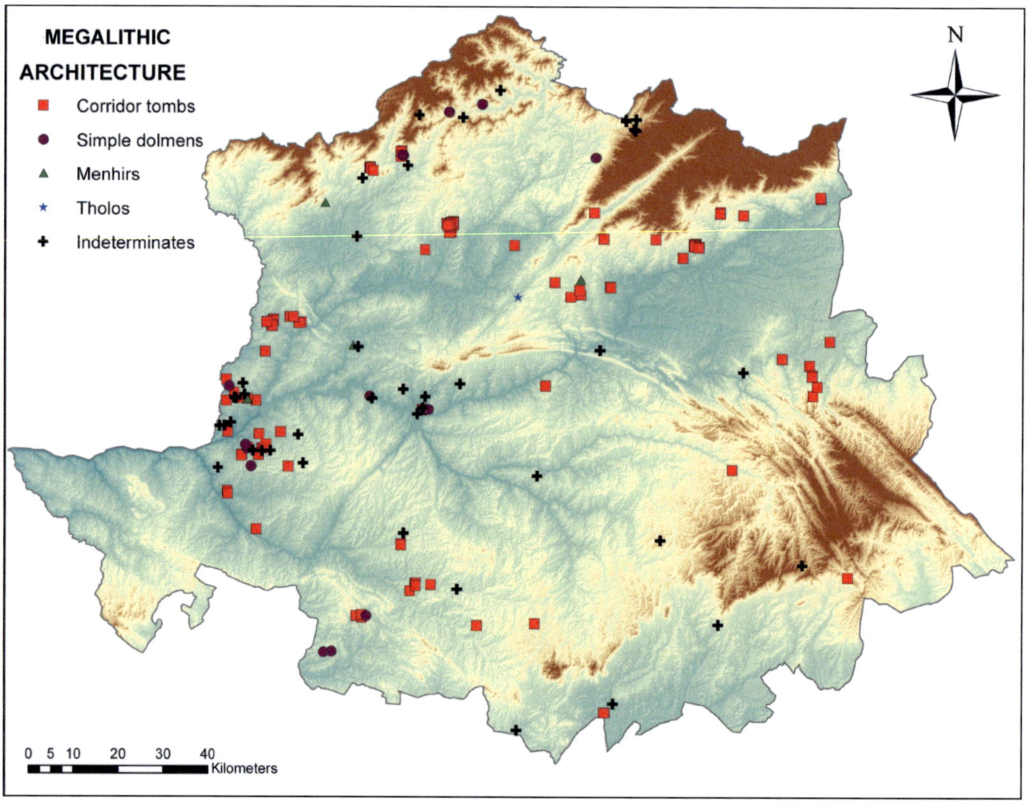

Figure 1: Megaliths of the study area.

megaliths (including dolmens and menhirs) for the province of Caceres, 165 of them are located in our study area. However, only 104 had been registered with geographic coordinates. Following a process of refinement, the sample has been expanded to 190 megaliths with the incorporation of data collected from fieldwork, academic papers and the Internet (Figure 1).

Monumental architecture in its geographic context. Understanding mobility in the past

Movement through space is a key factor in the structural organisation of human groups. Movement occurs routinely in the sense of habitus proposed by P. Bourdieu (1977). Therefore, we believe it is necessary to conceive movement as a sociological question (Llobera 2000), through which societies apprehend and organise space, as long as it is social space (Criado 1999). It is, therefore, necessary to analyse the interplay between mobility, physical space and the production of cultural materials. This is because movement is part of the socialisation process of individuals and the shaping of their identities. The structure of the movement is inherent in the process of social structuring (Gianotti 2014, 272).

Based on this premise, we have analysed the spatial distribution of megaliths in relation to mobility patterns and routes, using different statistical tests as a way of checking their mutual relationship.

The Kolmogorov-Smirnov test. A non-parametric test in order to understand the location of megaliths within their landscape

The primary hypothesis of this research is that the location of burial mounds in the landscape is related to the material expression of an ideological system, possibly related to issues of perception in the sense proposed by M. Marleu-Ponty (1975). Therefore, their presence in the field cannot be considered to be random.

Taking this hypothesis as a starting point, the decision was taken to perform the Kolmogorov-Smirnov test (Fletcher and Lock 2005, 111), a non-parametric test, in order to check whether a sample meets or comes from a population with a given probability distribution. This test allows us to measure the degree of agreement between the distribution of a dataset and a specific theoretical distribution.

In this case, we compared the distribution of the sample with a theoretical homonym in an attempt to establish the relative cumulative frequency in the working sample and in a hypothetical sample. The aim was to compare both of them in order to establish whether the level of significance responds to a random pattern or to a series of prior factors, which the test itself cannot specify. In other words, the goal is to indicate whether the data come from a population with a specified (random) theoretical distribution. If the level of significance is greater than 0.05, non-normality is accepted in the population analysed.

In order to carry out this test, three random samples were created, each consisting of as many points (190) as the working sample. The three random samples were compared to the working sample and among themselves in order to analyse the randomness of the sample.

Kolmogorov-Smirnov test	D	Prob
Megaliths vs Random 1	0.3809	3.958×10^{-9}
Megaliths vs Random 2	0.3227	1.153×10^{-6}
Megaliths vs Random 3	0.3651	2.083×10^{-8}
Random 1 vs Random 2	0.1085	0.393
Random 1 vs Random 3	0.1587	0.0625
Random 2 vs Random 3	0.1402	0.1342

Table 2: Kolmogorov-Smirnov test.

From this analysis, it was possible to interpret the results stating that the empirical sample (the burial mounds) responds to a location with a value of probability greater than 0.05, unlike the theoretical random samples (Table 2).

The results of this analysis are a key first step to the work, due to the fact that they support the hypothesis that the spatial distribution of the base population (the megalithic architecture) responds to particular, non-random patterns. These are as yet unknown; hence the interest in analysing different variables in order to gain an insight into the construction and location patterns of monuments in the landscape.

Cumulative Mobility Networks (MADO): lines, spots and stones

The use of mobility analyses to understand the spatial distribution of megalithic monuments is common in archaeological studies (Murrieta 2007; 2012; Wheatley and Murrieta 2008). These approaches focus mostly on creating optimal travel routes or least-cost pathways, which aim to establish the theoretical transit within a landscape by establishing a point of origin and a destination. A slightly different approach has been suggested by M. Llobera (2006), who seeks to define "natural corridors" as something related to movement patterns, rather than to specific paths.

In an attempt to analyse movement within the study area, we have continued the work begun by P. Fábrega (2006), which has been further developed in various publications in recent years (Fábrega and Parcero 2007; Llobera *et al.* 2011). These studies researched the idea of the "MADO", the representation of a cost accumulation model of displacement from a given origin without a specific point of destination (Fábrega 2006, 9) or vice versa (towards a given destination without a specific origin). The MADO represents the theoretical flow of "least effort displacement" on the ground.

In order to develop the model, a series of steps have been followed. Firstly, a friction map was produced (Figure 2), with 'friction' referring to the difficulty offered by the terrain for human movement. For the creation of the friction map, we have relied on Tobler's hiking function (Tobler 1993, as in Fábrega *et al.* 2011).

From this friction map, it is possible to easily compute the accumulated cost distance to access any specific point in the area. Based on this information, and by applying Hydrology Analysis Tools, it is again rather easy to create theoretical flow maps which are not related here to the theoretical flow of water on the ground, but to the theoretical flow of "least effort movement on the ground" towards the

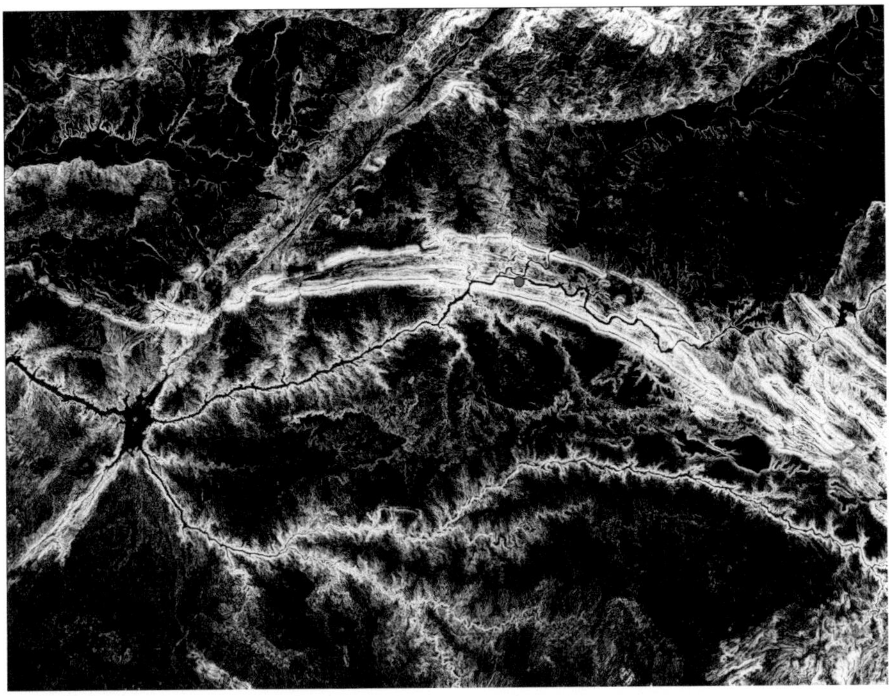

Figure 2: Sample of the friction map: lighter areas have higher friction values (i.e. they are harder to walk across).

specific point selected as a destination. A reclassification of this "flow model" leads to a "focal mobility network", a map highlighting the main paths leading to the destination point across the whole area, no matter what the origin may be (see Llobera *et al.* 2011 for a detailed explanation).

In this case study, this analysis was carried out for a series of different destination points, as detailed below. The aim of this study is to understand mobility in the study area and its relationship to the megaliths from different simulations. These simulations are differentiated from each other by the selection of different destination points.

A key factor in the creation of the models is the selection of the destination points. We started from the basis that there are a number of geographic points in the landscape which represent particular attractors for mobility, such as fords and passes (Llobera 2006, 117). Therefore, the decision was taken to develop three simulations, employing three samples of different destination points selected as follows:

- All passes and fords documented in modern maps. This model consists of a sample of 49 points.
- Historical fords and passes, meaning those with a historical record of relevant use from at least the Modern Age and, in many cases, from the Roman period (Menéndez Pidal 1851; Hernández Giménez 1967a; 1967b) (Figure 3).

- The final sample was produced from the morphological classification of the study area using the Landserf software and selecting the points classified as "passes". From this classification, a map of points was created with a total of 89 entities. Some of them correspond to the paths and fords listed above, although others do not.

Finally, the distance between the sample of archaeological entities and the three different networks of theoretical paths was analysed. The resulting values were compared with a random sample of points. When comparing the distance to any of the theoretical paths from monuments with the distance from random points, a clear difference emerges even from the basic descriptive statistics (Table 3).

As far as the paths produced with the Landserf passes are concerned, the average distance between the monuments and this network (1206 m with a SD of 1745 m), is less than half the average of the distance from the random points (3032 m, SD 5507 m). However, despite these notable differences, the Mann-Whitney significance test does not allow for the rejection of the null hypothesis, resulting in the fact that the distribution of both samples can be attributed to chance. The significance level P is above 0.05 (P=0.14). Consequently, the null hypothesis (Ho) cannot be rejected.

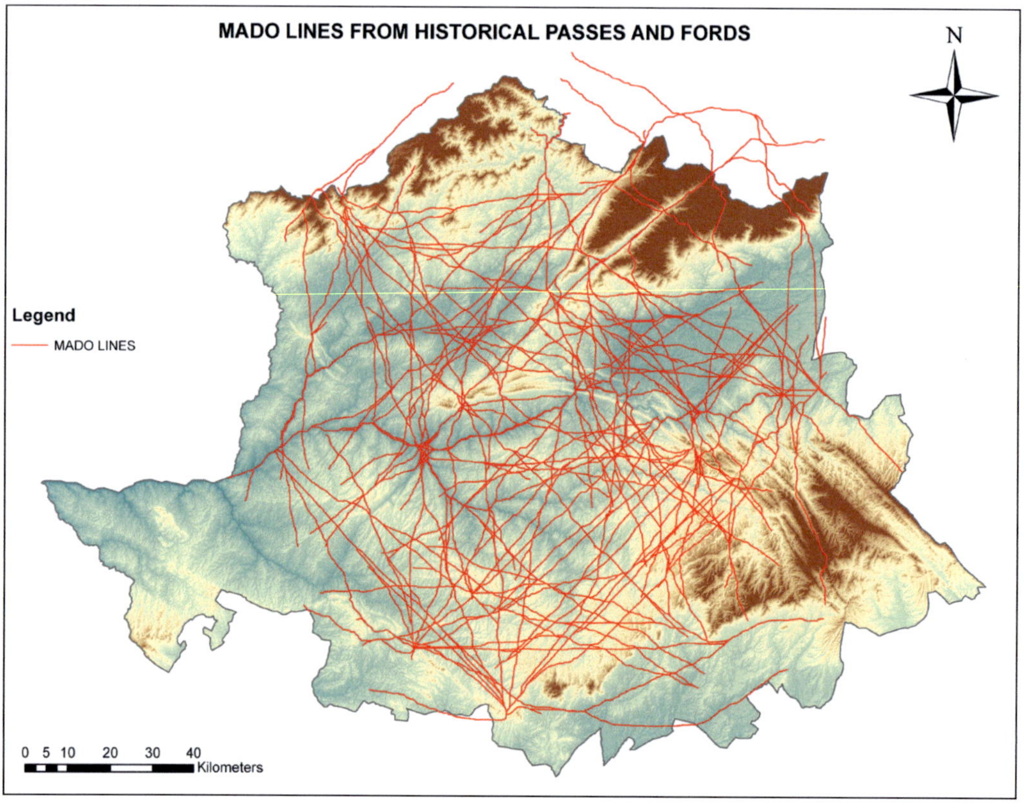

Figure 3: A MADO map calculated from historical passes and fords.

	Modern maps		Historical documents		Landserf classification	
	Monuments	Random points	Monuments	Random points	Monuments	Random points
Average distance	1185	2870	1697	5040	1206	3032
St. Dev. of distance	1618	4217	2134	6501	1745	5507
Significance (P value Mann-Withney test)	0.005	0.008	0.14			

Table 3: Distance between monuments and random points towards the 3 different sets of optimal corridors.

When using the passes and fords documented in modern maps, a significant difference is noticeable again. While the megaliths show an average distance of 1185 m (SD 1618 m), random points show an average distance of 2870 m (SD 4217 m). In this case, the results of the Mann-Whitney U test are completely different: the significance value of P is below 0.05 (P = 0.005). This value allows us to reject the null hypothesis (Ho) and to accept the alternative hypothesis (Ha) along with the non-randomness of the empirical sample. That is to say, the monuments are located at a significantly short distance from the optimal natural paths leading to the passes and fords documented in modern maps.

These results can be refined further if the typological differences in the architecture of the burial chambers in the monuments is taken into consideration. A striking difference emerges in the chambers with long corridors and false domes, which have the lowest values of the three models. In this regard, it must be added

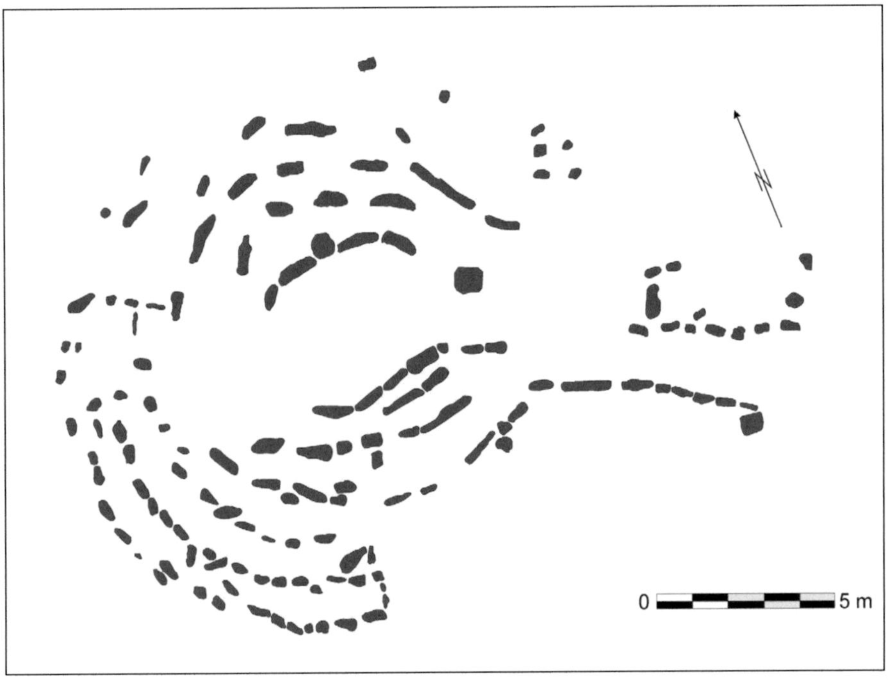

Figure 4: An archaeological map of the Guadalperal dolmen (adapted after Leisner and Leisner 1960, 78).

that chambers with long corridors account for 14% of the sample, whereas those with false domes account for 11%. These are the highest values within our sample. Special mention must be made of the megalithic group of Los Ibores. This group is located close to crossing points of the Tajo River. Some of these structures have a more modern chronology. Indeed, Bell-Beaker materials have been documented inside the Guadalperal dolmen (Leisner and Leisner 1956; 1960a; 1960b).

There seems to be a direct relationship between most modern chambers (chambers with false domes and tholos) and the optimal corridors. This hypothesis gains weight when below average values from different typologies are analysed. In the case of chambers with long corridors, it must be noted that one of the lowest values corresponds to El Guadalperal, where, as previously mentioned, Bell Beaker materials have been documented and the site is noted for its monumentality (Leisner and Leisner 1956). In the case of the tholos of Plasencia, Bell Beaker maritime-style ceramics have been documented (Castañeda and Matesanz 2006, 83). What is more, there are simple chambers with very low values, as documented in Plasencia (Sarasola 2006), for which absolute datings are not available, despite their recent excavation. However, the documented material is characterised by archer armlets, which could be framed within Bell Beaker contexts. The lowest values of the long corridor chambers again show materials from the Bell Beaker tradition.

Conclusions

At the beginning of this chapter, a hypothesis was proposed establishing the non-randomness of the location of megaliths in the landscape. Monumentalisation of space is the result of socialisation (Bender 1993) via experience (Ingold 1993; Tilley 1994). It is conditioned by human movement (Llobera 1996, 2005; Criado and Villoch 1998), among other factors.

With the aim of analysing and comparing the relationship between megaliths and mobility networks, the distribution of monuments known in the area has been compared with a random sample of points and the relationship of both with different models of optimal mobility corridors. These tests show a greater proximity between megaliths and mobility lines (MADOs) generated from paths and historical fords, which is statistically significant. This would allow us to accept the hypothesis relating megaliths with mobility patterns (Llobera 2015).

In short, a relationship between the monuments analysed in our study area and mobility routes originating from paths and historical fords can be contemplated. In addition, it can be stated that other parts of this analysis (not shown here) prove the continual proximity of monuments to cattle trails, over a long period of time, which suggests that this relationship between monuments and mobility corridors is deep and can open up future lines of research.

Acknowledgements

I would like to thank César González-García and César Parcero-Oubiña for their support in carrying out this study.

References

Almagro Basch, M. 1962a. *Megalitos en Extremadura*. Madrid: Servicio Nacional de Excavaciones Arqueológicas.

Almagro Basch, M. 1962b. *Megalitos en Extremadura*. Madrid: Servicio Nacional de Excavaciones Arqueológicas.

Bender, B. 1993 (ed.). *Landscape: politics and perspectives. Explorations in anthropology*. Providence: Berg.

Bourdieu, P. 1977. *Outline of a Theory of Practice*. Cambridge: Cambridge University Press.

Bradley, R. 1993. *Altering the Earth. The origins of Monuments in Britain and Ancient Europe*. Edinburgh: Society of Antiquaries of Scotland.

Bradley, R. 2000. *An Archaeology of natural places*. London: Routledge.

Bueno Ramírez, P. 1984. Tres nuevas estelas del Suroeste. *Revista de Estudios Extremeños* 40, 477-484.

Bueno Ramírez P. 1988. *Los dólmenes de Valencia de Alcántara*. Madrid: Excavaciones arqueológicas en España, Ministerio de Cultura.

Bueno Ramírez, P. and Balbín Behrmann, R. de 1992. L'art megalithique dans la Peninsule Ibérique. *L'Anthropologie* 96, 499-572.

Bueno Ramírez, P. and Balbín Behrmann, R. de 1994. Estatuas-menhir y estelas antropomorfas en megalitos ibéricos. Una hipótesis de interpretación del espacio funerario, in: Lasheras, J. A. (ed.), *Homenaje al Dr. Joaquín González Echegaray*. Madrid: Ministerio de Cultura. Dirección General de Bellas Artes y Archivos, 337-347.

Bueno Ramírez, P., González Cordero, A. and Rovira Llorens, S. 2000. Áreas de habitación y sepulturas de falsa cúpula en la cuenca extremeña del Tajo. Acerca del poblado con necrópolis del Canchal en Jaraíz de la Vera (Cáceres). *Extremadura Arqueológica* VIII, 209-242.

Castañeda Clemente, N. and Matesanz, R. 2006. Excavación arqueológica yacimiento número 2 Plasencia sur-Cañaveral este. *Extremadura Arqueológica* X, 73-90.

Cerrillo Cuenca, E. 1999. La cueva de El Conejar (Cáceres): avance al estudio de las primeras sociedades productoras en la penillanura cacereña. *Zephyrus* 52, 107-128.

Cerrillo Cuenca, E. 2005. *Los primeros grupos neolíticos de la cuenca extremeña del Tajo*. Oxford: Archaeopress.

Cerrillo Cuenca, E. 2006 (ed.). *Los Barruecos: primeros resultados sobre el poblamiento neolítico de la cuenca extremeña del Tajo*. Mérida: Junta de Extremadura, Dirección General de Patrimonio Cultural.

Criado Boado, F. 1989. Megalitos, espacio y pensamiento. *Trabajos de Prehistoria* 46, 75-98.

Criado Boado, F. 1993. Límites y posibilidades de la Arqueología del Paisaje. *SPAL* 2, 9-55.

Criado Boado, F. 1999. *Del Terreno al Espacio: planteamientos y perspectivas para la Arqueología del Paisaje*. Santiago de Compostela: Grupo de Investigación en Arqueología del Paisaje, Universidad de Santiago de Compostela.

Criado Boado, F. and Mañana Borrazás, P. 2003. Arquitectura como materialización de un concepto. La espacialidad Megalítica. *Arqueología de la Arquitectura* 2, 103-111.
Criado Boado, F. and Villoch Vázquez, V. 1998. La monumentalización del Paisaje: percepción actual y sentido original en el Megalitismo de la Sierra de Barbanza (Galicia). *Trabajos de Prehistoria* 55/1, 63-80.
Fábrega-Álvarez, P. 2006. Moving without destination: a theoretical, GIS based determination of routes (optimal accumulation model of movement from a given origin). *Archaeological Computing Newsletter* 64, 7-12.
Fábrega Álvarez, P. and Parcero Oubiña, C. 2007. Proposals for an archaeological analysis of pathways and movement. *Archeologia e Calcolatori* 18, 121-140.
Fernández Freire, C. 2008. *Paisajes agrarios pre y protohistóricos en la comarca de La Vera Alta (Cáceres): un enfoque arqueo-geográfico*. Madrid: Departamento de Prehistoria, U. Complutense de Madrid.
Fletcher, M. and Lock, G. 2005. *Digging numbers. Elementary statistics for archaeologists*. Oxford: Oxford University School of Archaeology.
Foucault, M. 2008 [1970]. *El orden del discurso*. Barcelona: Fábula Tusquets.
Gell, A. 1992. *The Anthropology of Time: Cultural Constructions of Temporal Maps and Images*. Oxford: Berg.
Gianotti, C. 2014. Procedimientos para el análisis de la movilidad prehistórica entre los constructores de cerritos mediante el uso de tecnologías geoespaciales. *Revista del Museo de Antropología* 7/2, 271-284.
Gil García, F. M. 2003. Manejos espaciales, construcción de paisajes y legitimación territorial: En torno al concepto de monumento. *Complutum* 14, 19-38.
Hernández Giménez, F. 1967a. Los caminos de Córdoba hacia Noroeste en época musulmana. *Al-Andalus* 32/1, 37-123.
Hernández Giménez, F. 1967b. Los caminos de Córdoba hacia Noroeste en época musulmana. (Conclusión). *Al-Andalus* 32/2, 277-358.
Hernando, A. 1999. Percepción de la realidad y Prehistoria. Relación entre la construcción de la identidad y la complejidad socio-económica en los grupos humanos. *Trabajos de Prehistoria* 56/2, 19-35.
Ingold, T. 2011. *Being Alive: essays on movement, knowledge and description*. London: Routledge.
Leisner, G. and Leisner, V. 1956. *Die Megalithgräber der Iberischen Halbinsel. Der Westen*. Berlin: Walter de Gruyter & Co.
Leisner, G. and Leisner, V. 1960a. El Guadalperal. *Madrider Mitteilungen* 1, 20-73.
Leisner, G. and Leisner, V. 1960b. *El Guadalperal: in memoriam Hugo Obermaier*. Heidelberg: Kerle.
Llobera, M. 1996. Exploring the topography of mind: GIS, social space and archaeology. *Antiquity* 70/269, 612-622.
Llobera, M. 2000. Understanding movement: a pilot model towards the sociology of movement, in: Lock, G. (ed.). *Beyond the Map: Archaeology and Spatial Technologies*. Ravello: IOS Press, 65-84.
Llobera, M. 2005. The Nature of Everyday Experience: Examples from the Study of Visual Space, in: Fisher, P. and Unwin, D. J. (eds.). *Re-presenting GIS*. Chichester; John Wiley & Sons Ltd., 171-192.

Llobera, M. 2006. Arqueología del Paisaje en el Siglo XXI. Reflexiones sobre el uso de los SIG y modelos matemáticos, in: Grau Mira, I. (ed.). *La aplicación de los SIG en la Arqueología del Paisaje*. Alicante: Universidad de Alicante: 109-124.

Llobera, M. 2015. Working the Digital: Some Thoughts from Landscape Archaeology, in: Chapman, R. and Wylie, A. (eds.). *Material Evidence. Learning from Archaeological Practice*. Oxon: Routledge, 173-188.

Llobera, M., Fábrega-Álvarez, P. and Parcero-Oubiña, C. 2011. Order in movement: a GIS approach to accessibility. *Journal of Archaeological Science* 38, 843-851.

Lock, G. and Molyneaux, B. L. 2006. Introduction: Confronting, in: Lock, G. and Molyneaux, B. L. (eds.). *Confronting scale in archaeology*. Boston: Springer, 1-14.

López-Romero González de la Aleja, E. 2005. *Arqueología del Paisaje y Megalitismo en el centro-oeste peninsular. Evolución de las pautas de poblamiento en torno a la cuenca del río Sever (España-Portugal)*. Madrid: Departamento de Prehistoria y Arqueología, Universidad Autónoma de Madrid.

Menéndez Pidal, G. 1851. *Los caminos en la Historia de España*. Madrid: Ediciones Cultura Hispánica.

Merleau-Ponty, M. 1975. *Fenomenología de la percepción*. Barcelona: Península.

Murrieta Flores, P. A. 2007. *Mobility, transhumance and prehistoric landscape: A GIS approach to the archaeological landscape of Almaden de la Plata in Andalucia, Spain*. Southampton: Department of Archaeology, University of Southampton. MSc in Archaeological Computing.

Murrieta Flores, P. A. 2012. Understanding human movement through spatial technologies. The role of natural areas of transit in the Late Prehistory of South-western Iberia. *Trabajos de Prehistoria* 69/1, 103-122.

Parcero Oubiña, C. 2002. *La construcción del paisaje social en la Edad del Hierro del Noroeste ibérico*. Ortigueira: Fundación Ortegalia.

Sarasola Etxegoien, N. 2006. Excavación arqueológica realizada en el subtramo Aldeanueva del Camino – Villar de Plasencia (Casas del Monte, Cáceres). *Extremadura Arqueológica* X, 15-28.

Shanks, M. and Tilley, C. 1987. *Social theory and Archaeology*. Cambridge: Polity Press.

Thomas, T., Sheppard, P. and Walter, R. 2001. Landscape, Violence and Social Bodies: Ritualized Architecture in a Solomon Islands Society. *The Journal of the Royal Anthropological Institute* 7/3, 545-572.

Tilley, C. 1994. *A Phenomenology of Landscape*. Providence: Berg.

Tobler, W. 1993. *Three presentations on geographical analysis and modeling*. Technical Report 93, 1. Santa Barbara, California: National Center for Geographic Information & Analysis (U.S.), University of California.

Vicent García, J. M. 1991. Fundamentos teórico-metodológicos para un programa de investigación arqueo-geográfica, in: López García, P. (ed.). *El cambio cultural del siglo IV al II milenios a.C. en la comarca noroeste de Murcia*. Madrid: Consejo Superior de Investigaciones Científicas, CSIC, 31-118.

Wheatley, D. W. and Murrieta Flores, P. 2008. *Grandes piedras en un mundo cambiante: la arqueología de los megalitos en su paisaje*. PH: Boletín del Instituto Andaluz del Patrimonio Histórico 67, 24-33.

Notes on contributor

Jose M. Señorán is a PhD student from the Complutense University of Madrid. His doctoral thesis is focused on the origins of the megalithic phenomenon in Alta Extremadura. He has also worked in the Institute of Heritage Sciences (CSIC), Santiago de Compostela. He is a member of the ARPA Research Team "Rock art, landscape and settlement in the High Atlas (Oukaïmeden, Morocco)", (Complutense University of Madrid). He is currently a co-director of the project "A postcolonial archaeology in Spain. Materialities and subaltern memories of the agricultural and industrial colonization in the twentieth century", with Dr. Xurxo M. Ayán Vila (UPV-EHU).

The answer is blowing in the wind

A method for measuring wind-protection as a criterion for settlement in the past

Marcos García García[1]

Use the Force!

'Don't be too proud of this technological terror you've constructed. The ability to destroy a planet is insignificant next to the power of the Force.' (Darth Vader, Star Wars, 1977).

In 2007, I attended the Second Edition of the course on Geographic Information Technologies (TIG in its Spanish acronym) back in the days when it was organised by the Archaeological Institute of Mérida. It was the first time that I had been involved in any kind of training in computer applications for Archaeology. My goal was to familiarise myself with this technology in order to learn new techniques for my PhD in Spatial Archaeology. Understandably, as a beginner, I was amazed by the characteristics of the software. I discovered some sophisticated programs, complex tools that delivered professional-quality maps and a huge amount of data difficult to manage. All of this constituted, in the words of Darth Vader, a true '*technological terror*'.

Looking back at that time, these courses can be considered to have been responsible for a large part of the innovation and success contained in recent theses defended in Spain. They have revealed the potential of such technologies (GIS, Remote Sensing, GPS-assisted surveying, etc.) to a younger generation of archaeologists. But, above all, during the course, we were warned against the risks of becoming fascinated by the machines, by the sophistication of the software, and by the novelty created by the illusion of doing something revolutionary. Rather, the important element was the thinking of the researcher, his or her ideas, his or her hypotheses, his or her intention of expanding our knowledge of the past. In other words, all of the aspects that, continuing with the comparison, we can call the '*Force*'.

---

1 Archaeologist, PhD in Prehistory marcggarc@gmail.com.

In this paper, I shall present an example of how theory can drive a technology to create an application which solves an archaeological question: Were prehistoric locations chosen according to whether they were protected from the wind?

Background: the question within its context

This question was one among many which I attempted to solve in my Doctoral Thesis (García García 2017). It was a study which, from the perspective of Landscape Archaeology, aimed to analyse the evolution of the societies which inhabited the central Duero valley between 2800-450 cal BC. The study area covered a total of 2483 km^2 encompassing three natural regions in the center of Castile and León (Spain): the plain of Tierra de Campos, the Torozos hillside and the lower Pisuerga valley (figure 1). The chronological scope of the analysis was the Metal Ages in the Duero valley: an interval of about 2,350 years, ranging from the Middle Chalcolithic (*c.* 2800 cal BC) to the end of the Early Iron Age (*c.* 450 cal BC ANE).

This research was set within the framework of Landscape Archaeology. In my work, I approached landscape as an object providing information to construct a discourse about past societies. Considering landscape as an expression of space and time, that is, as a dynamic and accumulative (stratigraphic) social construction, it is possible to study it from an archaeological perspective. This is the '*archaeo-geographic approach*' defined by Vicent (1991): a hypothetico-deductive method, the aim of which is to test hypotheses on what we do not know about social processes in light of what we do know. What we knew at the start of the research was the distribution of prehistoric sites and the existence of structuring factors in the present-day landscape which configure the dynamics of human occupation. The archaeological (time) axis of the research was subject to several methodologies of improvement, including two archaeological survey campaigns (2007-2009), which will be described in further detail at a later stage. Geographic factors (mainly relief and climate) were broken down into measurable variables (gradient, altitude, temperature and so on) so that they could first be translated into a mathematical and, then, into a computer language. Some of these variables were interpolated using GIS software, thereby defining a landscape factorial model. Such a model corresponds to a paleotechnic environment and, therefore, I took into consideration those factors which may have had an impact on a lifestyle the technological level of which we already have knowledge. Therefore, the landscape factorial model was based on today's landscape, while, at the same time, being informed by historical knowledge.

The other major methodological tool employed in our research was location analysis, defined as an analytical procedure intended to trace the decisions of past societies when choosing a particular spot to settle. As was stated at the beginning, the project was led by theory, so we established a consistent plan of questions, a sequence of model-proposal-testing, which, based on the knowledge we have (the position of sites and the factorial model) may help us to access what remains hidden. The historical hypotheses were laid out as a decision tree, thus enabling us to verify them in a systematic way and to even consider possibilities which, otherwise, may have been ignored. The questions making up the location analysis

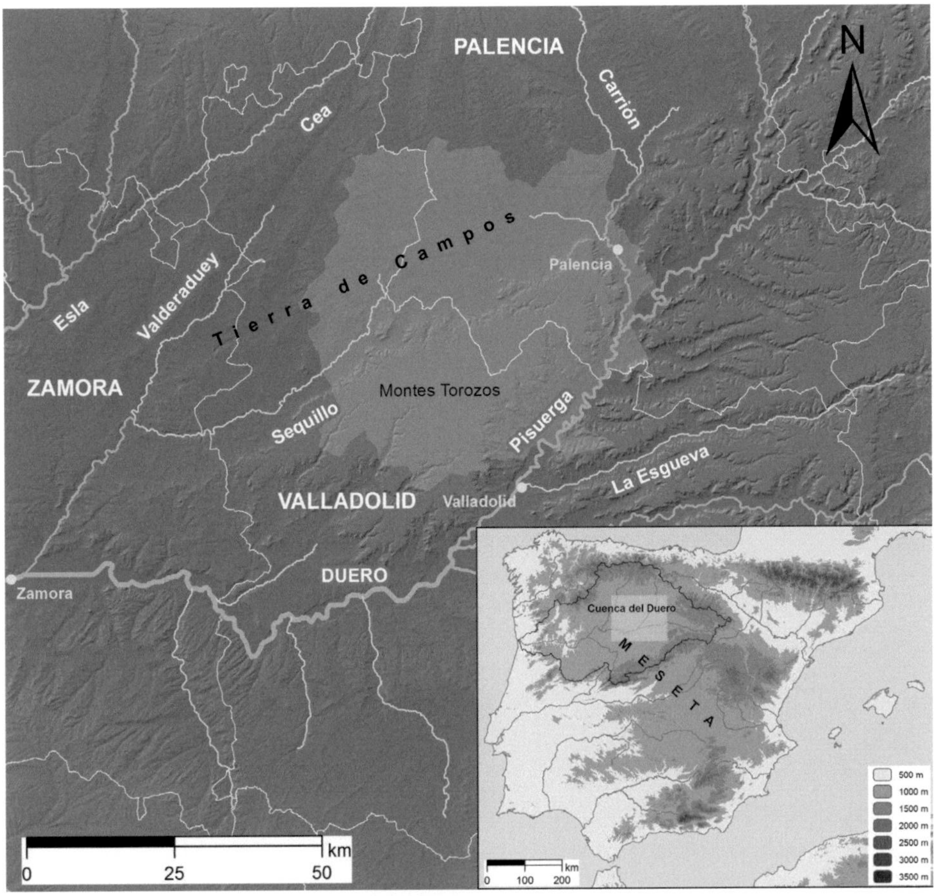

Figure 1: The study area is shaded in the main window. The inset shows the position of the study area in the Duero valley within the Iberian Peninsula.

were structured into three major categories: '*production*', '*location suitability*', and '*site size and distribution*'. The question which shall be dealt with in this paper was part of the '*location suitability*' group, which attempted to define the criteria leading to a given spot being considering appropriate for settlement, whether those aimed for a greater degree of comfort (not exposed to wind in this example) or those aiming for a better strategic position.

The fieldwork experience: a question arises

The necessity of carrying out fieldwork arose in order to review the chronological variables, since the archaeological information of our study area presents several problems. A first set of difficulties concerns the condition of the archaeological record, since, due to its compilation from surface remains, it is fragmentary and presents some voids. Another set of problems was posed by the structure of the Archaeological Inventory of Castile and León (IACyL). This inventory has a primarily administrative nature. Therefore, the sample of sites responds to managerial criteria and does not derive from a well thought out research plan. This involves an overrepresentation of the most visible types of site, either because they are at risk or because they were discovered by informants. Likewise, the IACyL is vague regarding the location of sites and makes excessive use of ambiguous chronological categories such as 'possible' or 'indeterminate prehistoric'.

In the light of these problems, we took the decision to review the archaeological information from the IACyL and to build a suitable database for our location analysis. For this reason, two archaeological survey campaigns (2007-2009) were carried out. Our objectives were to locate new sites, to better define the position and chrono-cultural attribution of some sites we already had knowledge of and to assess the prospecting strategies behind the compilation of the IACyL.

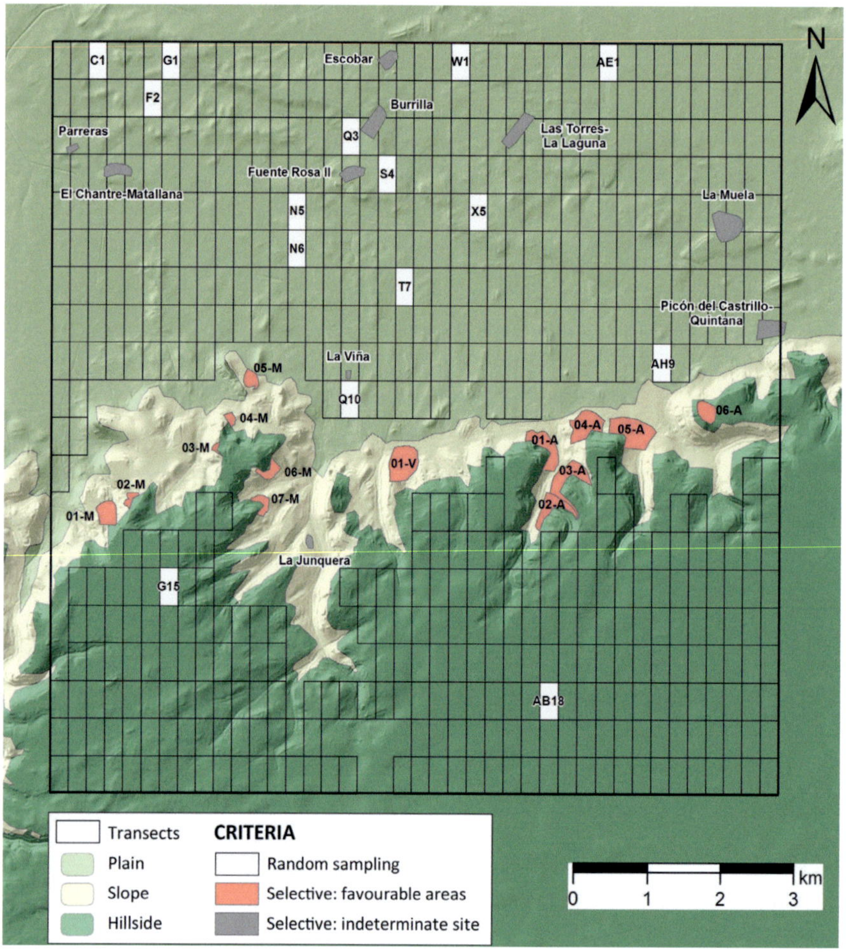

Figure 2: Areas surveyed during the first campaign, showing types of archaeological survey (random sampling and selective). The three zones into which the area was divided are also shown, along with the names of each surveyed area and the transect fishnet, which was later subject to a random selection.

In the first campaign (2007-2008) two different methodologies were used: firstly, a sample of 15 randomly selected transects of 500 x 250 m; and, secondly, selective prospection in sites labelled as 'undetermined prehistoric' and favourable areas according to the inventory criteria (figure 2). The results of the survey were quite satisfactory: those sites defined as 'undetermined prehistoric' were reduced by 31% and chronological and cultural attributions deemed to be certain increased by 17%.

However, the selective survey on favourable areas was, at first, disappointing: no prehistoric sites were found in these zones, despite the fact that the conditions looked promising: the smooth slopes of the Torozos moor descending to the fertile soils of the Tierra de Campos plain, at a maximum distance of 300 m from the closest river. I figured out a possible explanation for the absence of prehistoric remains in those more suitable areas: they might not have been such attractive locations if exposure to cold winds is taken into account. Indeed, my own experience during fieldwork facilitated this idea since, during the survey of these areas, we experienced strong cold winds blowing from the northeast.

I understood that the decision to avoid this kind of climatic annoyance could have been as desirable in Prehistory as it is today. From that moment on, I incorporated exposure to wind as a potential location criterion for settlements in the past. It then became necessary to design a methodology to measure this criterion quantitatively for the whole study area.

Dealing with the problem: GIS methodology

From the very first moment, it was clear that the tool to measure wind exposure should be implemented in GIS software, since every geographic and archaeological variable had been included in a database. Despite my decision, no specific algorithm could be found for this purpose, neither in the predesigned toolbox of *ArcGis* or in any published paper on the matter. This meant that I would have to design a tool from scratch which would have to combine the orientation of a given location and the effect of the wind upon it. Both elements can be simulated in GIS software, with the former being slightly easier than the latter.

I first used the 'aspect' tool to calculate the orientation of the cells of my DEM (Digital Elevation Model). I worked with a DEM of 10 m resolution. The 'aspect' tool returns a map with values ranging from -1 to 359: the -1 value represents flat areas; values between 0 and 359 represent sexagesimal degrees, with 0 representing north. After generating the aspect map, I reclassified the resulting values into the eight main wind directions (figure 3a). Wind is a climate factor which is difficult to generalise for a wide study area, due to its continuous changes in speed, direction and duration. It is also dependent on topography, so the relief surrounding measuring stations must be taken into consideration. In an attempt to solve this problem, I built a wind rose for the study area, using data collected over a 5-year period (2004-2009) from weather stations in very different locations: Medina de Rioseco (on the open plain of Tierra de Campos) and Villamuriel de Cerrato (a less exposed site in the Pisuerga valley), both located in the province of Valladolid. As can be observed (figure 3b), the mean of the collected data shows that both prevailing (more frequent) and dominant (faster) winds follow a

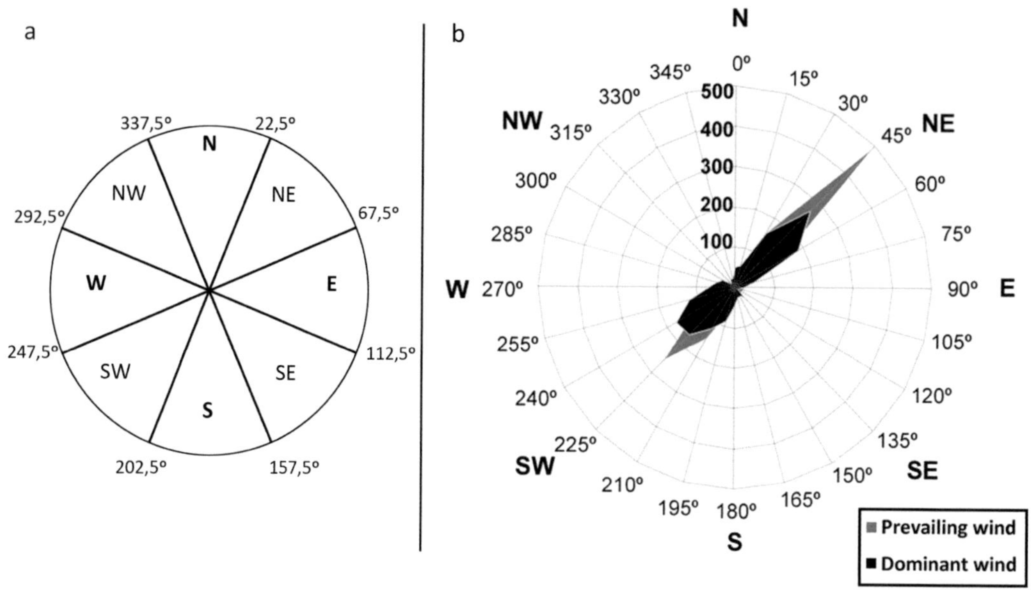

Figure 3: (a) Thresholds expressed in degrees for the reclassification of the aspect map into a map of the eight main wind directions. (b) A wind rose of the study area, built from data collected over a 5-year period (2004-2009) from two weather stations (Medina de Rioseco and Villamuriel de Cerrato, both located in the province of Valladolid); it represents the frequency in number of days (vertical axis) of wind directions (radial axis every 15°) for prevailing and dominant winds.

northeast-southwest axis. This wind line is consistent with the orientation of the Duero basin and the layout of natural regions. In addition, it implies that wind is a relatively uniform factor in the study area, thus making it easier to avoid when choosing a location to live, if it is necessary to do so.

The other variable necessary for measuring which areas are exposed to wind action was a map representing wind circulation in a given direction. There is no specific tool in GIS software for this task, so I looked for an algorithm which could simulate a continuous flux over the DEM. *ArcGis* uses the 'viewshed' tool to create solar radiation maps, which is also based on the continuous action of the Sun over certain areas of relief. Therefore, I worked around the lack of a specific tool by imitating the behaviour of GIS software: wind can be simulated using the algorithm for visible and invisible locations, which, in our case, were locations exposed, or not, to wind. The 'viewshed' algorithm calculates which cells of a DEM can be seen from a certain position, taking into consideration the height of the viewer and of the object to be seen. Translating this behaviour for our purposes required the creation of several 'viewshed' maps, at least one for each wind direction, which had to be combined with the orientation of relief, i.e. the aspect map which was explained above. It was also necessary to establish a height for the 'viewer' and the 'object': the first was the maximum height of the DEM (840.4 m) and the latter was 1.6 m, which was the mean height of the people of Iberia during the Bronze Age (Ruiz-Gálvez 1998: 126).

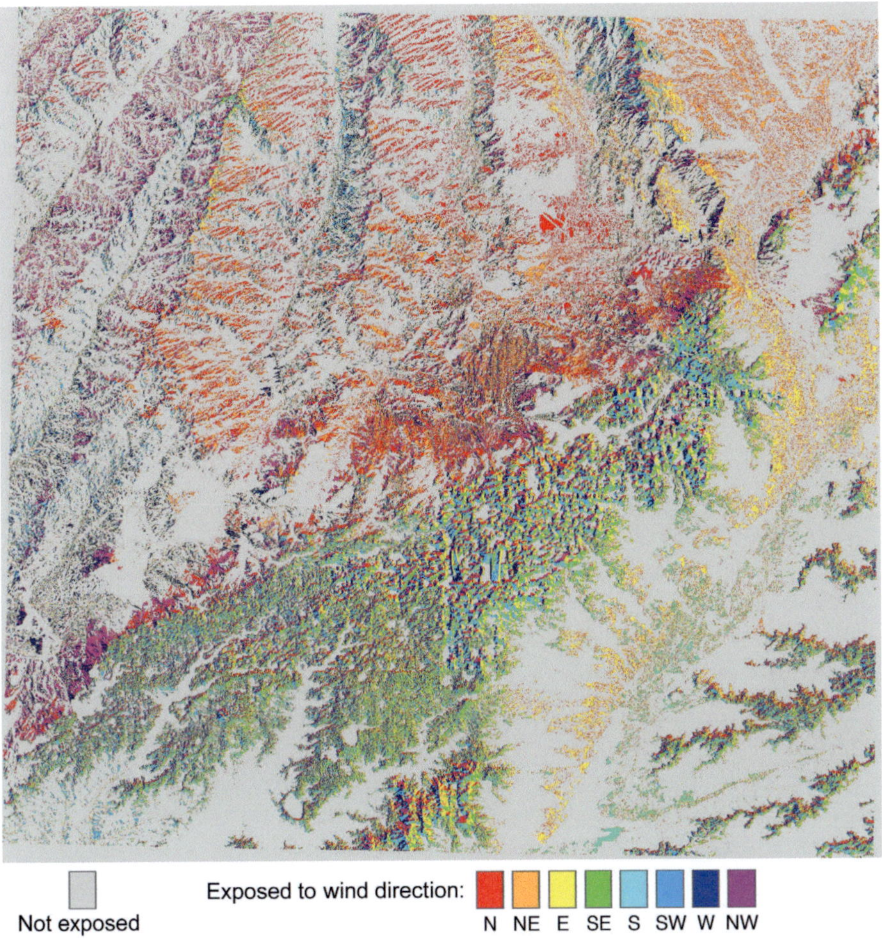

Figure 4: Map showing exposure to wind. The legend indicates whether the area is not affected by winds or, if affected, in which direction the wind blows.

The steps required in GIS software to create an 8-direction wind exposure map may be summed up as follows:

1. Apply 'aspect' tool to the DEM of the study area in order to obtain a map of orientations.
2. Reclassify the aspect map as follows: 0 for flat areas (-1 in the original aspect map), 1 to 8 for the eight main wind directions expressed in degrees (figure 3a). A more complicated reclassification is also possible (16 directions), depending on the climatic information available for the study region.
3. Then, several reclassifications are needed to individualise each orientation of the aspect map. In our example nine maps were created: a map of flat areas, another of areas facing north, a third one facing northeast, etc.
4. The creation of a multipoint layer of eight cardinal points at the edge of the DEM. For accuracy, we recommend using the XY position tool. In the table of attributes of the multipoint layer, a height field called

'OFFSETA' was created with a value equal to the maximum height of the DEM (840.4 m in our example).

5. Execute 'viewshed' for the multipoint layer, choosing the desired height for the observed object (OFFSETB). In our example, we considered that, as the wind affects humans, we should choose the mean height of prehistoric people (1.6 m in the Iberian Bronze Age). Eight viewshed layers would be obtained, one for each observer point, in other words, one for each cardinal point.
6. Each viewshed layer was combined with its corresponding orientation layer obtained in step 3 using the 'raster calculator'. For example, the viewshed layer from the southern observation point was multiplied by the layer representing the areas oriented to the south. The result was a map representing all the cells facing south which are exposed to southern winds. As for the flat areas, they must be combined with the eight wind directions, in other words, with the eight viewshed layers, since those cells are not protected from the wind in any direction.
7. Finally, the maps obtained in step 6 must all be joined together, thereby creating a single map of wind exposure (figure 4). It will have nine values, one representing areas that are not affected by the wind, while the other eight values indicate areas exposed to the eight wind directions.

Getting an answer

Once the tool was ready, it was time to ask the questions. On the wind-exposure map, 166 archaeological sites from the Copper to the Iron Age were subject to the same question: was the fact that a site was not affected by the wind a criterion to settle there? Results pointed clearly to an affirmative answer (figure 5): 70% of the prehistoric sites were in locations not exposed to any wind; if it is taken into consideration that prevailing and dominant winds come from the NE and SW (figure 3b), it can also be stated that 95% of prehistoric sites were not exposed to

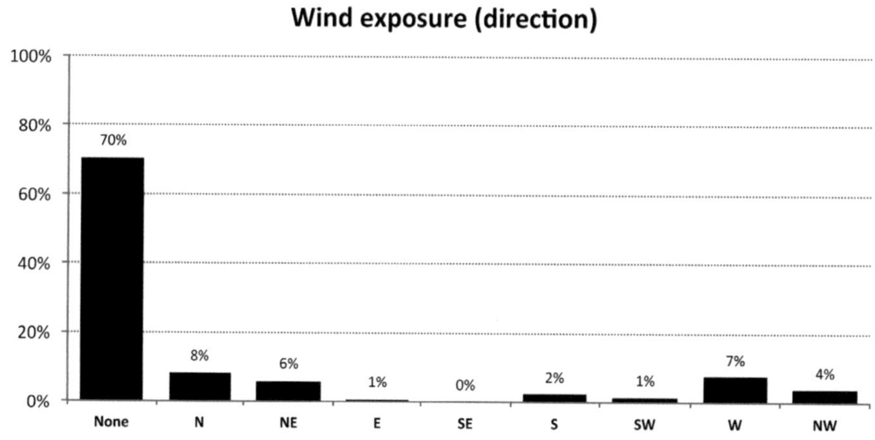

Figure 5: Results of wind exposure test (I): The number of prehistoric sites affected by each wind direction (horizontal axis). Prevailing and dominant winds coming from the NE and SW.

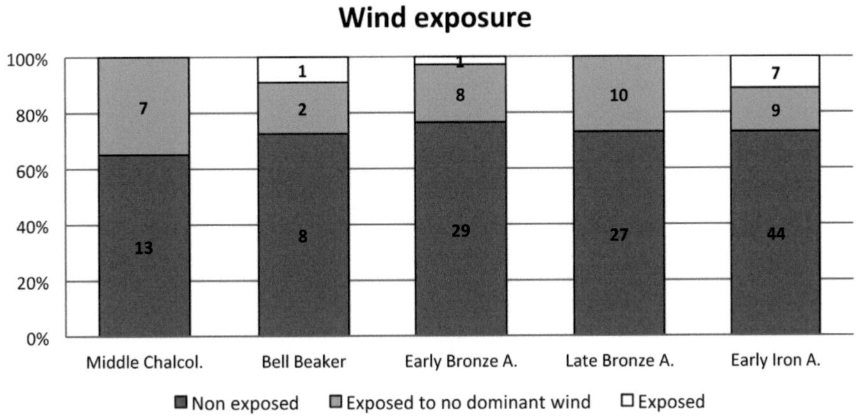

Figure 6: Results of wind exposure test (II): sites classified by archaeological periods and their exposure to wind. 'Exposed' means that sites are affected by prevailing and dominant winds (NE and SW).

any dominant wind. In the Middle Chalcolithic (*c.* 2800-2350 cal BC) and Late Bronze Age (*c.* 1450-1000 cal BC) percentages reach 100% (figure 5).

The values presented were above any random result, so it can be claimed that our perception during fieldwork was also shared in the past. As was observed in further tests included in my Doctoral Thesis, avoiding wind was preferred over other benefits, such as being in places with higher solar radiation or with good access to the surrounding area. Both criteria are related: a position sheltered from the wind tends to be an enclosed area, with more likely limitations for movement or for being exposed to sunlight.

'The answer, my friend, is blowing in the wind'

Here, a simple way of quantifying the incidence of the wind in the choice of the location of a settlement has been presented. From these results, it can be concluded that avoiding areas exposed to dominant winds was a key location criterion in the middle Duero valley during Prehistory.

However, the goal of this paper was, above all, to show a research path, a way of thinking for any archaeologist being introduced to computer methodology. Questions come from experience, curiosity, reading, etc. They fuel our theorising, thereby developing and shaping it. Only after questions are formed can we look for the tools to answer them. Therefore, technology aids theory. And the answer, in the words of Bob Dylan, 'is blowing in the wind', waiting for the right questions to be asked.

The courses in Mérida were meant to train archaeologists in GIS or Remote Sensing, something new for Spanish archaeologists back then. Ten years later, they have not only shown their relevance in the field but they have also provided new generations of archaeologists with the right attitude to create knowledge about past societies.

Acknowledgements

This paper explains the employment of a particular methodology and summarises part of my Doctoral Thesis. That research was funded by two scholarships: one from the Spanish Government (the FPU programme) and another from the Historical Heritage Foundation of Castile and León. It was begun and continued thanks to these "Mérida courses", so credit goes to their original organisers: Victorino Mayoral and Sebastián Celestino. And it was finished thanks to the huge support and care of my directors, Dr. Germán Delibes and Dr. César Parcero.

References

García García, M. 2017. La Edad de los Metales en el Duero medio: la evolución del paisaje y de las sociedades. Studia Archaeologica, 102. Valladolid: Universidad de Valladolid.

Ruiz-Gálvez, M. L. 1998. *La Europa atlántica en la Edad del Bronce. Un viaje a las raíces de la Europa Occidental.* Barcelona: Crítica.

Vicent, J. M. 1991. Fundamentos teórico-metodológicos para un programa de investigación arqueo-geográfica, in: P. López (ed.). *Cambio Cultural del IV al II Milenios a. C. en la Comarca NW de Murcia.* Madrid: CSIC, 29-119.

Notes on contributor

Marcos García García has a PhD in History from the University of Valladolid (Spain) and is a freelance researcher on Prehistory and Archaeology. His main interest is social change in the Iberian Age of Metals, which he has addressed with a Landscape Archaeology approach. He has studied settlement patterns in the Duero valley and the significance of causewayed enclosures in the same region. In his studies, methods such as GIS, Remote Sensing and archaeological surveying are always present. He is now part of a research project focused on causewayed enclosures in Southern Iberia in collaboration with the University of Málaga.

Section 3

Patterns, behaviour, decisions

Analysing the archaeological evidence

The Application of GIS to flint management studies during the Pleistocene to Holocene transition: the case of Baltzola (Dima, Bizkaia, Spain)

Maite García-Rojas,[1-2] *Alejandro Prieto,*[2]
Aitor Sánchez,[2] *Cristina Camarero*[2]
and Lydia Zapata (†)[2]

Introduction

Over the course of the last decade, archaeological studies involving Geographic Information Systems (GIS) have increased greatly in the context of the Iberian Peninsula. At the same time, there has been an emphasis on the research of lithic raw materials and the understanding the catchment areas of lithic resources; with all of this being entangled with the articulation of economical territories and patterns of mobility of Palaeolithic groups (Terradas 2002; Mangado 2006; Djidjian 2009). Overall, these studies suppose an advance in the accurate location and petrological characterisation of flint catchment sources in many geographic areas of the Iberian Peninsula (Sarabia 2000; Bustillo and Peréz-Jiménez 2005; Mangado 2005; Tarriño 2006; Tarriño *et al.* 2015; Fuertes-Prieto *et al.* 2014; Sánchez de la Torre 2015; Molina Hernández 2015). Despite the multiple applications offered by GIS, *Cost Distance Analysis* among them, they have barely been used to explore the correlation between flint outcrops and sites. Furthermore, the majority of researchers still

1 Dipartimento di Studi Umanistici Sezione di Scienze Preistoriche e Antropologiche. Universita di Ferrara.
2 Departamento de Geografía, Prehistoria y Arqueología. Área de Prehistoria. Universidad del País Vasco/Euskal Herriko Unibertsitatea (UPV/EHU).

present these relationships in a linear, geographically abstract, manner with no concern for the geographic constraints which would have influenced the gathering of lithic resources.

The use of GIS to help understand the mobility of human groups has a considerable background in Iberian archaeological research, particularly with regard to the Holocene (e.g. Fairén 2004; Llobera *et al.* 2011; Murrieta Flores *et al.* 2009; Murrieta Flores 2010). Furthermore, there has been some research aimed at the study of settlement patterns (García 2010; Turrero *et al.* 2013; Ríos and García 2015; Fontes 2016) and land management of faunal resources during the Palaeolithic (Coward 2004; Marín 2008).

To date, lithic catchment studies have been more limited with some good work being carried out in the *Foz Côa* valley in Portugal (Aubry *et al.* 2012), in the Trubia valley in Asturias (Fernández 2010) or the *Asón* river in Cantabria (Rissetto 2009; 2012). Within that line of research, we have used the concept of *Cost Distance* to approach flint catchment patterns of groups from the Upper Palaeolithic in the Western Pyrenees (García-Rojas 2014; Prieto *et al.* 2016; Sánchez *et al.* 2016).

In brief, we aim here to understand the unequal importance of different flint outcrops by analysing the distribution of their raw materials through the use of *Cost Distance* calculations and to understand the patterns of catchment used over time by the prehistoric groups living in the different sites analysed here. In addition, this study seeks to relate geographic information with the techno-typological analysis of the lithic industry at the site of *Baltzola*. With this correlation, we attempt to analyse the management of flint by prehistoric humans in this cave during the transition from the Pleistocene to the Holocene.

Context and Materials

The cave of Baltzola (Coordinates UTM 30N: X:522468 Y:4774409) is located in the natural park of Urkiola in the municipality of Dima (Biscay, Spain). It is a great cavern bordered by the Indusi river, an affluent to the Arratia river within the Ibaizabal river basin. This site is located on the drainage divide at an altitude of 370 metres above sea level.

The first archaeological studies of this site were carried out by R. Jaggor in 1866 and by Gavez Cañero in 1912. Subsequently, Jose Miguel de Barandiarán carried out another excavation in 1932, in which he described three levels belonging to the Magdalenian, Azilian and Neolithic periods (Barandiarán 1932). The last excavation was carried out by Lydia Zapata between 2006 and 2010 with the aim of updating the stratigraphic sequence and collecting samples for diverse analyses. This intervention revealed a repeated occupation of the cave by human groups during the later episodes of the Upper Pleistocene and the beginning of the Holocene, in a chrono-cultural range from the Upper Palaeolithic, Azilian and Neolithic (Zapata *et al.* 2007; 2008; 2009; 2010; 2011). In this study, we shall analyse the industrial lithic assemblages from this last excavation, which belong to the Azilian level 9 (10,431-9,825 cal BC 2 sigma) and Magdalenian levels 6 (10,910-10,752 cal BC 2 sigma) and 7 (13,014-12,253 cal BC 2 sigma).

A total of 741 lithic implements have been analysed in the three archaeological levels according to the following distribution: 481 in level 9, 207 in level 6 and 53 in level 7. The flint found in Baltzola was extracted from 4 sources: Flysch Kurtzia (Barrika, Biscay), Treviño (Sierra de Araico-Cucho, Burgos), Urbasa (Sierra de Urbasa, Navarre) and Loza (north of the Sierra de Cantabria, Álava). The types of flint identified, as well as the location of the source outcrops, have recently been analysed by Tarriño *et al.* (2015).

Methodology

For this study, some of the spatial analysis tools offered by ArcGIS 10.2 were used. The main data layer was the Digital Elevation Model (MDT25, according to its Spanish acronym) created by the Spanish National Geographic Institute (IGN 2016). All of the data included in this study were georeferenced following the UTM-ETRS89 (zone 30N) system of projected coordinates. We acknowledge the emphasis other studies have made on submarine topography, agents such as vegetation cover, lithology, the fluvial network or weather as elements of the landscape (Leorri *et al.* 2013; Iriarte *et al.* 2007-2008), all of which have had a great influence on the mobility of the groups. However, the necessity of accurately modelling their effects on each period hinders the systematic inclusion of these elements in the study. Although they will be taken into consideration when drawing conclusions, we avoided including them in the data analysis process.

Our methodology will allow us to relate the archaeological sites with flint outcrops through geographic variations in two ways. Firstly, by taking the source outcrops as reference points in the analysis in order to understand the spatial distribution of the flint extracted from them. Secondly, by taking the archaeological settlement sites as reference points in the analysis in order to approach the issue of the management of flint by prehistoric groups. In this article, we have focused on the second case in order to understand how the people in Baltzola managed their flint resources.

Taking into account the bases and biases mentioned above, we proceeded as follows: Firstly, after merging the different portions of the MDT25 into a single layer (Figure 1.1), we calculated a slope map (Figure 1.2). This map was later reclassified in order to convert slope values into movement cost values (Figure 1.3), according to a model of cost assignment used in previous studies by our group (Prieto *et al.* 2016; Sánchez *et al.* 2016). These studies follow the proposals of Marcos Llobera (Llobera 2000) and López Romero (López Romero 2005) developed from the experiments of A. E. Minnetti (Minnetti 1995). Other formulas relating terrain slope and movement costs exist, such as those of Pandolf (Pandolf *et al.* 1977), Langmuir (Langmuir 1984), Tobler (Tobler 1993) or even more recent adaptations (Kramer 2010; Martín Arroyo 2009). These are, at present, a matter of contrast and debate (Kantner 2012), with no clear conclusions having been drawn as yet. Therefore, we took the decision to continue with our model, which had already been tested in the area. This model takes, as the unitary value of effort, the cost generated by moving at a constant speed of 5 kph over flat terrain. As a result, we obtained a new raster map in which pixel values correspond to the cost of walking on that section of terrain (Figure 1.4) based on the slope.

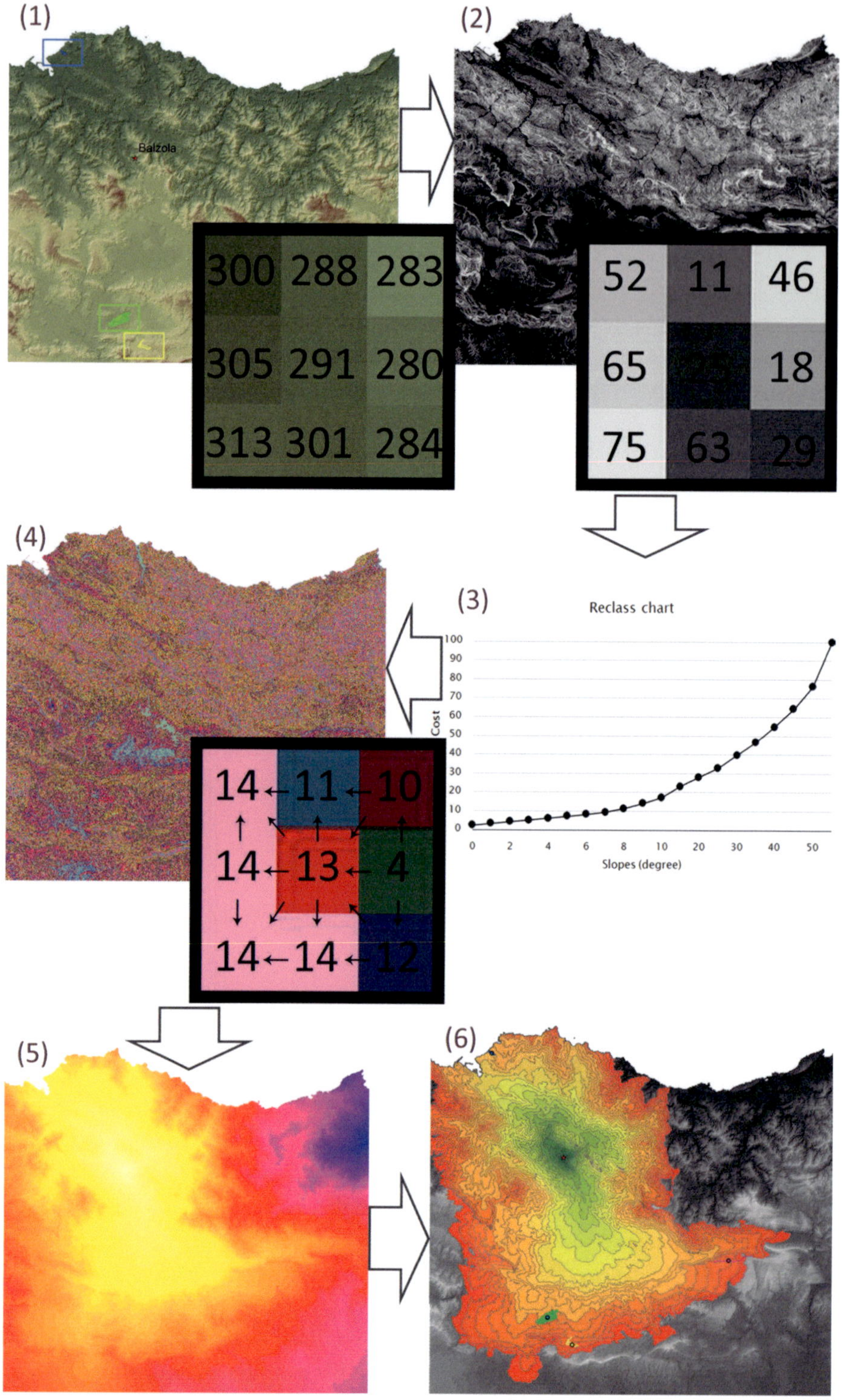

Figure 1: Methodological diagram showing the steps in calculating cost maps: 1) DEM with the location of Balzola cave and flint outcrops; 2) Slope map; 3) Cost reclassification function; 4) Friction surface; 5) Accumulated cost surface from Balzola; 6) Accumulated cost surface reclassified in discrete intervals.

The location of the flint outcrops was mapped on the basis of the geological units they are assigned to, from the information in the geological maps of the series MAGNA 1:50,000 provided by the *Instituto Geológico y Minero de España* (Spanish Geological and Mining Institute) (IGME, 2010a, IGME, 2010b, IGME, 2010c). The next step was to process a *Cost Distance* calculation from the location of the *Baltzola* cave, using the aforementioned Cost surface as a friction input in order to obtain an *Accumulated Cost Distance* map for *Baltzola*. This map shows the cost involved in moving from *Baltzola* to any position in the surrounding terrain, including the cost of reaching the different flint outcrops (Figure 1.5).

In order to simplify the representation of the accumulated costs, we have reclassified the resulting map by arbitrary discrete intervals of 15,000 cost units. This helps us to organise and visualise the results and to compare them with other studies. (Prieto *et al.* 2016) (Figure 1.6).

Subsequently, the same sequence was applied for the flint outcrops in order to obtain a series of maps representing the costs of movement from every single outcrop. Since, at present, we have no specific information on which specific outcrops were used as flint sources, we have established a conventional centre point (centroid) for the 4 polygons which delimit the extension of the flint bedrock areas in the National Geological Map of Spain (sectors of Kurtzia Flysch, Treviño, Urbasa and Loza) (Prieto *et al.* 2016).

Finally, we related the former data with generic techno-typological categories with the aim of understanding the flint management of these prehistoric societies. To that end, we used Analytic Typology, which focuses on the modal and morphological structures identified in the retouched tools (Laplace 1987; Fernández Eraso and García-Rojas 2013), the technical structures for raw reduction products (core trimming flakes, cores, blanks and retouched products) (García-Rojas 2010; 2014) and the petrological structure to identify different types of flint.

Results

As was mentioned previously, the analysis of the materials consisted of the techno-typological definition and identification of every source of raw material used. In relation to the techno-typological definition, we established four typological groups associated with the remains of the different stages of the technical process: 1) non-retouched blank (including blades and flakes); 2) core trimming flakes; 3) cores; 4) retouched products (Table 1). Each of the levels displayed a different distribution, according to the different economic activities recognisable at each moment (Garcia-Rojas 2014).

Table 1 details the distribution of the different types of flint in the three levels of Baltzola under analysis. The three stratigraphic levels are similar, with the Flysch type being the most common. In order to geographically relate the site with the different flint outcrops we have carried out a *Cost distance* geoprocess. In this way, we have been able to determine the effort involved in moving from the site to the sources of catchment of raw resources (Figure 2). As a result, we determined that all of the flint outcrops were located at a similar cost distance (between 16 and 23 CU). The average total cost would have been 19.75 units, with the Flysch and

Level	Lithic productions	Flint types									
		Flysch		Treviño		Urbasa		Loza		Indet.	Σ
9	Blanks	186		22		10		3			221
	Core trimming flakes	30		3		2		0			35
	Cores	7		0		0		0			7
	Retouched products	129		28		8		0			165
	Σ	352	82.24%	53	12.38%	20	4.67%	3	0.70%	53	481
6	Blanks	135		2		4		1			142
	Core trimming flakes	6		1		0		0			7
	Cores	3		2		0		0			5
	Retouched products	32		9		0		0			41
	Σ	176	90.26%	14	7.18%	4	2.05%	1	0.51%	12	207
7	Blanks	28		4		0		0			32
	Coree trimming flakes	5		0		0		0			5
	Cores	0		0		0		0			0
	Retouched products	12		0		0		0			12
	Σ	45	91.84%	4	8.16%	0	0.00%	0	0.00%	4	53

Table 1: Data analysed

Treviño outcrops being below average, while the Urbasa and Loza outcrops are above average.

Crossing these data with the percentages of flint types identified per level, we noticed an inverse correspondence between the cost involved in the displacement to the flint outcrop and the amount of flint types. In other words, higher accumulated cost values are linked to smaller amounts of the raw material in the stratigraphic levels and *vice versa* (figure 3). However, there is a disproportion between the amount of *Flysch* type flint and the other types and the CU progression. The pattern is a trend which is repeated throughout the three levels, both as far as the Magdalenian and the Azilian are concerned.

When flint types are distributed according to their relative frequencies and compared to cost, it can be seen that the difference increases, with *Treviño* being the outcrop behaving differently to the rest if we were to expect a linear relationship between proximity and use of sources. This means that even though the cost of reaching Treviño is below the average, the amount of flint remains from that source is underrepresented in the three archaeological levels (figure 4).

In order to gain a better understanding of the dynamics of flint catchment and management, the levels were divided into techno-typological categories. In this way, it could be determined whether there is any correlation between the three levels or, on the contrary, whether each one of the chrono-cultural periods displays its own characteristics (Table 1). The same analysis as was carried out in the previous case was repeated on this occasion, although, this time, splitting the flint remains into levels and techno-typological categories. The result is that, although each of them is adjusted to the main identified dynamics, some new trends emerge.

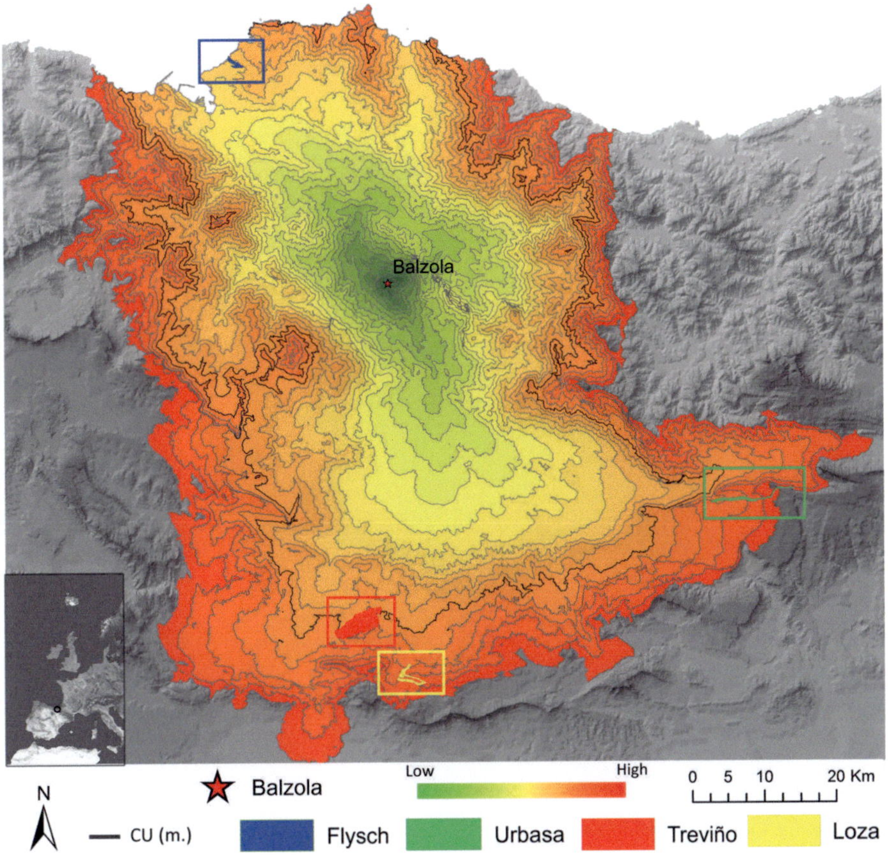

	FLYSCH	TREVIÑO	URBASA	LOZA	TOTAL (CU)	CU *m.*
CU (mdt25)	16	19	21	23	**79**	**19,75**

Figure 2: An accumulated cost map projected from the Baltzola cave, classified by Cost Units (CU), with reference to the flint outcrops (Flysch, Treviño, Urbasa and Loza). The numerical value of the cost unit where flint outcrops are found is included at the bottom.

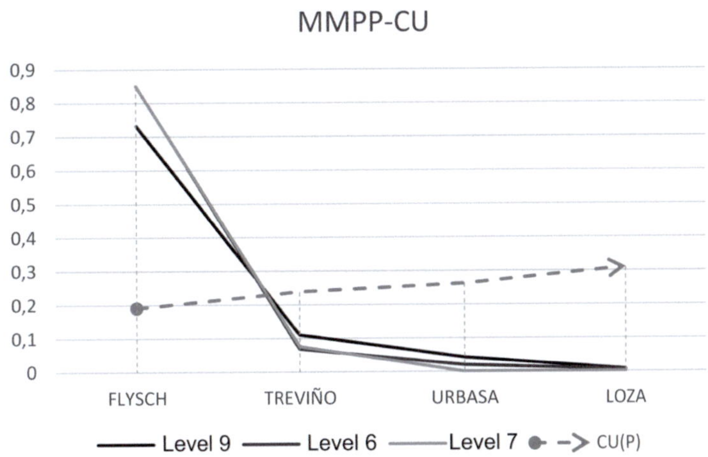

Figure 3: The relationship between Cost Units and the relative frequency of flint types identified in the Baltzola cave.

In level 9, there is an overrepresentation of flaking products, trimming flakes and cores within the Flysch type, while the retouched artefacts are below the average. The Treviño flint type shows an opposite trend to the Flysch type, while the Urbasa and Loza types are noticeable by their scarce overall representation (Figure 5).

Level 6 displays a different distribution. Only flaking products are above the average within the Flysch flint type and both cores and retouched products are underrepresented. The Treviño type, again, shows its own peculiarities on the

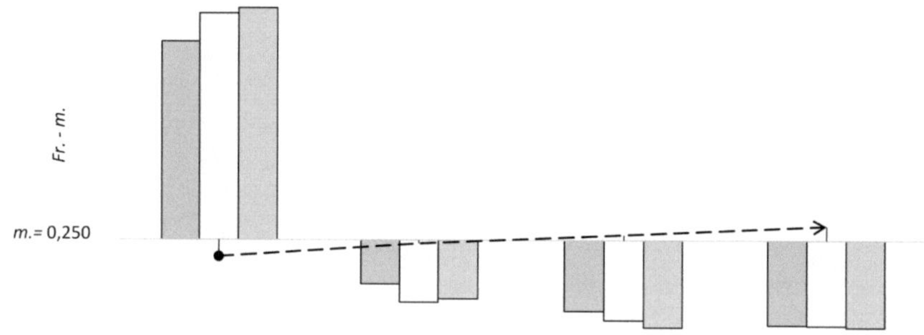

	FLYSCH	TREVIÑO	URBASA	LOZA
Level 9	0,572	-0,126	-0,203	-0,243
Level 6	0,653	-0,178	-0,229	-0,245
Level 7	0,668	-0,168	-0,250	-0,250
Fr.-m (CU)	-0,047	-0,009	0,016	0,041

Figure 4: Chart representing each type of flint per stratigraphic level based on the relative frequencies of each category in relation to the mean. The dotted line shows the representation of Cost Units based on the relative frequencies of each category in relation to the mean.

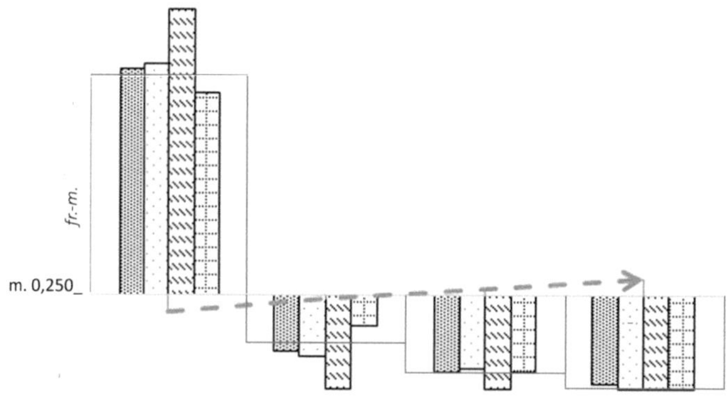

	Flysch	Treviño	Urbasa	Loza
Banks	0,592	-0,150	-0,205	-0,236
Core trimming flakes	0,607	-0,164	-0,193	-0,250
Cores	0,750	-0,250	-0,250	-0,250
Retouched products	0,532	-0,080	-0,202	-0,250
CU	-0,047	-0,009	0,016	0,041

Figure 5: Chart representing types of flint and technological categories from level 9 based on the relative frequencies of each category and type in relation to the mean. The dotted line shows the representation of Cost Units based on the relative frequencies of each category in relation to the mean.

shape of overrepresented cores, followed by retouched products and an almost complete absence of flaking products and trimming core flakes. Lastly, the Urbasa and Loza types are only present as flaking products (Figure 6).

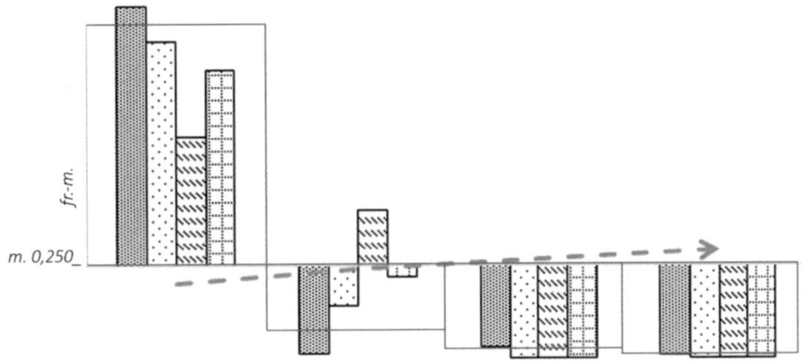

	Flysch	Treviño	Urbasa	Loza
Blanks	0,701	-0,236	-0,222	-0,243
Core trimming flakes	0,607	-0,107	-0,250	-0,250
Cores	0,350	0,150	-0,250	-0,250
Retouched products	0,530	-0,030	-0,250	-0,250
CU	-0,047	-0,009	0,016	0,041

Figure 6: Chart representing flint types and technological categories on archaeological level 6 based on the relative frequencies of each category and type in relation to the mean. The dotted line shows the Cost Units based on the relative frequencies of each category in relation to the mean.

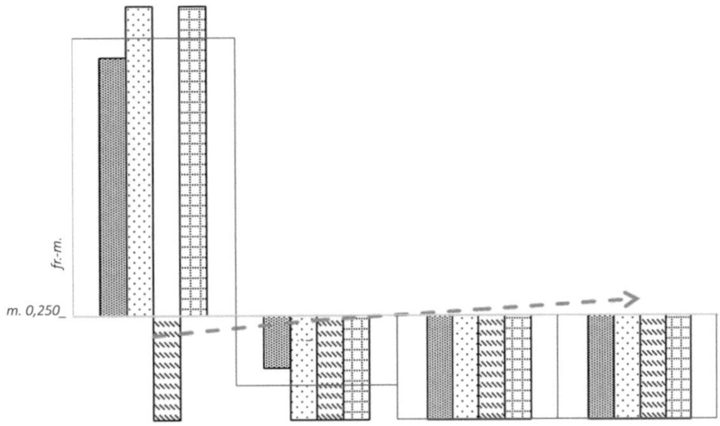

	Flysch	Treviño	Urbasa	Loza
Blanks	0,625	-0,125	-0,250	-0,250
Core trimming flakes	0,750	-0,250	-0,250	-0,250
Cores	-0,250	-0,250	-0,250	-0,250
Retouched products	0,750	-0,250	-0,250	-0,250
CU	-0,047	-0,009	0,016	0,041

Figure 7: Chart representing flint types and technological categories on archaeological level 7 based on the relative frequencies of each category and type in relation to the mean. The dotted line shows the Cost Units based on the relative frequencies of each category in relation to the mean.

On level 7, only the Flysch and Treviño types were found. The former has a remarkable underrepresentation of cores and flaking products as opposed to an overrepresentation of retouched products and trimming flakes. On the other hand, the Treviño type only has flaking products (Figure 7).

Discussion

Exploring, determining and understanding the socio-economical mechanisms developed by prehistoric societies are some of the most intriguing challenges offered in the field of research on Prehistory. In order to shed some light on these matters, we propose a methodology based on the integral analysis of lithic industry from typological, geological and geographic points of view. The interaction between these different factors may allow us to propose hypotheses on the social and economic mechanisms these societies developed for the catchment and management of flint.

The systematic use of analytic typology for the coding of data coming from the field record has provided high-resolution data from the formative geological and geographic perspectives of the objects. The first of these perspectives enables us, through a hierarchical organisation of traits, to understand the formatting and production stages of the transformation of the object by the artisan. The second enables us to recognise which flint types are managed, according to the previous definition of their geological distribution. Finally, the geographic perspective helps us to understand the place from which each object originates by considering topography and movement costs as measurable elements which may be used to compare the provenance of the lithic materials on a quantitative basis.

The use of Least Cost Paths as minimum effort routes is a widespread tool for analysing movement in archaeological contexts. In our case, we decided not to use them for several reasons. The alteration of the physical environment (erosion, deposits, and changes in the coastline) and the lack of accurate data to systematically model them are major drawbacks. Furthermore, these kinds of routes simplify the approach to prehistoric mobility by applying an excessively simplistic criterion whilst excluding other elements. These elements, such as territoriality, recurrent mobility and the delimitation of crossing areas which, nowadays, may be considered irrelevant landmarks or the catchment of other equally important resources such as faunal or vegetable resources, are extremely interesting to take into consideration.

As far as the case of *Baltzola* is concerned, our analysis helps us to gain a better understanding of the management of flint as a raw material in a synchronic and diachronic way.

From a synchronic point of view, each of the levels studied displays its own characteristics, although, diachronically, they show common links in terms of the management of this type of raw resource. Based on the petrological study, four types of flint have been identified in the cave of *Baltzola*, which are related to the geographically located outcrops of Flysch, Treviño, Urbasa and Loza. When the petrological data were combined with the Cost Units obtained via GIS, it turned out that the overall percentage of flint objects corresponds roughly with the degree of accessibility from the site to the outcrops (table 1). The exception was the flint from Treviño, which, although it is located in a relatively accessible place, has negative presence indexes (figure 3).

The techno-typological analysis displays a larger divergence between the three stratigraphic levels, with each one of them showing different patterns of exploitation of lithic resources from a technical point of view, as demonstrated by the quantitative analysis. Upon more detailed examination, the remains of almost all the stages of the lithic *chaine operatoire* were found at the Azilian level (level 9), whereas in the Magadalenian levels (levels 6 and 7) only a few stages of the lithic *chaine operatoire* were found (table 1).

This divergence within the techno-typological structure allows us to gain a better understanding of the catchment and management of lithic resources. If we were to make an exclusive reading through the petrological structure, we might reach the conclusion that the management of flint had been relatively similar in the three periods. However, when introducing the techno-typological approach, we observe certain diachronic changes in management (figure 4), which are visible on the distribution of the stock between the technological strategies identified and related to flint types.

These management dynamics detected in each chrono-cultural episode seem to represent different strategies related with the mobility and economic activity carried out in each place (García-Rojas 2014). In this way, an initial Magdalenian occupation of hunter-gatherer communities may be interpreted with an important hint on mobility, in which different hunting activities would have been carried out. Later, a second, more intensive, occupation and, finally, an occupation in which the settlement would have been stabilised by these communities.

Finally, when putting together the information obtained from the techno-typological analysis of the different lithic assemblages along with the flint catchment areas and the radiocarbon dates, an expansion of the areas of influence of the human groups which inhabited *Baltzola* can be observed. In this regard, level 7 (the oldest one) only reveals the management of two outcrops, Flysch and Treviño. These are the most accessible ones and depict a north to south axis of movement, although taking into account the quantity and morpho-technical traits represented in the reduction products, the Flysch type was clearly more used. On the other hand, the Treviño flint, despite being present, had a minor presence. Therefore, it may be said that in this period, the hunter-gatherer groups had a preference for mobility and the exploitation of coastal resources.

Moving forward in time, the analysis of the lithic assemblages from the upper archaeological levels shows the inclusion of the Urbasa and Loza as new flint types. This implies the incorporation of new catchment areas and a modest enlargement of the space managed by these groups, expanding their horizons inland, to the current plateau of Álava and the valley of the Ebro River. This data reveal a change of the transit axes between levels 6 and 9, expanding the north-south trajectory with another route to the east.

The quantitative data resulting from the geographic relationship between the site of *Baltzola* and the flint outcrops, established from the cost distance analysis and simplified through the CU, provide data which we have directly related with the amount of raw material brought to the site. Additionally, the cost distance analysis offers an insight into the complexity of prehistoric human mobility, the possibility of using different routes of the same cost and the likely directions of their mobility. All of this data helps us to understand the economic dynamics and mobility of the prehistoric groups inhabiting the site of *Baltzola*.

Conclusion

The use of GIS enables better understanding of the geographic setting as a productive source of information. Furthermore, with the use of *Cost Distance Analysis* tools, we have been able to measure the effort involved in moving over the landscape, synthesised in the concept of *Cost Unit*.

- This quantification allows for the equal comparison of the raw material found in the cave of *Baltzola*.
- The techno-typological analysis of lithic industry has enabled us to determine the degree of exploitation of the lithic masses found on site.
- The combination of data provided by each different analysis enables us to examine the different patterns of management developed by the hunter-gatherer societies during the occupation of the site of *Baltzola* at a higher degree of resolution. This method does not only point out the quantity of the raw material brought to the site and exploited there, but also allows us to know the stages of the *chaine operatoire* of the flint found on site.

Finally, and looking forward to future research, we consider that the use of GIS, along with the inclusion of petrological, use-wear and techno-typological analyses, provides new perspectives on the management of raw materials by prehistoric groups. This higher degree of resolution also allows mobility patterns to be studied in detail by bringing to the fore a wider spectrum of the archaeological record, both in terms of biotic and abiotic resources. In this way, understanding of the management and exploitation of resources by prehistoric groups in a particular territory can be improved.

Acknowledgements

The first four authors would like to express their gratitude to Lydia Zapata, who sadly passed away recently, not only for her management of the Baltzola Cave project but also for her confidence in their work and her contribution to their education and life.

The results presented in this paper have been partially funded by the Spanish Science Ministry research project HAR2014-53536-P (La ruta occidental del poblamiento de la Península Ibérica durante el Paleolítico medio y superior) and the Spanish Science Ministry research project HAR2015-67429-P (El complejo minero prehistórico de sílex de Araico-Cucho (Cuenca Vasco-Cantábrica) and the Research Team in Prehistory of the University of the Basque Country (IT-622-13).

A.P., C.C. and A.S. are funded by fellowships from the pre-doctoral programme of the Department of Education, Linguistics and Culture of the Government of the Basque Country (BFI-2012-121, PRE-2013-1-948 and BFI-2012-205). M.G.R is funded by a fellowship from the post-doctoral programme of the Department of Education, Linguistics and Culture of the Government of the Basque Country (POS-2017-2-0045).

References

Aubry, T., Luís, L., Mangado, J. and Matias, H. 2012. We will be known by the tracks we leave behind: Exotic lithic raw materials, mobility and social networking among the Côa Valley foragers (Portugal). *Journal of Anthropological Science* 31: 528-550.

Bustillo, Mª.A. and Pérez-Jiménez, J.L. 2005. Características diferenciales y génesis de los niveles silíceos explotados en el yacimiento arqueológico de Casa Montero (Vicálvaro, Madrid). *Geogaceta* 38, 243-246.

Coward, F.S. 2004. Transitions, Change and Identity: the Middle and Upper Palaeolithic of Vasco-Cantabrian Spain. Southampton: Ph.D. dissertation, University of Southampton.

De Barandiarán, J.M. 1932. Una visita a la cueva de Balzola. *Anuario de Eusko Folklore* 12, 111-114.

Djindjian, F. 2009. Le concept de territoires pour les chasseurs cueilleurs du Paléolithique supérieur européen, in Djindjian, F., Kozlowski, J. and Bicho, N., (eds). *Le concept de territoires dans le Paléolithique supérieur européen. Actes du XVe Congrès mondial (Lisbonne, 4-9 septembre 2006)*, BAR International Series 1938. Oxford: Archaeopress, 3-25.

Fairén, S. 2004. ¿Se hace camino al andar? Influencia de las variables medioambientales y culturales en el cálculo de caminos óptimos mediante SIG. *Trabajos de Prehistoria* 61/2, 25-40. doi:10.3989/tp.2004.v61.i2.41.

Fernández Eraso, J. and García-Rojas, M. 2013. Tipología Analítica, in: García Díez, M. and Zapata, L. (eds.). *Métodos y técnicas de análisis y estudio en arqueología prehistórica: de lo técnico a la reconstrucción de los grupos humanos.* Bilbao: Servicio Editorial de la Universidad del País Vasco, 479-497.

Fernández, J.F. 2010. *Una aportación desde la arqueología del paisaje al conocimiento del primer poblamiento humano del Valle del Trubia. Estudio geoarqueológico y análisis SIG del territorio.* Oviedo: Ediciones de la Universidad de Oviedo, 198.

Fontes, L.M. 2016.The Initial Magdalenian mosaic: New evidence from Urtiaga cave, Guipúzcoa, Spain. *Journal of Anthropological Archaeology* 41,109-131.

Fuertes-Prieto, M.N., Neira-Campos, A., Fernández-Martínez, E., Gómez-Fernández, F. and Alonso-Herrero, E. 2014. "Mucientes Chert" in the Northern Iberian Plateau (Spain). *Journal of Lithic Studies* 1/1, 117-135.

García, A. 2010. *Patrones de asentamiento y ocupación del territorio en el Cantábrico Oriental al final del Pleistoceno. Una aproximación mediante SIG.* Santander: Departamento de Ciencias Históricas, Universidad de Cantabria.

García-Rojas, M. 2014. *Dinámicas de talla y gestión de las materias primas silíceas a finales del Pleistoceno en el País Vasco.* Bilbao: Servicio Editorial de la Universidad del País Vasco.

García-Rojas, M. 2010. Propuesta de descripción y clasificación de los productos de debitado desde la tipología analítica. *Zephyrus* 66, 93-107.

Iriarte, M.J., Pérez Díaz, S., Ruiz Alonso, M. and Zapata, L. 2007-2008. Paleobotánica del epipaleolítico y mesolítico vascos, in: Fernández Eraso, J. and Santos, J. (eds.). *Homenaje a Ignacio Barandiarán Maestu.* Veleia Revista de Prehistoria, Historia Antgua, Arqueología y Filología clásicas 24-25. Vitoria-Gasteiz: Universidad del País Vasco/ Euskal Herriko Unibertsitatea.

Kantner, J. 2012. Realism, reality and routes. Evaluating cost-surface and cost-path algorithms, in: White, D. and Surface-Evans, S (eds.). *Least Cost Analysis of Social Landscapes: Archaeological Case Studies.* Salt Lake City: The University of Utah Press, 225-238.

Kramer, P.A. 2010. The effect on energy expenditure of walking. *American Journal of Human Biology* 22, 497-507.

Langmuir, E. 1984. *Mountaincraft and Leadership.* Glasgow: Scottish Sports Council.

Laplace, G. 1987. Un exemple de nouvelle écriture de la grille typologique. *Dialektikê. Cahiers de Typologie Analytique* 1985-1987, 16- 21.

Leorri, E., Cearreta, A., García-Artola, A., Irabien, M.J. and Blake, W.H. 2013. Relative sea-level rise in the Basque coast (N Spain): Different environmental consequences on the coastal area. *Ocean & Coastal Management* 77, 3-13.

Llobera, M. 2000. Understanding movement: a pilot model towards the sociology of movement, in: Lock, G. (ed). *Beyond the Map. Archaeology and Spatial Technologies.* Amsterdam: IOS Press, 65-83.

Llobera, M., Fábrega-Álvarez, P. and Parcero-Oubiña, C. 2011. Order in Movement: a GIS approach to accessibility. *Journal of Archaeological Science* 38, 843-851..

López Romero, R. 2005. Cálculo de rutas óptimas mediante SIG en el territorio de la ciudad celtibérica de Segeda. Propuesta metodológica. *SALDVIE* 5, 95-111.

Mangado, J. 2006. El aprovisionamiento en material primas líticas: hacia una caracterización de los comportamientos paleoeconómicos. *Trabajos de Prehistoria* 63/2, 79-91.

Mangado, X. 2005. *La caracterización y el aprovisionamiento de los recursos abióticos en la Prehistoria de Cataluña: las materias primas silíceas del Paleolítico Superior Final y el Epipaleolítico.* British Archaeological Reports (International Series) 1420. Oxford: Archaeopress.

Marín, A.B. 2009. The use of optimal foraging theory to estimate late glacial site catchment areas from a central place: The case of Eastern Cantabria. *Journal of Anthropological Archaeology* 28/1, 27-36..

Marín, A.B. 2008. Patrones de movilidad y control del territorio en el Cantábrico Oriental durante el Tardiglaciar. *Trabajos de Prehistoria* 65/1, 29-45.

Minnetti, A.E. 1995. Optimum gradient of mountain paths. *Journal of Applied Physiology* 79/5, 1698-1703.

Molina Hernández, J. 2015. *El sílex del prebético y cuencas neógenas en Alicante y sur de Valencia: su caracterización y estudio aplicado al Paleolítico Medio.* Unpublished doctoral thesis

Murrieta-Flores, P.A. 2010: Travelling in a Prehistoric Landscape: Exploring the Influences that Shaped Human Movement, in: Frischer, B., Webb Crawford J. and Koller D.(eds.). *Making History Interactive. Computer Applications and Quantitative Methods in Archaeology (CAA), Proceedings of the 37th International Conference, Williamsburg, Virginia, United States of America, March 22-26, 2009* British Archaeological Reports S2079 (International Series). London: Archaeopress, 258-276.

Murrieta-Flores, P.A., Wheatley, D. and García Sanjuán, L. 2009. Movilidad y vías de paso en los paisajes prehistóricos: megalitos y vías pecuarias en Almadén de la plata, in: Mayoral Herrera, V. and Celestino Pérez, S. (eds.). *Tecnologías de información geográfica y análisis arqueológico del territorio. Actas del V Simposio Internacional de Arqueología de Mérida.* Mérida: Instituto de Arqueología-Mérida. CSIC-Junta de Extremadura-Consorcio de Mérida, 1-16.

Pandolf, K.B., Burse, R.L. and Goldman, R.F. 1977. Role of physical fitness in heat acclimatization, decay and deinduction. *Ergonomics* 20, 399-408.

Prieto, A., García-Rojas, M., Sánchez, A., Calvo, A., Domínguez-Ballesteros, E., Ordoño, J., García-Collado M.I. 2016. Stones in Motion: Cost units to understand flint procurement strategies during the Upper Palaeolithic in the south-western Pyrenees using GIS. *Journal of Lithic Studies* 3/1, in press.

Ríos, J. and García, A. 2015. Middle Paleolithic Mobility Patterns and Settlement System Variability in the Eastern Cantabrian Region (Iberian Peninsula): A GIS-Based Resource Patching Model, in: Conard, N. and Delanges, A. (eds.). *Settlement Dynamics of the Middle Palaeolithic and Middle Stone Age.* Tübingen: Publications in Prehistory, 329-360.

Rissetto, J.D. 2012. Using least cost path analysis to reinterpret Late Upper Paleolithic hunter-gatherer procurement zones in Northern Spain, in: White, D.A. and Surface-Evans, S.L. (eds.). *Least Cost Analysys of Social Landscapes. Archaeological Case Studies.* Salt Lake City: University of Utah Press, 11-31.

Rissetto, J.D. 2009. *Late Pleistocene Hunter-Gatherer mobility patterns and lithic exploitation in Eastern Cantabria (Spain*). Albuquerque: Ph.D. Dissertation. Department of Anthropology, University of New Mexico, 336.

Sánchez de la Torre, M. 2015. El sílex en su contexto geológico: Un corpus de datos para el Pirineo centro-oriental. *Journal of Lithic Studies* 2/2, 167-187.

Sánchez, A., Dominguez-Ballesteros, E., García-Rojas, M., Prieto, A., Calvo, A. and Ordoño, J. 2016. Patrones de aprovisionamiento de sílex de las comunidades superopaleolíticas del Pirineo occidental: el "coste" como medida de análisis a partir de los SIG. *Munibe Antropologia-Arqueología. 67, 235-252.*

Sarabia, P.M. 2000. *Aprovechamiento y utilización de materias primas líticas en los tecnocomplejos del Paleolítico en Cantabria.* Tesis doctoral. Santander: Universidad de Cantabria.

Tarriño, A. 2006. *El sílex en la cuenca vasco-cantábrica y Pirineo navarro: caracterización y su aprovechamiento en la Prehistoria.* Monografías del Museo Nacional y Centro de Investigación de Altamira 21. Madrid: Ministerio de Educación y Cultura.

Tarriño, A., Elorrieta, I. and García-Rojas, M. 2015. Flint as raw material in prehistoric times: Cantabrian Mountain and Western Pyrenees data. *Quaternary International* 364, 94- 108.

Terradas, X. 2002. *La gestión de los recursos minerales en las sociedades cazadoras-recolectoras.* Treballs dÉtnoarqueología 4. Madrid: CSIC.

Tobler, W. 1993. *Three presentations on geographical analysis and modeling.* Technical Report 93, 1. Santa Barbara, California: National Center for Geographic Information & Analysis (U.S.), University of California.

Turrero, P., Domínguez-Cuesta, M.J., Jiménez-Sánchez, M. and García-Vázquez, E. 2013. The spatial distribution of Palaeolithic human settlements and its influence on palaeoecological studies: a case from Northern Iberia. *Journal of Archaeological Science* 40(12), 4127-4138.

Zapata, L. Gallaga, I., García-Rojas, M.; Regalado, E., Ruíz Alonso, M., San Pedro Z. 2008. Baltzola. *Arkeoikuska: investigación arqueológica* 07, 223-224.

Zapata, L. Gallaga, I., García-Rojas, M., Regalado, E., Ruíz Alonso, M., San Pedro, Z. 2009. Balzola. *Arkeoikuska: investigación arqueológica* 08, 249-250.

Zapata, L., Gallaga, I., García-Rojas, M., Regalado, E., Ruíz Alonso, M., San Pedro Z., 2011. Cueva de Baltzola. *Arkeoikuska: investigación arqueológica* 10, 238-240.

Zapata, L., Gallaga, I., Regalado, E., Ruiz Alonso, M., San Pedro, S. and Savanti, F. 2007. Cueva de Balzola (Dima). *Arkeoikuska: investigación arqueológica* 06, 142-145.

Zapata, L., Regalado, E., García-Rojas, M., Gallaga, I., Pérez, A. 2010. Cueva de Baltzola. *Arkeoikuska: investigación arqueológica* 09, 220-221.

The Archaeology of Rock Art as Archaeology of the Mediterranean Landscape

María Sebastián López[1]

Introduction

This paper argues that rock art sites, when characterised geographically on a regional scale, can reflect the basic dynamics of the construction of the landscape and its use. In this case study in the Mediterranean basin, it can be seen how the spatial analysis of sites with rock art enables the social and economic situation to be modelled for the beginning of anthropisation. It is in this period that the erosion rates in the Mediterranean landscape in the Neolithic Age increase. This article proposes that Levantine rock art is the first visible sign of the acculturation of the landscape. Rock art and its location in specific places in the countryside are analysed via a multiproxy study.

To this end, the project is tackled from a geographic perspective in which the land is seen as a convergence between natural and cultural elements, a space perceived through various aspects of human experience and activity. Moreover, it is understood that the space acts reciprocally by expressing a network of relationships between the people and places providing the context in which daily activities and habits take place.

Therefore, the Neolithic landscape is understood as a part, and at the same time a product, of social action (Bernabeu *et al.* 1993). We do not suggest that they should be studied separately but that an analytical strategy of breaking up the elements comprising and intervening in the landscape should be used by carrying out an in-depth study of each of the components. We shall begin by analysing the distinctive geographic features of our rock art sites. Then, we shall analyse their spatial distribution before going on to offer an interpretation of the results.

---

1 Department of Geography and Spatial Planning. University of Zaragoza; msebas@unizar.es.

Having reached this point, we should emphasise the fact that we do not aim to make a positive reconstruction of the conditions of the past based on the direct correlation among elements which were originally related. Our working method fits perfectly with that defined by J. Vicent: *"the archaeology of the agricultural landscape demands an unconventional archaeological approach, whose aim is not to make a positive reconstruction of facts, but to compare hypotheses on aspects not directly observable in the process (...) based on those that are"* (Vicent 1993, 37). Therefore, we have adopted a methodological sequence based on the elaboration of models and a comparison which enables us to analyse the most obvious material (functional) aspects of objects, moving on to subtler (symbolic) displays from their creators (Parcero, 2002). Although rock art has mostly been studied from a cognitive and cultural perspective, this article analyses rock art sites as the main proxy for understanding the rates of acculturation of the Mediterranean landscape in the long term.

We assume that the distribution of painted caves is not random but that it arises from specific criteria, which were not initially know. We define these criteria as landscape elements (for example, the choice of sites with a good view of the surroundings, river gorges, cut-off ravines, etc.). These will be used as a basis for returning to the criteria which may have shaped the distribution of archaeological evidence. In this way, it can be determined to what extent decisions or criteria on location fits with our initial hypotheses by exploring the priority given to some factors over others. However, identifying these working guidelines in the field does not guarantee direct explanations on decision-making but does link certain geographic factors to possible patterns. For example, a correlation has been found between two factors (caves with schematic rock art and building reservoirs). This fact does mean that there is a direct causal connection between the two (obvious in this case) but that there is a relationship with a third factor, river gorges, which provide access to such a vital and necessary resource as water, particularly in summer.

When the spatial distribution of rock art is combined with ethnographic reasons, the potential use that nearby land could have had may be explained. However, it should not be forgotten that ethnographic information should be used with care as "*...modern or early modern ethnohistoric parallels are merely a source of models, not a blueprint for claiming, for example, a continuity of land use and mode de vie in a given regional landscape*" (Barker and Bintliff 1999, 207).

The study area: homogenous geographic areas

Rock art, i.e. the practice of marking natural places with graphic representations of differing complexity, has been practised since the Upper Palaeolithic. Examples can be found all over the world wherever suitable rocks exist. The earliest examples are thought to be reliable indicators of the emergence of modern human behaviour. As a vehicle for the transmission of aesthetic values and the marking of identity, it is part of the cultural heritage of contemporary society. Furthermore, it is of extraordinary scientific value as it provides us with the only testimony of the symbolistic behaviour of many prehistoric societies.

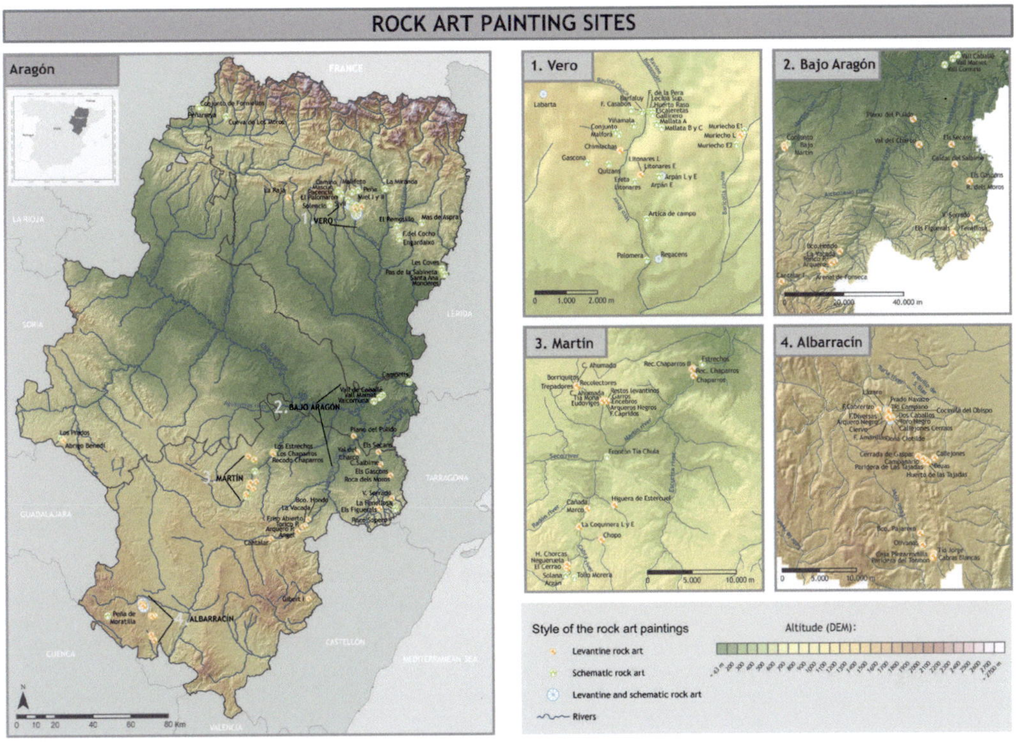

Figure 1: Location of the study areas.

Therefore, the UNESCO World Heritage list contains many rock art sites, to which ARAMPI (Rock-Art of the Mediterranean Basin on the Iberian Peninsula) was added in 1998, with more than 370 rock art sites. Aragon, with its extraordinary examples of rock art sites, is located within ARAMPI.

Due to its many painted shelters (176) and the contrasts in its landscape, Aragon is an excellent pilot area to carry out a landscape study. This region has a surface area of 47,755 km^2. Our aim was to implement the method over a wide area in order to verify the results in several characteristic ecosystems of the Mediterranean region (flat areas, mountains, depressions, etc.) and to work on different scales. In addition, equal access to the archaeological information studied was preferred and restricting the analysis to current administrative areas (even though they do not correspond to geohistorical areas) meant that we could locate all the rock art sites and then check them in situ.

In general, Aragon is far from being a homogenous natural region; the physical environment is characterised by huge degree of diversity leading to a wealth of resources and great potential for use. The basic feature of the land in Aragon is the conformation of lowlands between two mountain areas. With the Ebro valley (a contrast between flat and mountainous areas with the high ground on the border), the topographic framework explains the physical features of the region to a large extent. There is great diversity of vegetation in the landscape: from Euro-Siberian in the Pyrenees to the more arid Mediterranean areas in the centre of the depression. The climate is determined by opposing Atlantic and Mediterranean influences, a mix of continental winds trapped in the Ebro valley. The extreme aridity in the central, low-lying, sector extends to the foothills of the river valleys (Peña *et al.* 2004).

This wide variety of geographic environments is what makes the terrain of Aragon an extremely interesting case to study, as this one area contains very contrasting landscapes. There are four main regions (fig. 1):

Exterior Mountain Ranges in Huesca

These lie in the interior, in the midst of the Pyrenean foothills over 200 km from the coast, in the westernmost area in which rock art from the east coast is found. As its name indicates, it has a large number of ranges consisting mainly of tertiary materials (limestone and marls from the Eocene period) and, to a lesser extent, from the Triassic (chalk and marl) and Cretaceous periods (hard limestone of marine origin which comprises the main elevations of the mountain ranges).

Only seven caves with Levantine paintings have been documented in the province of Huesca, accounting for 15% of rock art in the area, with the remaining 85% being schematic art found mostly in the basin of the River Vero.

Lower Aragon

This area consists of three large geomorphological units, the Iberian system, the Ebro depression and the westernmost foothills of the coastal-Catalan range. The river system, with the Guadalope as the main axis crossing the district in a south-east to north-east direction, together with its tributaries (Alchozas, Guadalopillo, Bergantes and Mezquín) have chiselled the high ground to open up a way through the various geological materials (limestone and deposits from the Quaternary period).

The Levantine art style in this area is characterised by its many types (with up to five different types of humans) (Utrilla and Bea, 2007) and the abundance and density of sites and paintings (over 50 human figures are found in Civil).

Bajo Martín

Closely linked to the previous region, is the third zone of the river Martín. This is an area which is mainly characterised by the river and its tributaries carving out a south-east to north-east path through the mostly limestone rock. This has created steep canyons crossing the Iberian mountain ranges, whose structure of folds in a north-east to south-east direction is located in the middle stretch of the Martín River.

The Levantine art style predominates in this area, which is notable both for its quantity as well as for some almost exclusive themes of great interest (running archers, wild boar hunting).

Albarracín

The last region occupies the south-west sector of the province of Teruel and comprises the Albarracín and Montes Universales ranges, together with the valleys beyond at the head of the Tajo and Gallo rivers. It forms part of the vast orographic system of the Montes Universales, with more than one range being a complex of ridges, plateaux, hills and moors separated by deep passes, ravines and gorges with steep slopes, with a clear predominance of tabular topography.

As far as the art is concerned, the strong Levantine nucleus is outstanding and has its own personality, shown by the unique white figures on red buntsandstein sandstone, the predominance of large bovids, frequent depictions of horses (which are extremely rare in classic Levantine art), the re-painting of figures and very few human forms.

Methodology: spatial and inferential statistical analysis

The working methodology proposed was similar to the theory of system analysis, differentiating among the significant environmental components in order to explain the distribution of the first traces of human activity and to search for connections or relationships between the chosen components to explain the behaviour of the system.

All of the methods and instrumentation used in this research fall within the design and implementation of a global model, which implies the creation and analysis of a geographic information system (Fernández-Freire and Uriarte 2011). The use of this tool provides a large amount of relatively objective information, with statistics being the best method to process the data obtained with these programs. They should be seen as an extension of our capacity to observe (Wheatley and Gillings 2002, 125) but not a tool to produce complete archaeological interpretations, especially when dealing with the large datasets obtained from our analysis. The variables we have used are as follows:

Elevation (metres above sea level)

The digital elevation model was generated from information available from the Historical and Geographical Science Laboratory at the University of Zaragoza. It is based on 129 pages from MTN in a vector format at a scale of 1:25,000 containing accurate 10 m contours.

Relative altitude and visual dominance

There are several methods for calculating this (Cruz-Berrocal 2004; Fairén and García-Atiénzar 2005; Fábregas 2005).

We calculated the average altitude based on an MDT at a resolution of 25 m, according to Parcero, 2002. This method is based on the following equation:

$$Ar = (A\ absolute - m)/DT$$

Where Ar is the relative altitude resulting from the division between the absolute value of each cave (Aabs) and the average altitude (m) found in the chosen search area (circular area of 1000 and 3000 m). DT is the deviation of values in the surroundings analysed.

We calculated their values within a radius of 1000 m (immediate surroundings) and 3 km (average distance), respectively. Therefore, a value of approximately 1 would indicate that a point is prominent in the surrounding area, and a value of approximately 0 would represent the lowest point in the terrain.

Aspect

Variables derived from the MDT can work continuously or be reclassified. We reclassified due to the fact that it gave us an idea of different potential uses. Therefore, we divided the territorial analysis by aspects: north, north-east, east, south-east, south, south-west, west, north-west and flat.

Slope

The degree of slope on which the caves are located is not only an essential factor for determining accessibility but it also offers significant clues to the type of activity of their inhabitants. The type of crop or simple human need can nuance the classification; it is generally thought that slopes of more than 20º are only suitable for livestock and forestry, whereas those between 0º and 12º are best suited for agriculture (García Sanjuán 1999).

Litho-geology

We have treated the information from the Territorial Information Service of Aragon by adding lithological types according to the most common categories and their degree of resistance to erosion. Thus, we differentiated between the following types: sandstone, limestone, conglomerates, clays, surface formations (terraces, alluvial cones, slope deposits, river beds), chalk, granite, quartzite and slate, dolomites and marls.

Soils

The layer we used is from the Soil Map of Spain 1:1.000.000 (Gómez-Miguel 2006), based on the USDA (United States Department of Agriculture 2003) soil classification. Four common soil types were chosen, all of which are typical in Mediterranean environments: Inceptisol xerept (soils with a certain edaphological development and some presence of organic material), Entisol orthent (soil with scant edaphological development and high amounts of lithological substrate), Entisol fluvent (not very developed, characteristic of alluvial areas) and Aridisol calcid (typical soil in arid Mediterranean areas, which is poor in organic material and rich in carbonates).

Classification of the river system

We classified the river system where the rock art was located in accordance with Horton's classification, which was later modified by Strahler (Horton 1945; Strahler 1957). This method forms a "natural" grouping of the various systems and indicates the most useful for general studies. Rivers belonging to a specific order tend to form groups whose constituents can be examined together. In addition, we analysed their degree of sinuosity and paid special attention to the presence of river gorges.

Thermalism

We geographically located geothermal traces corresponding to current upwellings of water and subsequently related them to the rock art sites in Aragon, taking into account their distance at thresholds of 5, 10 and 15 km.

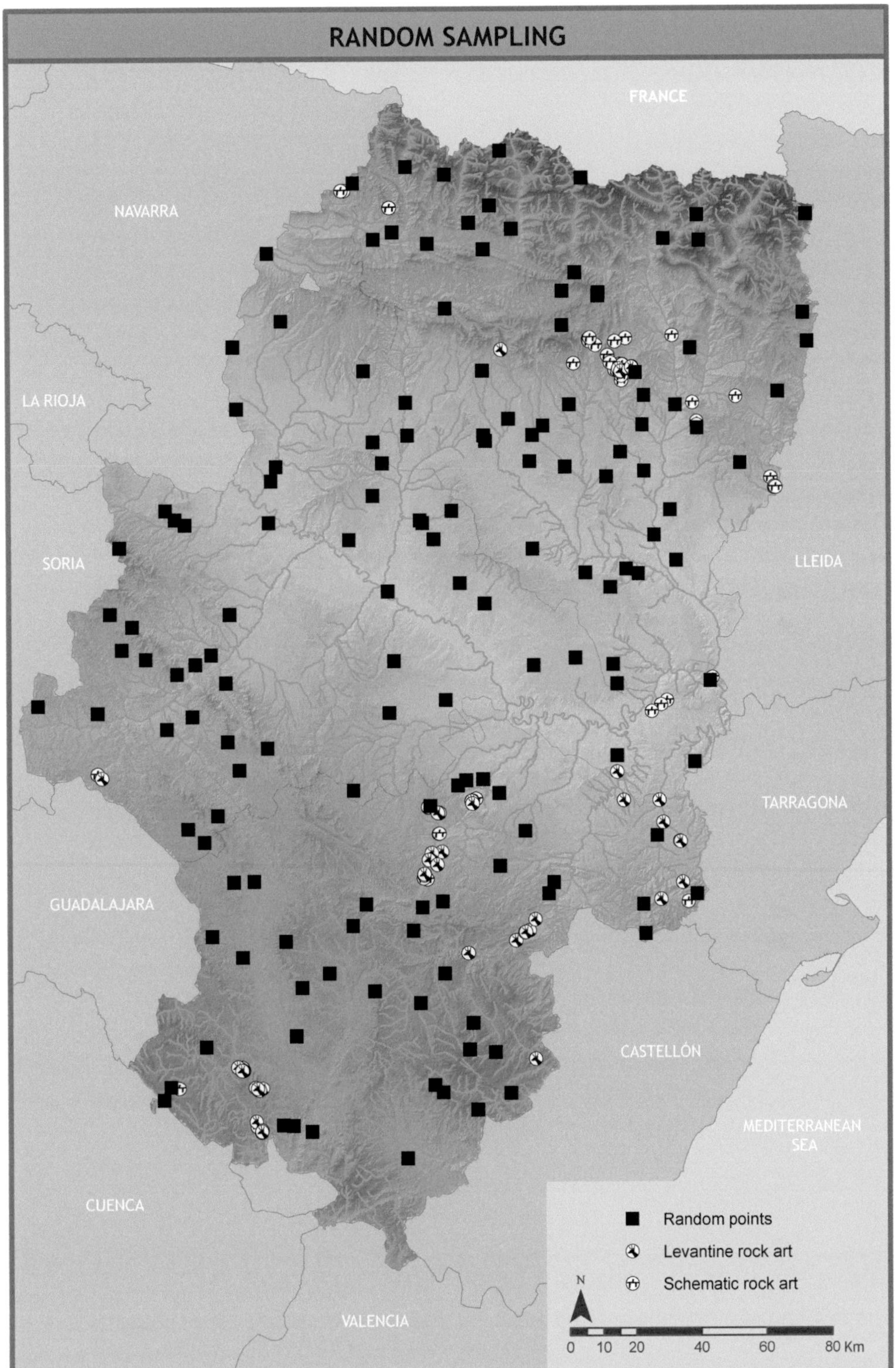

Figure 2: A random sample and shelters with rock art in Aragon. The random sample includes all the territory analysed.

Once the variables for sites with Levantine and schematic rock art had been analysed and calculated, we chose a null hypothesis as the sites were randomly located (fig. 2). The interpretation of this alternative hypothesis led us deduce whether or not there were location patterns, not only through spatial observation but also by the objectification of the observed variables and indicating significant differences or otherwise (Fernández-Freire 2009).

The variables described above were calculated for 176 painted rock art sites. Then, a random sample was created and the same landscape variables were calculated for them. We compared our samples of rock art sites in Aragon with the random points generated, using the dependent variables from our model which are always geographic.

Geographical variables	Rock art sites	Random points
Mean elevation (m)	794.88	440.43
Mean relative altitude (m)	23.17	11.55
Terrain aspect (% area)		
north	2.45	13.00
north-east	3.33	11.65
east	33.48	12.79
south-east	2.58	13.23
south	30.45	9.76
south-west	13.18	11.18
west	4.33	9.33
north-west	9.00	14.00
flat	1.2	5.06
Slope (mean º)	12.50	4.89
Litho-geology (% area)		
sandstone	23.45	10.45
limestone	34.65	9.54
conglomerates	16.98	7.53
lutite	1.89	5.09
clays	2.00	7.56
surface formations	0.87	8.24
chalk	1.45	9.22
granite	8.34	11.00
quartzite and slate	5.23	13.31
dolomites and marls	5.14	18.06
Soils (% area)		
inceptisol xerept	51.99	23.49
entisol orthent	40.76	33.06
entisol fluvent	4.25	25.45
aridisol calcid	3.00	18.00
Hydrology (% of 1st & 2nd order rivers)	67.21	12.19
Termalism (% area)	54.89	34.47

Table 1: A summary of variables used in the analysis

The samples in each analysis were compared using Chi-squared or Anova testing, depending on the type of variable (discrete or continuous). The Chi-squared test measures whether there is a significant difference between an expected distribution and an observed distribution (García Ferrando 1985, 184; Cruz-Berrocal 2004, 260). The Anova test (variance analysis) is a test to compare averages, based on calculating their variance (García Ferrando 1985, 311; Cruz-Berrocal 2004, 260).

The null hypothesis is corroborated when the result is greater than 0.05. If it is less than 0.05, the null hypothesis is to be discarded and the alternative hypothesis accepted. This means that conclusions will be confirmed at a 95% margin of confidence. In addition, the distributions were compared. We first descriptively examined the differences in frequency value using bar graphs. Secondly, the Kolmogorov-Smirnov test was used to check if the differences were statistically significant.

The result (Table 1) was a model of landscape factors (Vicent 1991, 40-47). Data provided by the statistics show that the rock art sites are not randomly located but that they appear in very specific areas with different spatial characteristics. In comparison with the random sample, they returned positive results for the hypothesis of a previous selection of the painted caves. The analysis also shows that variations in the location of the rock art sites respond, to a great extent, to regional differences and, therefore, we should be fairly cautious in confirming clear differences between both styles.

Significant and defining aspects of the terrain in which painted rock art sites are located

Following the spatial analysis and fieldwork, it can be confirmed that this is a structured landscape, in which certain differences can be seen between the territories where Levantine art is found compared with areas containing mainly schematic art. These contrasts are much more noticeable on a regional scale and are slightly disseminated on a macro-scale for the region of Aragon as a whole. Once again, these territorial differences allow us to broadly corroborate (with certain exceptions) two slightly different landscape models. Although most of the caves with rock art are characterised by the fact that they follow a longitudinal distribution along river axes (ravines and gorges), we could discern variations according to the different styles.

Thus, by making a brief summary of the geographic features characterising the rock art sites, we can state that:

- The sites are located in **mountain areas**, very often in smaller ravines, as these usually consist of medium-sized river systems, and almost never near the Ebro River, except for the complexes of Vall Mamet, Vall Caballé, Vall Comuna and Campells, which are located on main rivers, although their chronology is, according to Baldellou, extremely uncertain, (Baldellou, "personal communication").

Moreover, within these **biotopes** (as we believe the ravines to be geographically), the general tendency to **choose specific points** in the landscape is significant:

- At the **entrance** and **exit** to **defiles**: the caves of Los Estrechos and Los Chaparros on the Martín River mark the beginning and end of a defile. In the area of Ladruñán, the Arenal de Fonseca cave marks the start of a canyon; on the Vero River, the Chimiachas deer seems to point to the start of a ravine.
- The **confluence** of two **ravines**: the Gallineros schematic paintings at Lecina seem to mark out the confluence of the ravine of La Choca with the Vero River; the Huerto Raso cave is situated at the confluence of the La Choca ravine and Besender, the recently discovered cave of Benedí etc.
- An **upwelling** of **water:** at Obón, the Cerrao paintings stand at the mouth of a ravine and at the rise of an upwelling; the same occurs with the Fenellosa de Beceite cave which seems to mark a natural spring located opposite it; the Canatal I cave is located close to a small stream originating from a spring; El Mortero de Alacón, where nine sites are grouped around a spectacular waterfall at the start of the ravine; El Cerro Felío, with the San Miguel spring that rises at the foot of the caves; La Cañada de Marco is near the Pinarosa spring and the schematic sun in Barranco de Solencio appears to mark a stunning waterfall in the rainy season. Finally, there are still waters and a large spring at Prado del Tormón.
- **River gorges** (reservoirs): in the case of schematic art, there was a very strong preference to mark the narrowing of rivers and confluences of ravines, places where water would accumulate in the dry season and which are used for reservoirs nowadays. The caves of La Vacada, El Torico and El Arquero in the Santolea reservoir; the Plano del Pulido cave (located in the only wet area in the mainly arid terrain), the Caspe reservoir; the Remosillo cave next to the power station; Cañada de Marco next to the Foradada reservoir; the sites at Baldellou and Les Coves, above the Santa Ana dam.
- **Predominant altitudes** are above the average, both for schematic and Levantine art. This means that they are located at higher altitudes than might be expected for a random location. Furthermore, there is a clear tendency for high altitudes to be chosen for schematic art. Thus, on the Vero River, the caves in the cliffs of Mallata and Lecina are spectacular as is that of Muriecho (Levantine and schematic) from which the Portal of Cunarda can be seen. On the Ésera River, the paintings of Remosillo are located in extremely high vertical cliffs. On the Martín River, the cave of Los Estrechos stands at 30 m above the river bed.
- The **slope** of the terrain near the sites is usually middling, although slightly higher than that of the data from a randomly taken point in the landscape (average values between 10º and 15º).
- **Orientation:** places in the east and south were preferred, except for the north-facing Pudial de Ladruñan, from which the natural bridge of Fonseca could be seen.
- The **geological substrate** in which the sites are located was chosen extremely carefully in all of the regions studied. The statistical work shows that the sites are restricted to limestone and sandstone (perhaps due to how they erode and form caves). In addition, during field studies, we observed

that on occasion, the sites were located in places characteristic of local outcrops of other material (as in Rodeno in the district of Albarracín) or in contact zones.

- **Soils**: as far as this variable is concerned, we noted a remarkable preference for locating rock art sites on Inceptisols and Entisols.
- **Thermalism**: in the spatial analysis, we found that there was some preference for choosing places close to thermal waters, thus the majority of schematic sites stand at 5-15 km from a thermal spring.

The landscape of the schematic art *versus* the Levantine structure

The geographic analysis carried out previously discards the idea that rock art sites are found in degraded areas with a poor landscape. On the contrary, the mountains of the Exterior Sierras of Huesca represent a concentration of economic resources which are difficult to deny. In our view, the location of rock art sites in these wet mountainous areas is linked to the keeping of livestock and a pastoral life, as Colón Diaz (1998, 145) once noted. In addition, the choice of the karst system in Aragon, far from being a disagreeable place for its inhabitants, is an ideal biotope for the most productive human groups as they are stable and wet geosystems with a wealth of raw materials.

It is also much more convenient to keep semi-corralled livestock in the surrounding area (Fairén and García-Atiénzar 2005). Unfortunately, livestock archaeology is very poorly developed, even within the historical context. However, among other factors, Prescott (1995, 168) defines the natural potential for keeping livestock as an identifying element. In accordance with this idea, we believe that, from the pre-historic point of view, there are certain theoretical and practical conditions which allow us to suggest a possible link between schematic art and livestock.

More specifically, in Aragon, it is highly significant that the Chaves cave is closely linked to schematic art through the iconography of its painted boulders. This excellent Neolithic site bears witness to the domestication of ovicaprids (Castaños 2004), although little is known about their farming practices. The absence of wheat and barley in the records is surprising, despite there being 36 illustrations with a ceremonial patina (Cava 2000). On the other hand, the dense series of large black dots at both ends leads us to take the size of corrals (the 3000 m^2 lit entrance allowing animals to be stabled there) into consideration.

Throughout the region of the Exterior Sierras of Huesca, there are documented signs of early, long and short distance, transhumance. Three types of livestock movements are known for the area. First of all, transhumance to higher ground: this consisted of short movements upwards within the same valley or district. The livestock wintered in low areas in stables or was semi-corralled and, in summer, they were moved up to alpine pastures in the mountain passes. Secondly, summer or ascending transhumance: longer distances were covered from districts outside the region. The livestock was kept near the winter pastures but in spring and autumn the animals could graze in intermediate pastures. Thirdly, winter or descending transhumance: the shepherds living in the summer pastures moved the livestock to winter pasture (Fernández-Otal 1993).

In fact, the successive occupation and abandonment of the Olvena cave, near the schematic art cave of Remosillo, at times ranging between the Early Neolithic and Late Bronze Age, has been interpreted as a sign of a halt in transhumance to the high ground of the Sierra Ferrera, in search of the summer pastures used by the population of Puyascada and Forcón (Utrilla 1996).

The Pyrenees have two very different zones: the colder high ground at an altitude of over 1000 m, and the more hospitable lower ground used in winter. Therefore, the high valleys of the Pyrenees which were wet and had large passes and hay meadows, were the traditional places for transhumance for most people. However, in the pastures of the pre-Pyrenees, which were drier, with no passes or meadows, transterminance was practised and made an important contribution to the subsistence of the local and neighbouring populations. We consider that this fact is what gives schematic rock art in Aragon its longitudinal character (fig.3).

Thus, the preference for small ravines was not accidental. Rather, it was a choice which made them more attractive for these communities due to their convenience and capacity to create humid micro-climates, as well as for their suitability for handling livestock, as some authors have pointed out (Cruz-Berrocal 2004). The ravines where rock art is mostly found are classed as "ravines easy to close off to

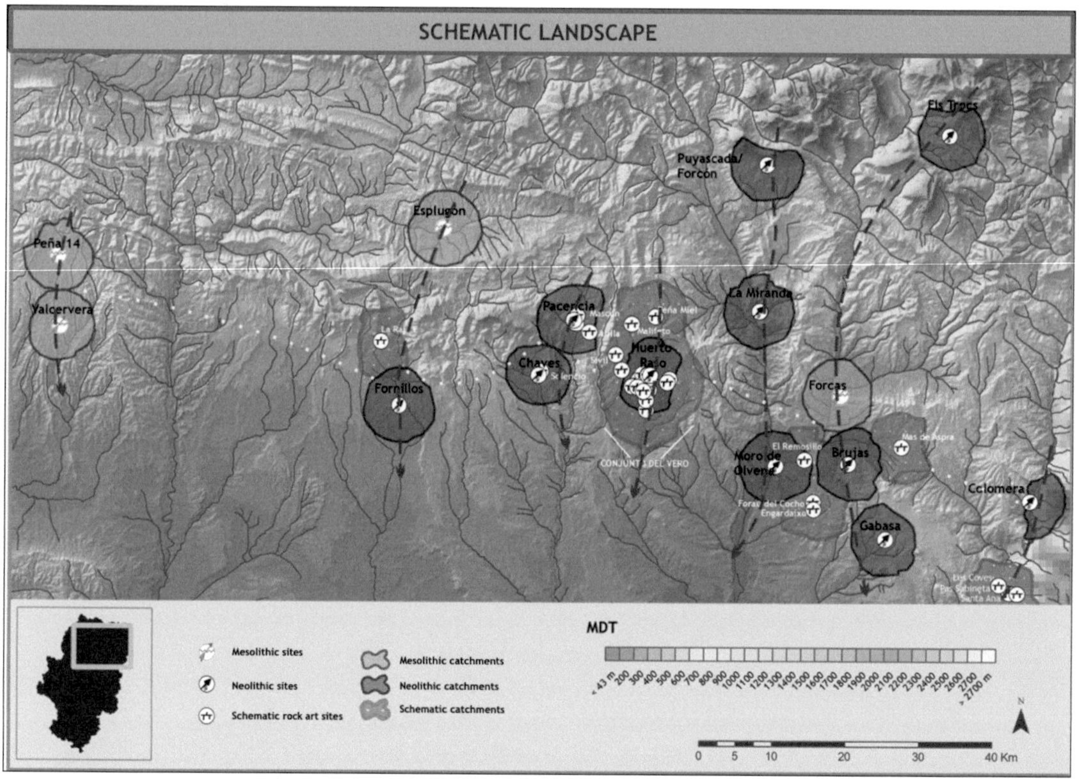

Figure 3: A map of the schematic art in the pre-Pyrenees landscape in which its longitudinal character can be appreciated.

hunt and enclose prey" (Beltrán 1993, 72). Indeed, in our field studies, we often had a strong feeling of being in ravines-enclosures, or rock formations in general which could have been used for this purpose.

This situation would be quite similar to the semi-husbandry proposed by Davidson in the Upper Palaeolithic Age for the Marchuquera valley in Valencia, with all the inhabited sites (Parpalló, Barranc Blanc, and the slightly more distant Mallaetes, etc.) controlling the closed valley. There is no such impressive list of occupied sites here but there are plenty of ravines. For example, there is Huerto Raso on the Vero River, positioned in the only area of the river wide enough to have a small vegetable plot. It is also the case of the cave of Pacencia in Rodellar (both have simple schematic paintings based on dots and bars, which are marking signs) and the Neolithic Remosillo cave on the Ésera River. In this respect, there may be a depiction of a barrier semi-corralling the animals in a schematic painting in the Palomera cave in Alquézar (the arrival point when canyoning down the Vero River and the only exit to the ravine). It shows schematic animals appearing to encounter a vertical barrier.

On the other hand, schematic rock art caves are characterised by being immediately visible. Thus, although they are found at altitudes which are relatively higher than those of Levantine art; they have visual dominance over the whole ravine. This visual control is highly significant in Barfaluy and Mallata on the Vero River, and in Santa Ana in the Ribagorza district. They are also obligatory crossing places due to their enclosure as is the case of the Alagón River. These days, they are popular places for the sport of canyoning. It is not by accident that this activity in the Vero starts at the Mallata paintings in Lecina and ends near the Palomera cave in Alquézar (fig. 4).

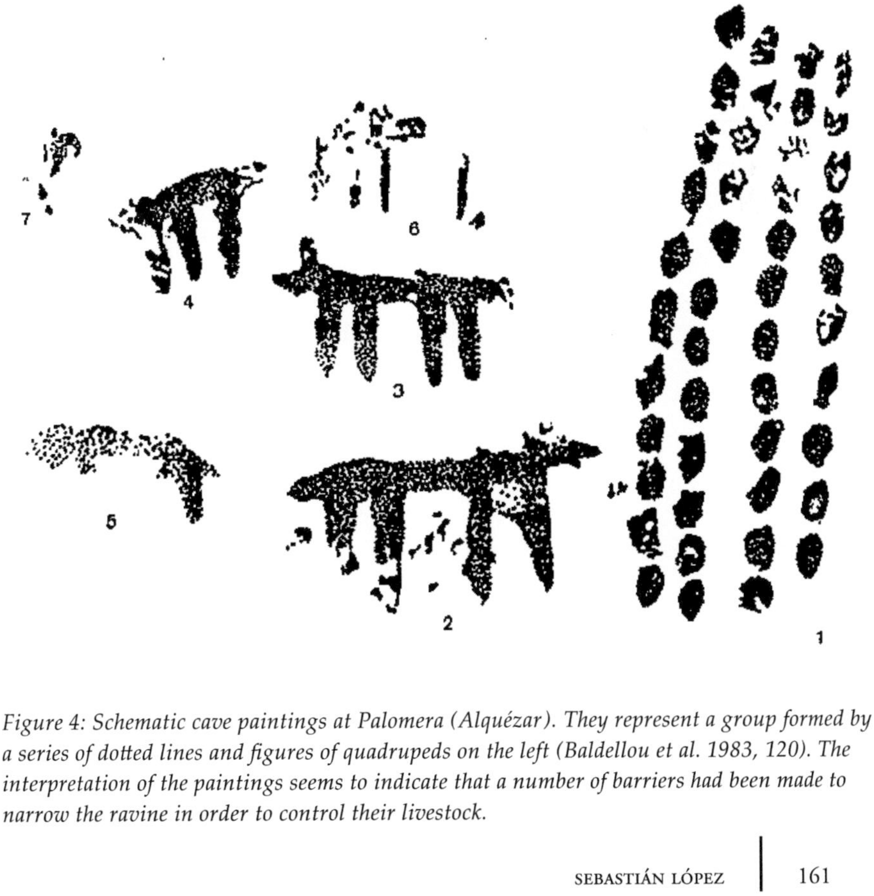

Figure 4: Schematic cave paintings at Palomera (Alquézar). They represent a group formed by a series of dotted lines and figures of quadrupeds on the left (Baldellou et al. 1983, 120). The interpretation of the paintings seems to indicate that a number of barriers had been made to narrow the ravine in order to control their livestock.

Finally, the sites marking the narrowing or confluence of ravines may also have marked places in which it was possible to collect water during the summer. Therefore, people would have come down with their animals to obtain such a vital resource.

Moreover, after analysing the potential land use, we found that around the schematic rock art sites, generally in the higher areas, there are flat areas where subsistence, although not intensive, farming could have taken place to complement livestock production. This is the case with the flat areas in the present-day villages of Lecina (with regard to Barfaluy, for example) and Olvena and La Puebla de Castro (with regard to Remosillo).

However, is it reasonable to think that the people of the High Aragon in the Neolithic Age would have kept their livestock in ravines, such as those of the Vero or Ésera Rivers, which are prone to flash flooding in May? Would they not have lost all their animal resources? There are two possible answers: a positive one: yes, they could have if they moved their livestock to much wetter high ground during the hot summer; and a negative one: the schematic art would not have marked out the ravines for domestic livestock use, but rather to pen in wild prey, as hunting was still carried out in the Neolithic Age, albeit possibly more for sport, as it was not required for subsistence.

Only one thing can be proved for certain: schematic art is mainly found in places with canyons or gorges, such as the Vero in the High Aragon, the Ésera and the Noguera in Ribagorza, and the Martín in Lower Aragón, with the Los Estrechos gorge in Albalate and the canyon of the Cabra River in Obón (Tollos de la Morera). In Jaraba, the canyon of the Mesa River is the site of the schematic cave art of Los Prados. In other words, where there are no narrow canyons, as in Bajo Matarraña, Bajo Guadalope and Albarracín, hardly any schematic art can be found. One has to go up to the Beceite passes (and to the ravine of Parisal) to find the sole site with schematic art along the river (La Fenellosa). Once again, the Mequinenza complexes are the exception, although this could be due to a highly doubtful chronology, dating perhaps from the Middle Ages.

On the other hand, the landscape of Levantine art is much more open and circular (fig.5), as can be seen in the map of Lower Aragon, the area containing the highest number of Levantine rock art caves. The map below shows the area of caves within an hour's walk. Along the edge of the semi-circular layout are caves with Levantine art (Val del Charco, Plano del Pulido, Caídas del Salbime, Roca dels Moros and Gascons) surrounding the inhabited caves (Costalena, Pontet, Botiquería, Sariñena) with Ahumada cave, the largest and deepest and therefore the most likely to hold the main dwelling, at the centre. The terrains are strictly at a tangent, and some are secant, such as Els Secans with respect to Botiquería. However, we wish to emphasise that both sites merged in times of occupation (Secans between Botiquería 4 and 6).

The high ground of Matarraña has two different types of cave. One is the Vidre cave, in a "pure" Neolithic style similar to that of Chaves, with bone spoons, long point drills, retouched segments with a double bevel, Cardial pottery with cockle-shell motifs, and, therefore, not at all accultured (similar to the caves in Bajo Matarraña: Botiquería, Secans, Pontet, and Costalena). The other is the cave painted with schematic art at La Fenellosa de Beceite, at a tangent to the former,

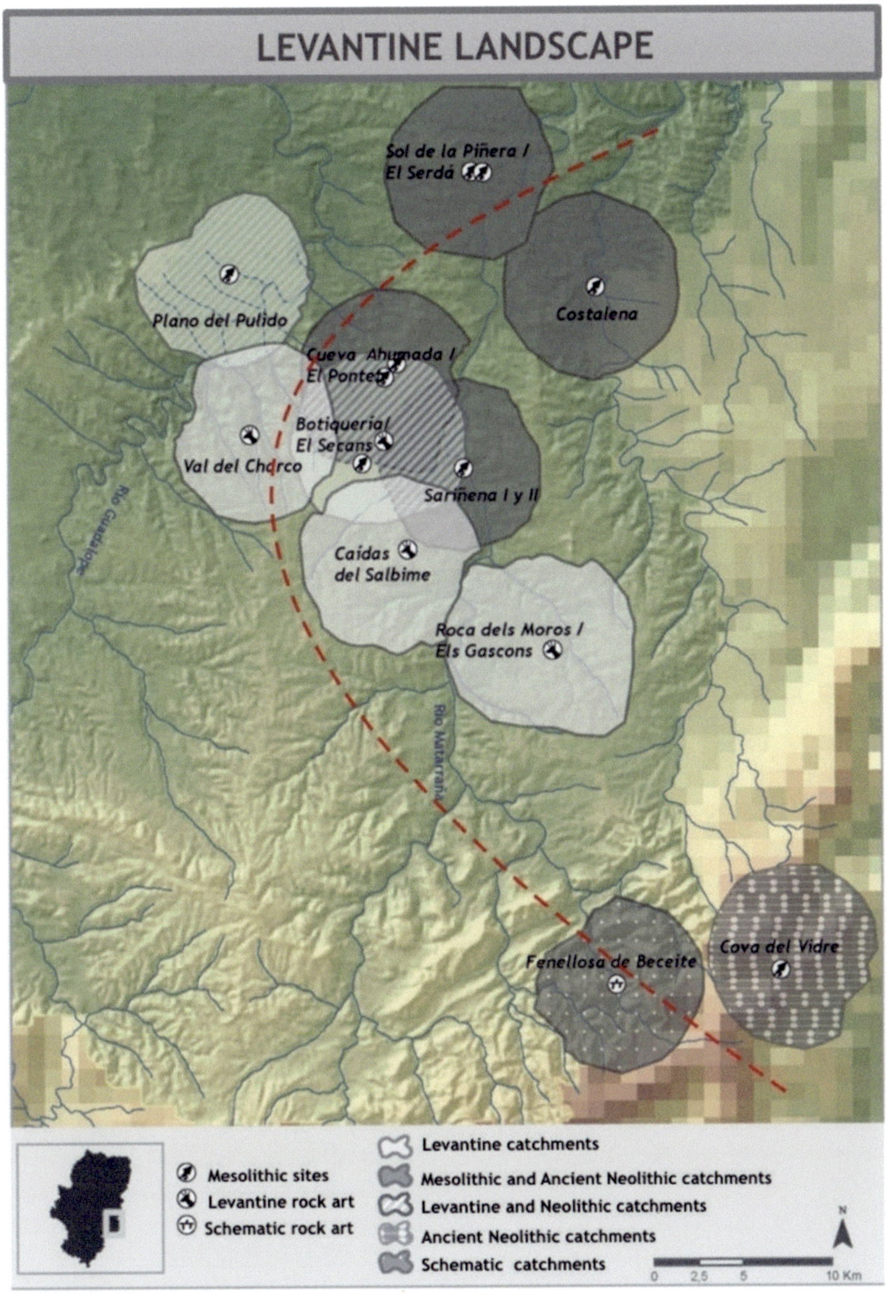

Figure 5: Circular distribution of the landscape of Levantine art.

which may be linked to the Cova del Vidre. The subject matter is people riding animals and could not be more symbolic or schematic, as is the case with similar examples in Los Estrechos in Albalate. It is possible that the Cova del Vidre was the neo-pioneering focal point, a key settlement of Neolithic expansion in the area, from which Neolithic culture and technology emanated to Lower Aragon.

The grass in these areas would have withered early in the summer. The pasture formed part of the undergrowth protected by the trees. Therefore, it is significant that almost half of the land in the area (43% in the area of Albarracín and Ladruñán, 33% in the Matarraña) was given over to forestry, while pasture land, strictly speaking, covered only 13% of the whole area. Nevertheless, the vegetation was better on the higher ground, especially in places where there was a great deal of shade. However, the main activity in the area was certainly hunting, with agriculture coming a poor second due to the climate (extreme cold in winter and too hot and dry in summer (Bacaicoa *et al.* 1993, 17)).

In this respect, it is significant that the most common animals in high terrain with Levantine art (Castellote/Ladruñán and Albarracín) are bovids, which do not fit in with a landscape of canyons and narrows. In the lowlands of Matarraña, Guadalope and Martín, deer (together with wild boar in Valdelcharco) are the most commonly depicted animals, which would denote a forest landscape.

On the other hand, mountain slopes were very useful for hunting goats (which are plentiful nowadays around Ladruñan in the Maestrazgo and Beceite passes), as the herds could be observed from several places without disturbing them. A similar practice could have been an advantage in exploiting wild goats in Prehistoric times. It could also be suggested that, at that time, there was sufficient ground cover to leave the deer in low lying areas where goats were no longer kept. Thereby, the tactics for hunting deer may have been similar to those for goats; observe and do not disturb.

These factors lead us to one of the main characteristics of the Levantine landscape; visibility. This was usually much wider than in landscapes with schematic art, despite the relative altitude of schematic art sites. This is the case of the Raja cave at Nueno, which controlled almost the whole of the Hoya de Huesca, and had the highest degree of accumulated visibility. The visibility from Roca Benedí in Jaraba allowed it to control access from the head of the ravine, in an area in which several caves were discarded in the ravine in favour of visibility. El Plano del Pulido in Caspe looks out over the entire plain as far as Val del Charco. Although it does not have a wide degree of visibility, the Muriecho cave does control the only route to the Cunarda pass. The Barranco Hondo in Castellote is positioned on a spur (not a cave) which looks out over a natural amphitheatre with excellent acoustics. In the same way, the Levantine caves of Pudial in Ladruñan are atypical due to their unusual orientation to the north. However, this characteristic can be explained by the excellent visual dominion over all the land they possessed.

It must also be pointed out that inter-visibility between the caves is immediate (in the area of Alacón and Albarracín) and that when visual communication was not possible due to the twisting landscape (between Coquinera I and Cañada de Marco or the complexes of Albalate), it was possible to whistle or shout, perhaps to alert companions to the arrival of a herd to hunt. We, along with our colleague Carlos Gracía Benito, have proved this in our experimental work on acoustic communication.

Conclusions

The study of rock art from the point of view of its distribution is an excellent way to approach "landscape archaeology". Unlike settlements, the spatial distributions of rock art sites over time may explain the functionality of the land, relating to social or cultural trends or even the environment. Rock art allows us to observe a complete landscape (Cruz Berrocal and Vicent 2007; Cruz Berrocal *et al.* 2014) in which the low and high lands are linked to form a regional territory, consisting of a series of connected local areas. Therefore, rock art reveals the interaction between macro and micro, the connectivity which characterises the fragmentation of the Mediterranean landscape.

The correlation between the distribution of Neolithic paintings and the traditional pastoral landscape can be interpreted more as a continuous process in the recurrent use of economic resources on the high ground of the Mediterranean for thousands of years. There are several papers on how this relates to the archaeology and ethno-archaeology of grazing in the Mediterranean (Creighton and Seguí 1998; Mientjen 2004; Christie *et al.* 2007), as well as studies focusing on change and continuity in a given area (Ejarque and Orengo 2009). However, due to their nature and use, there is scant evidence of settlement in the Mediterranean mountains in the Holocene Age.

As proved, the archaeology of rock art also displays a wider social context through the presence or absence of patterns, and therefore spatial modification is seen as a result of the social change that gave rise to its creation.

In the case in hand, it also points to "popular knowledge" of the landscape, and to continuity in the use of places, even at different social and economic times. Unfortunately, we do not have a great deal of information on this type of social memory and knowledge.

Therefore, our two proposals only serve to indicate that maps and the geographic analysis of rock art sites reflect the structural characteristics of the Mediterranean mountain ranges, such as topography, the seasonal availability of water and the coexistence of multifunctional spaces, among others. The vast majority of structural characteristics described are not prone to change and, therefore, it is probable that they have remained stable for centuries. At the same time, human activity in the landscape and especially ground cover (which reached its maximum complexity in the Mediterranean in the Holocene Age), seem to be found at the same time as rock art came into being in the Neolithic (Carrión *et al.* 2001). Patterns of change in vegetation reached their peak complexity between the middle and end of the Holocene Age. Thus, for example, although the trend towards greater aridity has been well established over the last five thousand years, the moment in which the forest opened uniformly cannot be explained by solely paleo-environmental change (Carrión 2002). Among the influencing factors, the role of anthropogenic disturbances (burning and grazing, among other activities) are decisive in this change (Carrión *et al.* 2007; López-Sáez *et al.* 2009).

Thus, we argue that acculturation of the Mediterranean landscape began in the Neolithic Age and was linked to the use of certain resources, either for hunting or grazing. We were unable to find a positive definitive relationship between rock art and land for hunting or grazing, as has been argued by several authors. However, we propose rock art as a "proxy" to understanding the underlying structure of the landscape, since rock art sites are found where geographic features are relatively uniform. Furthermore, these sites have been used throughout the course of history.

Therefore, we believe that, if society and nature form a complex system, just as environmental markers are used to reconstruct social patterns, landscape analysis should be complemented by the use of historical markers in order to reconstruct, or at least to imagine, what the landscape was like in the Holocene Age.

Studies such as this one, with a clear human bias, are needed for the integral analysis of the Mediterranean landscape and its acculturation. This is due to the fact that each landscape directly relates to a way of life, through mutual dependence between man and the environment created over thousands of years. Indeed, if these sites are abandoned, it can lead to a change in the landscape and the destruction of the Mediterranean landscape. Over the last few decades, the Mediterranean landscape in Spain has been abandoned, mostly in the low lands suitable for agriculture, thus causing a dramatic change in land use. Giving up sustainable economic practices in these areas is the most probable reason for the catastrophic change and degradation of the Mediterranean landscape today. Such an argument has also been made in the past (Hobbs *et al.* 1995; Naveh 1998; Rundel *et al.* 1998; González-Hidalgo *et al.* 2007; García-Ruiz 2010). As the land is being lost, its potential to feed the population, as well as thousands of years of work invested, acquired knowledge and its history are lying in ruins.

References

Bacaicoa Salaverri I. 1993. Albarracín-Cuenca-Molina. *Cuadernos de la Trashumancia 8, ICONA.*

Baldellou, V., Painaud, A., Calvo, Mª. J. 1983. Las pinturas esquemáticas de Quizans y Cueva Palomera (Alquézar, Huesca). *Zephyrvs* XXXVI, 117-122.

Burillo, F. 1997. Prospección arqueológica y geoarqueología, in: Armada, D. (coord.). *La Prospección Arqueológica, Actas II Encuentros de Arqueología y Patrimonio.* Granada, 119-132.

Butzer, K.W. 2005. *Arqueología, una ecología del hombre: método y teoría para un enfoque conceptual.* Barcelona: Bellaterra arqueología.

Carrión, J.S. 2001. Dialectic with climatic interpretations of Late-Quaternary vegetation history in Mediterranean Spain. *Journal of Mediterranean Ecology* 2, 145-156.

Carrión, J.S. 2002. Patterns and processes of Late Quaternary environmental change in a montane region of southwestern Europe. *Quaternary Science Reviews* 21, 2047-2066.

Carrión, J.S., Fuentes, N., González-Sampériz, P., Sánchez Quirante, L., Finlayson, C., Fernández, S. & Andrade, A. 2007. Holocene environmental change in a montane region of southern Europe with a long history of human settlement. *Quaternary Science Reviews* 26, 1455-1475.

Castaño, A. 2004. Estudio arqueozoológico de los macromamíferos del neolítico de la cueva de Chaves (Huesca). *Saldvie. Estudios de Prehistoria y Arqueolgoía* 4, 125-171.

Cava, A. 2000. La industria lítica del Neolítico de Chaves (Huesca). *Saldvie. Estudios de Prehistoria y Arqueolgoía* 1,77-164.

Colón Díaz, M. 1998. *Montaña y karst mediterráneo. Especificidad, antropización y gestión ambiental.* Cádiz. Universidad de Cádiz.

Cruz Berrocal, M. 2004. La investigación del arte rupestre desde la geografía: la pintura neolítica del ámbito mediterráneo de la Península Ibérica. *Trabajos de Prehistoria* 61/2, 41-62.

Cruz Berrocal, M. & Vicent, J. 2007. Rock art as an archaeological and social indicator: the neolithisation of the Iberian Peninsula. *Journal of Anthropological Archaeology* 26, 676-697.

Cruz-Berrocal, M., Sebastián, M., Uriarte, A., López-Sáez, A., 2014. Landscape contruction and long-term economic practices: an example from de Spanish Mediterranean uplands through Rock Art Archaeology. *Journal of Archaeological Methods and Theory* 21, 589-615.

Davidson, I. 1989. *La economía del final del paleolítico en la España oriental.* Valencia: Diputación provincial de Valencia.

Fábrega, P. 2005. Tiempo para el espacio. Poblamiento y territorio en la Edad del Hierro en la comarca de Ortegal (A Coruña, Galicia). *Complutum* 16,125-148.

Fairén, S., García-Atiénzar, G. 2005. Arte rupestre y territorio. Contribución de los Sistemas de Información Geográfica al análisis del paisaje neolítico en el interior de la Marina Alta (Alicante), in: Arias Cabal, P., Ontañón Pereda, R., García Moncó C. (eds.). *Actas del III Congreso sobre Neolítico en la Península Ibérica.* Santander, 2003. Santander: Universidad de Cantabria, 569-578.

Fernández-Freire, C. 2009. *Paisajes agrarios pre y protohistóricos en la comarca de la Vera Alta*. Madrid: Universidad Complutenses de Madrid (unpublished PhD Thesis).

Fernández Otal J., 1993, *La Casa de Ganaderos de Zaragoza. Derecho y trashumancia a fines del siglo XV.* Zaragoza: Institución Fernando el Católico.

Fernández-Freire, C., Uriarte, A., 2011. Modelización del paisaje mediante SIG para la investigación de sociedades agrarias paleotécnicas, in: Mayoral, V., Celestino, S. (eds.). *Tecnologías de Información Geográfica y Análisis Arqueológico del Territorio. Actas del V Simposio Internacional de Arqueología de Mérida.* Mérida, 2007. Mérida: Junta de Extremadura-Consorcio de Mérida, 449-458.

García Ferrando, M. 1985. *Socioestadística: introducción a la estadística en sociología.* Madrid: Alianza.

García-Ruiz, J.M. 2010. The effects of land uses on soil erosion in Spain: A review. *Catena* 81/1, 1-11.

García San Juan, L. 1999. *Los orígenes de la estratificación social: patrones de desigualdad en la Edad del Bronce del Suroeste de la Península Ibérica (Sierra Morena Occidental c.1700-1100 a.n.e/ 2100-1300 A.N.E).* BAR International Series (XXIV) 307. Oxford: Archeopress.

Gómez-Miguel, V. 2006. *Mapas de Suelos de España 1:1000000.* Madrid: Instituto Geográfico Nacional.

González-Hidalgo, J.C., Peña-Monné, J.L. & De Luis, M. 2007. A review of daily soil erosion in Western Mediterranean areas. *Catena* 71: 193-199.

Hobbs, R.J.; Richardson, D.M. & Davis, G.W. 1995. Mediterranean-type ecosystems: opportunities and constraints for studying the function of biodiversity, in G.W. Davis and D.M. Richardson (eds.). *Mediterranean-type ecosystems. The function of biodiversity.* Ecological Studies 109. Springer, Berlin.

Horton, R. 1945. Erosional development of streams and their drainage basins: Hydrophysical application of quantitative morphology. *Geological Society of América Bulletin* 56, 275-370.

Leori-Gourhan, A. 1964. *Le geste et la parole.* Paris: Albin Michel.

López Sáez, J.A., López-Merino, L., Alba Sánchez, F. & Pérez Díaz, S. 2009. Contribución paleoambiental al estudio de la trashumancia en el sector abulense de la Sierra de Gredos. Hispania. *Revista Española de Historia* 231, 9-38.

Naveh, Z. 1998. From biodiversity to ecodiversity. Holistic conservation of the biological and cultural diversity of Mediterranean landscapes, in: P. Rundel, G. Montenegro and F. Jaksic (eds.). *Landscape disturbance and biodiversity in Mediterranean-type ecosystems.* Ecological studies 136. Berlin: Springer,

Parcero, C. 2002. Using GIS for the historical analysis of archaeological landscapes. *Archaeological Computing Newsletter* 59, 4-10.

Peña, J.L., Julián, A., Chueca, J., Echeverría, M.T., Ángeles, G. 2004. Etapas de evolución Holocena en el valle del río Huerva: Geomorfología y Geoarqueología, in: Peña, J.L. (ed.). *Geografía Física de Aragón. Aspectos generales y temáticos.* Zaragoza: Universidad de Zaragoza e Institución Fernando el Católico, 289-302.

Prescott, C. 1995. Aspects of early pastoralism in Sogn, Norway. *Acta Archaeologica* 66, 163-190.

Rundel, P., Montenegro, G. & Jaksic, F. (eds.). 1998. *Landscape disturbance and biodiversity in Mediterranean-type ecosystems.* Berlin: Springer.

Sánchez-Navarro, J.A. 2000. *Las aguas termales en Aragón: estudio hidrogeotérmico.* Serie de Investigación 23. Zaragoza: Consejo de Protección de la Naturaleza de Aragón.

Soil Survey Group. 2003. *Keys to soil taxonomy.* United States Department of Agriculture.

Vicent, J.M. 1991. Fundamentos teórico-metodológicos para un programa de investigación arqueo-geográfica, in: López, P. (ed.). *El cambio cultural del IV al II milenios a.C. en la comarca noroeste de Murcia.* Madrid: CSIC, 29-117.

Wheatley, D., Gillings, M. 2002. *Spatial Tecnology and Archaeology.* London: Taylor y Francis.

Utrilla, P., Rodanés J. M. 2004 *Un asentamiento epipaleolítico en el valle del río Martín: el abrigo de los Baños (Ariño, Teruel).* Monografías Arqueológicas 39. Zaragoza: Universidad de Zaragoza.

Utrilla, P., Bea, M., Benedí, S. 2010. Hacia el Lejano Oeste. Arte levantino en el acceso a la Meseta: la Roca Benedí (Jaraba, Zaragoza). *Trabajos de Prehistoria* 67/1, 227-243.

Notes on contributor

María Sebastián López is Assistant Lecturer at the University of Zaragoza (Spain). Her research has been devoted to the integration of interdisciplinary knowledge derived from geomorphology, rock art archaeology and geo-archaeology for the study of the Mediterranean Holocene Landscape. Her research is concerned with the characterisation of geographic and climatic processes; the analysis of the effects of erosion processes and sediment transport in rock art sites; the relationships between rock art sites and landscape dynamics; and a large-scale approach to landscape dynamics in Mediterranean areas.

Building landscapes: a landform approach for the Iron Age sites in the Upper Duero River

Raquel Liceras-Garrido,[1] Enrique Cerrillo-Cuenca[2] and Alfredo Jimeno-Martínez[1]

Introduction: where we have come from and where we are going

This approach understands landscapes from a holistic perspective and always in the plural, as they are not something which is unique but are, rather, the result of numerous processes, practices or reasons, a set of nested images created by humans. According to B. Olsen (2007, 291), landscapes and artefacts are not quietly waiting to embody socially constructed meanings, but possess their own materialities and competences which have an influence on our lives. Thus, we propose an analysis of landforms in order to consider the motivations and social strategies taken into account by Iron Age people when choosing a location for their habitats.

Our study is focused on northern-central Spain, at the headwater of one of the country's main rivers, the Duero, which runs across Iberia from east to west towards the Atlantic Ocean. It is a natural region which includes the upper-mid basin of this river where the topography is flatter and crossed by many more tributaries than in the headwater region. Our study area lies between two of the Iberian Peninsula's major mountainous systems, the Iberian System to the north and east, and the Central System to the south. The Duero River lies between both, in a valley crossed by several mountain ranges and foothills resulting in a rugged and undulating landscape.

The archaeological context of our analysis is the Iron Age, which lasted over 700 years in this region and is divided into three periods with differences in material culture, settlement patterns, landscapes and identities. The Early Iron

---

1 Department of Prehistory, Complutense University of Madrid. Edificio B C/ Profesor Aranguren, s/n Ciudad Universitaria 28040 Madrid (Spain) rliceras@ucm.es aljimen@ghis.ucm.es.

2 Archaeologist, PhD in Prehistory enrique.cerrillocuenca@gmail.com.

Age ranges from 700 to 400 BCE and is characterised by small communities which inhabited the landscape in not very large sites. During the Late Iron Age, a process of synoecism took place, so that the existing communities gathered around centralised sites which we refer to as cities or *oppida,* with complex systems of social organisation and a territory to rule. In the 2nd century BCE, these cities were involved in wars against Rome, which resulted in the Roman conquest of Celtiberia by the end of the century. Finally, the 1st century BCE, a short period full of changes, resulted in a progressive adoption of Roman ways of life. This society is the outcome of the redistribution of land following the conquest to the indigenous peoples who had helped the Romans, until the time of the Sertorian War (75-72 BCE), when several of these settlements were again destroyed.

Settlement pattern analysis has been one of the mainstream areas of research in this area (see Taracena 1954; Jimeno and Arlegui 1995; Jimeno 2011; Heras 2000; Romero 1991; Romero and Lorrio 2011). Numerous researchers have partially studied the settlement of certain regions, which has resulted in a marked regionality which must be overcome by approaching Iron Age landscapes as a whole in both time and space. This is the main motivation for our analysis; to comprehend the settlements of the entire region through the 1st millennium BCE and to understand modifications in the landscape as a result of the social dynamics which took place

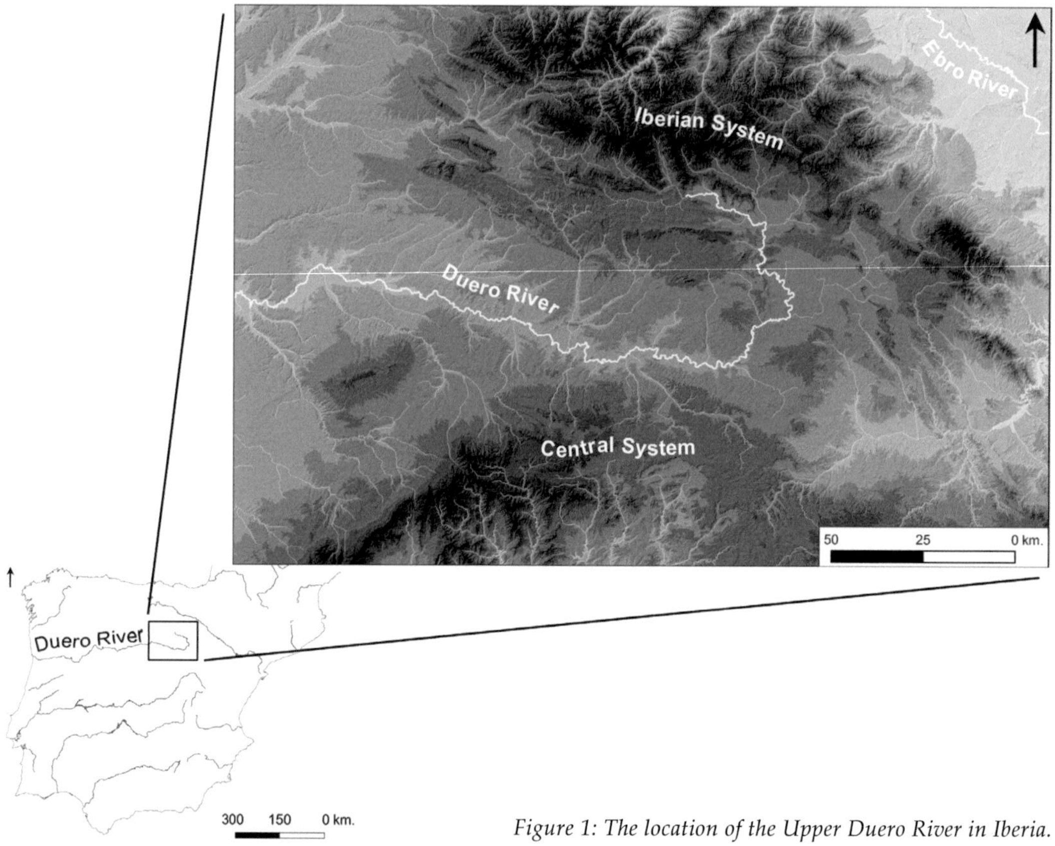

Figure 1: The location of the Upper Duero River in Iberia.

in the different periods of the Iron Age. In order to achieve this, we have carried out a geomorphological assessment of the places where the sites were built through a variable which has remained largely constant in time, landform, with the aim of exploring the chronological variability of these places as a key component of cultural diversity. It is an exploratory technique which enables us to evaluate both the specificity of the sites and their environments.

In order to carry out this analysis, we have gathered all the bibliographical information related to the location, chronology and cultural features of the 327 known sites in the study area (see Taracena 1941; Revilla 1985; Borobio 1985; Pascual 1991; Morales 1995; Romero 1991; Heras 2000; Alfaro 2005; Sacristán 2007; Barrio 1991). We have also included some sites from neighbouring natural regions in order to observe the differences between archaeological traditions.

Methodology: exploring the landforms

The method which we have employed follows three steps: 1) a classification of landform units, 2) an analysis of particular site locations relating to specific landscape units and, 3) the evaluation of settlement patterns considering the representativeness of certain types of landforms around the sites.

Landform classification

GIS-based terrain characterisation has become one of the aims of Geomorphology. Several methods for classification have been published since the popularisation of GIS in this discipline (Pike 1988, among others). We have applied a classification method designed for the geomorphological evaluation of landscape proposed by Andrew Weiss (2001). This procedure is based on two parameters: the Topographic Position Index (TPI), and the Slope Position as the best way of understanding variability in land morphology.

According to J. Jenness (2006), the TPI is: "the difference between a cell elevation value and the average elevation of the neighbourhood around that cell". Positive TPI values represent locations which are higher than the average of their surroundings, thereby, they are related to landforms such as hills or ridges. On the contrary, negative TPI values are related to locations which are found in a lower position than their neighbours, corresponding to valleys or depressed areas. However, values close to 0 may present some problems as they include flat or relatively flat terrain which may characterise different topographic forms: from mesas or upper slopes to flat lower areas.

The TPI neighbourhood is calculated on two scales: 300 and 2,000 metres which makes it possible to record both the local features, where the sites are located, and the background topography. The combination of these two scales (the small and large TPI scales) allows us to differentiate a variety of nested landforms. This results in a thorough and detailed classification of the morphology of the sites, where the 300 m neighbourhood identifies the local forms, such as small hills, isolated mounds or small streams, while the 2,000 m neighbourhood reflects the landscape features on a wider scale, showing their characterisation on a regional level.

Class	Landform
1	Canyons. Deeply incised streams.
2	Midslope drainages. Shallow valleys.
3	Upland drainages. Headwaters.
4	U-shape valleys.
5	Plains.
6	Open slopes.
7	Upper slopes. Mesas.
8	Local ridges. Hills in valleys.
9	Midslope ridges. Small hills in plains.
10	Mountain tops. High ridges.

Table 1: The landform classes from the classification.

However, TPI calculation requires a correction of certain values according to the slope position. Thus, when both TPI values and the slope are close to 0, they are classified as flat areas or plains, whereas when the slope values are above certain thresholds (see Weiss 2001), they are defined as middle or steep slopes. In this way, the slope position values consider the variability of elevation values within that neighbourhood. The final result is a landform classification map in which every pixel of the DEM is assigned one of the ten different landform classes listed in Table 1.

For the purpose of this paper, we used the digital elevation model (DEM) produced by the PNOA (Spanish National Plan for Aerial Orthoimage, Spanish National Geographic Institute). This DEM is freely available and has a resolution of 5 metres, which is more than adequate for representing and characterising the minor landform units.

A significant particularity of our work is that the landform classification process was not performed within desktop GIS software. Instead, computer libraries and scripts in the Python programming language were used to generate the landform maps for each of the sites making up our sample. A script automatically merged and cropped the DEMs employing a circular buffer of 3000 metres around the absolute central coordinate of each site, whilst another script created the landform classification maps, obtained after applying the aforementioned method created by Weiss (2001).

Evaluation of sites and landform types

In this paper, we estimate the relationship between the sites and the landscape classification in two different ways. The first reflects the direct relationship between the sites and the specific landform class to which they are ascribed. Although this approach can serve as a highly descriptive manner for finding trends among the sites and along the cultural sequence, it also becomes essential in the evaluation of the surroundings of the sites to shed some light on additional aspects, such as land use or the strategic occupation of the landscape. Thus, we took into consideration an area with a radius of 2,500 metres from the centre of the site, within which we computed the surface occupied by each landform class. The information for the whole dataset was gathered along with the chronological attribution for the sites in a text file for later processing.

In order to find trends, we analysed the resulting dataset with two independent but complementary statistical techniques: K-means and Principal Components Analysis (PCA). K-means partitions a set of *n* observations from a *d* dimensional space into *k* classes fixed *a priori*. Thus, we fixed a number of three or four classes, depending on the complexity of the data in order to classify the representativeness of each type of landform within the given buffer. A number of 20 iterations was fixed for the calculation of K-means.

PCA reduces the dimensionality of the data by projecting the values from the original cases into a new and more meaningful vector space, in which they remain linearly related to the original dataset. One advantage of PCA is that the re-projection of the values maintains the information with the least possible loss. By means of PCA, the cases can be represented in a new 2D or 3D space in which it becomes easier to understand the representativeness of each landform class in the pattern of settlement for each period. The results consist of sets of components which gather the different percentages of variance. For the present case study, we only selected components which jointly gather more than 70% of the explained variance (the first and the second components for all the cases). The values from eigenvectors allow us to determine which landform classes would have a stronger effect in the choice of certain locations in the landscape. The combined graphic representation of K-means and PCA allows us to understand whether or not there are significant trends in the distribution of the sites in the landscape.

Results and discussion: landforms and Iron Age settlements in the Upper Duero River

A direct relationship between sites and landform classes

The first outcome of our analysis is the allocation of a landform class to each site, which shows the main trends in the settlement distribution across the Iron Age. This is summarised in Table 2. These data allow for an initial exploratory analysis for the identification of trends in the pattern of distribution of settlements.

Landform class	Early Iron Age		New sites in the Late Iron Age		Late Iron Age		New sites in the 1st century BCE		1st century BCE	
	Sites	%	Sites	%	Sites	%	Sites	%	Sites	%
1	0	0	1	0.8	1	0.6	0	0	0	0
2	0	0	0	0	0	0	0	0	0	0
3	0	0	0	0	0	0	0	0	0	0
4	0	0	1	0.8	1	0.6	2	6.2	2	3.8
5	9	7.75	17	14.9	20	12.5	12	37.5	16	30.7
6	4	3.4	11	9.6	11	6.9	4	12.5	5	9.6
7	26	22.4	21	18.4	30	18.8	5	15.6	13	25
8	1	0.8	0	0	0	0	0	0	0	0
9	17	14.6	11	9.9	16	10.06	5	15.6	5	9.6
10	59	50.8	52	45.6	80	50.8	4	12.5	11	21.1
Total	**116**	**100**	**112**	**100**	**159**	**100**	**32**	**100**	**52**	**100**

Table 2: Number of sites and percentages for each landform class and period.

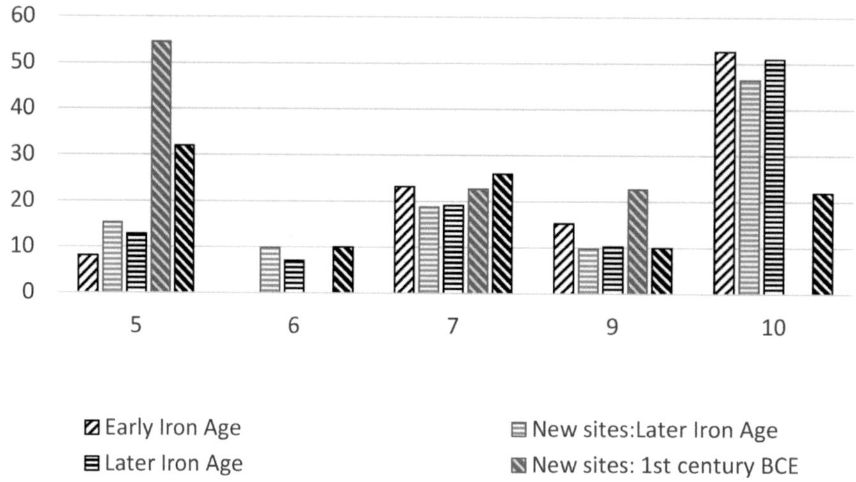

Figure 2: Bar graphic representing the percentage of sites per class (≥5 cases).

During the Early Iron Age, the preferred sites for the construction of settlements were on the highest locations; mountain tops or high ridges (class n. 10), upper slopes (n. 7) or midslope ridges or small hills located on plains (n. 9). The trend during the Late Iron Age is quite similar, class 10 remains as the preferred location, although there is a slight rise in class 5 corresponding to plains. However, while we can see a similar affinity between both periods, more than 70% of the sites were established during the 4th century BCE, that is, in the Late Iron Age. Finally, people in the 1st century BCE, influenced by Roman policies, reversed the traditional inclination to build sites on the higher areas of landscape, with a sharp decline in class 10, which had been the main location in the previous periods, in favour of plains (class 5) which are the main landform of this period.

The Early Iron Age duality

At the beginning of the Iron Age, we can observe, for the first time, the emergence of stable occupation of territories in which small communities built their villages scattered over the landscape. In the previous section we have suggested the significance of settlements on hilltops, which represent more than 80% of the total number of settlements in this period.

The results of the PCA analysis reveal that principal components (PC) 1 and 2 contain more than the 70% of the explained variance: 76.1%, which is enough to understand the principal causes for the variability of the locations. As derived from the analysis of the eigenvectors, landform classes 1, 5, 6, and 7 seem to have a significant influence on the differences between the cases for PC1, whereas classes 4 and 10 seem to be meaningful for describing the variability in PC2.

K-means results show three well-distinguished types of landscapes (G1, G2 and G3), each one with a similar number of cases (Figure 3B). G1 contains 44 sites and is related to mountainous areas. These sites are located on rough and steep areas with sharp differences in elevation, combining deep valleys with high peaks. Figure

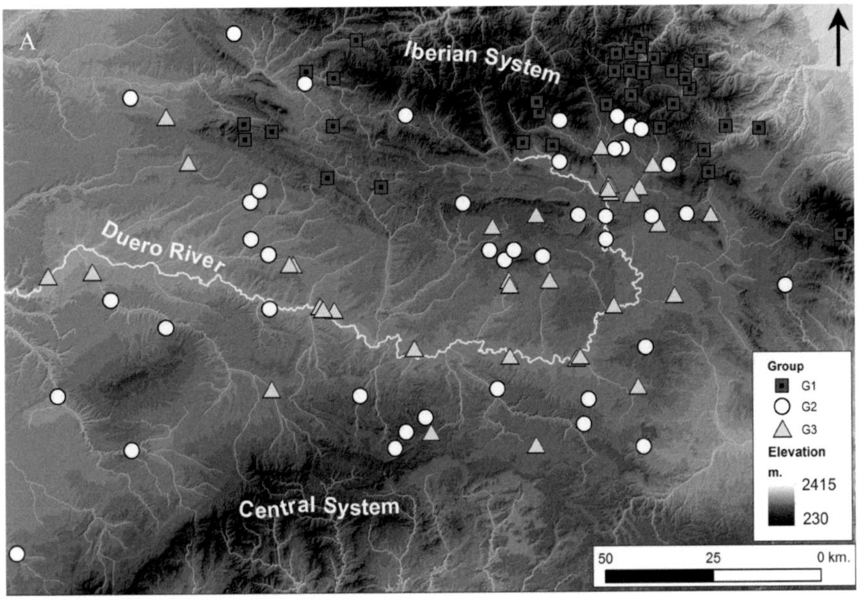

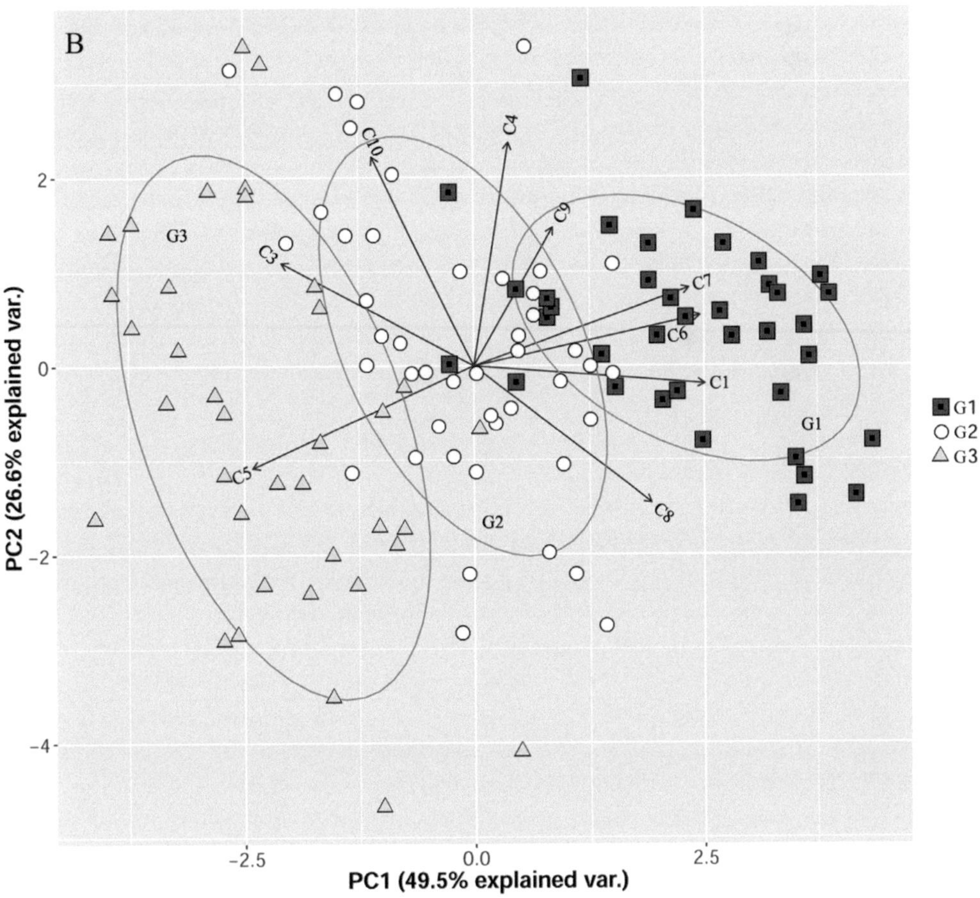

Figure 3: Groups in the Early Iron Age: A. Distribution map for the groups. B. Scatter plot representing the projection of the cases (n=116) for the two first components and the K-means groups.

3B shows that classes 6, 7 and 1 have a distinctive weight in shaping this group. On the other hand, G3 includes 37 sites and is strongly related to class 5. These sites are the settlements in the Duero valley and its tributaries which generally correspond to dominant places in undulating areas, usually buttes or prominent headlands. The final set, G2, includes 37 sites. This group shows the transition between two distinct types of environments, the high mountains and the valley. It can be seen in the chart how this group is located between the other two groups and is related to the secondary foothills.

Through this classification of sites, we can suggest a duality of places between the sites located in mountainous areas and those in the valley. This dichotomy has previously been pointed out by M. Revilla and A. Jimeno (1986-1987) in their regional study of the Almazán area, suggesting some geographic, cultural and economic differences. As far as the outcomes of our method are concerned, the spreading of the analysis to the limits of the whole natural region shows that such duplicity of types of settlements is still maintained.

From the 1920s, researchers have focused on the particularity of the defensive systems of certain settlements, especially for the group called "*Cultura de los Castros Sorianos*", which is located on the southern slope of the Iberian System facing the Duero valley (Taracena 1933, 1954; Romero 1991; Romero and Lorrio 2011; Lorrio 2005). According to our analysis, G2 encompasses most of the defended sites, including these *castros* (hillforts). It extends over mountainous areas, especially in the heart of the Iberian System and the secondary foothills. These sites usually present the most complex and diverse defences with a combination of ramparts, ditches, towers and *chevaux de frise*. There are only a few examples of open settlements in this group, which are closely related to fortified sites by direct visibility, such as El Castillo del Avieco (Molinos de Razón), La Coronilla Negra (Yanguas) and Valdegén (Villaseca Bajera). These sites could perhaps be interpreted as subordinated villages relating to an increase in population in some hillforts which required an efficient exploitation of resources from the bottom of the valleys.

G2 and G3 are mainly composed of open sites relating to valley environments, with no defensive structures or artificial boundaries, at least in the light of current data. However, we have found several sites with mixed features, usually on intermediate areas, such as slopes, foothills or spurs.

Therefore, it can be observed how people in the Early Iron Age looked mostly for high and prominent places in the landscape to build their settlements, with the clear aim of changing their surroundings. Thus, where landscapes were rough and mountainous, people built massive defences, whereas when the site surroundings were gentle and undulating, they chose prominent locations. Therefore, their strategies of landscape modification were different depending on the landforms and the environments of each region along the Upper Duero River.

On the basis of this analysis, it can be seen how the debate about the polarisation of the settlement is not only pertinent to the traditionally studied smaller regions, but to the area as a whole. Two main groups can be distinguished which chose different strategies in terms of landscape appropriation depending on their ideologies, cosmologies and understanding of the world. Landforms are just one way of characterising their identities but if we look at the material culture, we can see other relevant differences: in the open sites people lived in roundhouses and Celtiberian warrior cemeteries can be found, while in the hillforts they lived in rectangular houses and no burial grounds have been found. These different forms of identities and power would have had an influence in the Late Iron Age.

Cities in the Late Iron Age

From the 4th century BCE, a deep transformation of the landscape occurred as a result of an increase in population accompanied by social and technological changes (Jimeno 2011; Jimeno and Arlegui 1995; Alfaro 2005). The pre-existing small communities grew in size and underwent a process of synoecism which concluded with the organisation of the territories around cities such as Numantia, Termes and Uxama. By referring to cities we understand what has been suggested by M. Fernández-Götz and D. Krausse (2013, 480): "a numerically significant aggregation of people permanently living together in a settlement which fulfils central place functions for a wider territory".

The number of sites increased and, consequently, the landforms diversified, as can be seen in Figure 2. The tendency of locating settlements on hilltops continued and class 10 (mountain tops or high ridges) was still the preferred landform type. However, most of the Early Iron Age sites were abandoned and continuous occupation sequences, such as that of El Castillejo (Fuensaúco), are extremely rare. Over 70% of the settlements were new foundations, with a total of 114 sites.

Figure 4: Numantia seen from the air.

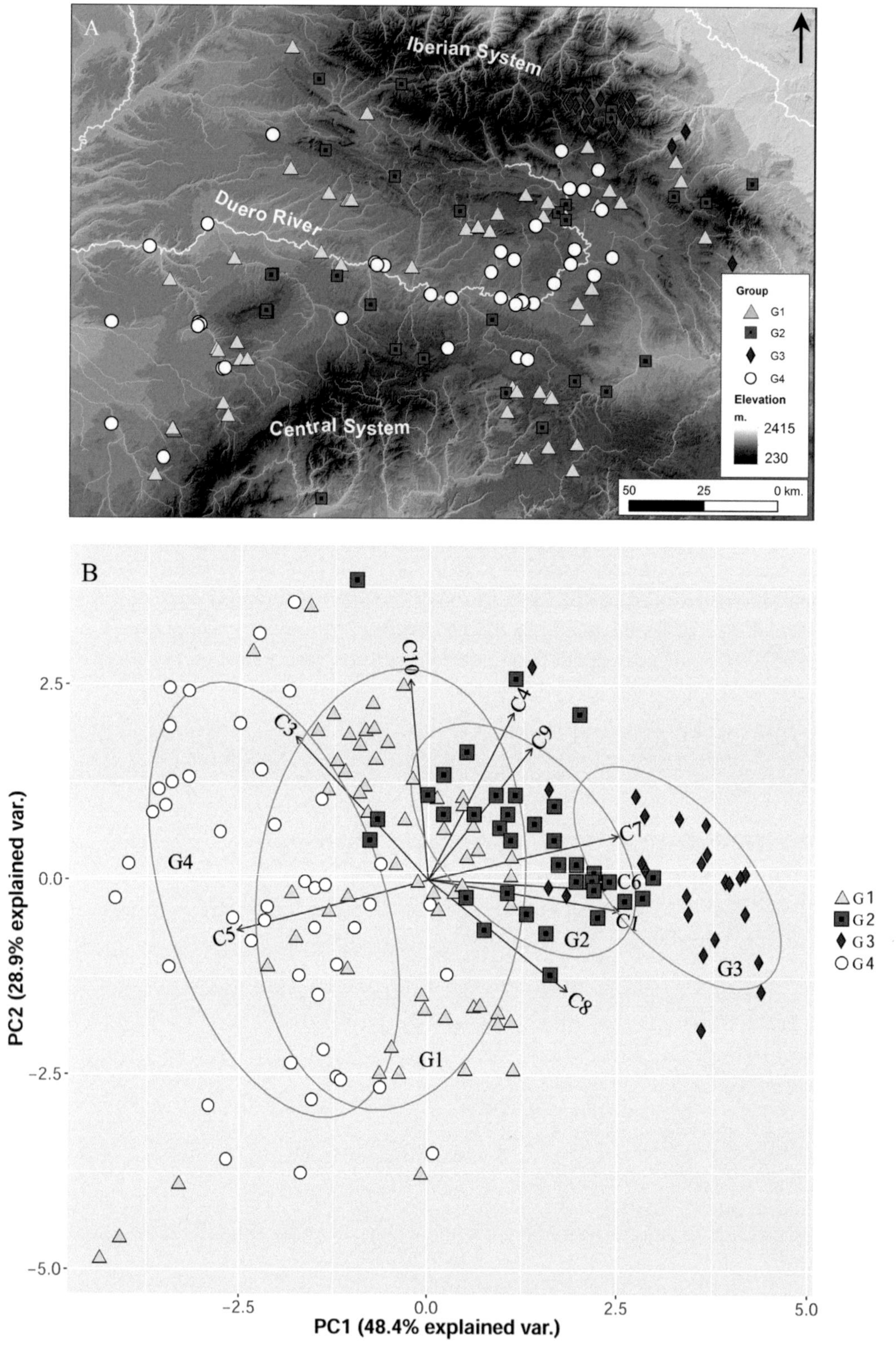

Figure 5: Groups in the Later Iron Age: A. Distribution map for the groups. B. Scatter plot representing the projection of the cases (n=159) for the two first components and the K-means groups.

As far as PCA is concerned, PC1 and PC2 include 77.27% of the explained variance. Landform classes 1, 6 and 7, as well as 5, are truly representative of the variance of PC1. Landform class 10 has a strong influence on the variance of PC2, and less influence is detected for classes 3 and 4.

Due to this apparently larger variability, we set the K-means algorithm to output four different groups (G1, G2, G3 and G4). G3 is a clearly distinct group by its location which includes 24 sites on the mountains of the Iberian System, in which classes 6, 7 and 1 are especially significant to shape the group as a result of the rugged environments. On the opposite side of the graph, G4 contains 44 sites with a close relationship with plain environments and a key role of class 5. Between these two sets, there are two more groups. G2 includes 35 sites where classes 1, 6, 7, 9 and 4 have a similar weight. The largest group (G1) contains 58 sites, which represent, along with G2, the intermediate landscape units between the high mountains and the valleys. Therefore, if we consider the map of the site distribution by groups, an obvious polarisation of the landscape into the two groups can be observed, with G4 in the valley and G3 in the mountains and G2 and G1 in between.

In terms of the distribution of the groups among the different landscape units, G3 sites are concentrated in the Iberian System, whereas in the Duero valley groups G2, G1 and G4 are intertwined. As the Duero River flows towards the west, G2 sites disappear and only G1 and G4 sites exist. To the south, in the Central System, G2 and G1 sites are again related, while G4 is absent. Thus, it may be suggested that the mixture of sites in different landscape units could be due to the diversification of the economy, resources and specialisation of craft activities.

As has already been pointed out, we have included several sites from the peripheral regions in our analysis in order to observe any possible differences or any transition zone to other landscapes and cultural groups. In this case, a clear differentiation can be found to the west and the south. The western group represents the changes between the Upper and Middle Duero River landscapes where the environments are more open and flatter and the rivers wider. The settlements are quite different and the people were clustered in large cities, such as Pintia, Cuellar and Cauca and there are no dependent sites (*cf.* Sacristán 2011, 1989). However, the southern group is related to the beginning of the Tajo and Jalón Rivers where the settlement pattern is dispersed and apparently not linked to cities.

There is a dichotomy between how people built in the landscapes and how the natural features influenced the constructions, which led to several outcomes in the same region depending on the physical characteristics, the communities which created them, and the messages they wanted to communicate through them, thus forging a profound relationship between the landscapes and the people. Thus, it can be seen how the Middle Duero River and the Upper Tajo and Jalón Rivers present differences in both the material culture and environments with the studied area. The results of our analysis may support an interpretation of those differences in terms of identity changes.

To sum up, in the Upper Duero River, two different cultural groups can be found which settled in different environments, which can roughly be defined as the mountains and the valleys. This dichotomy of settlements has been noted

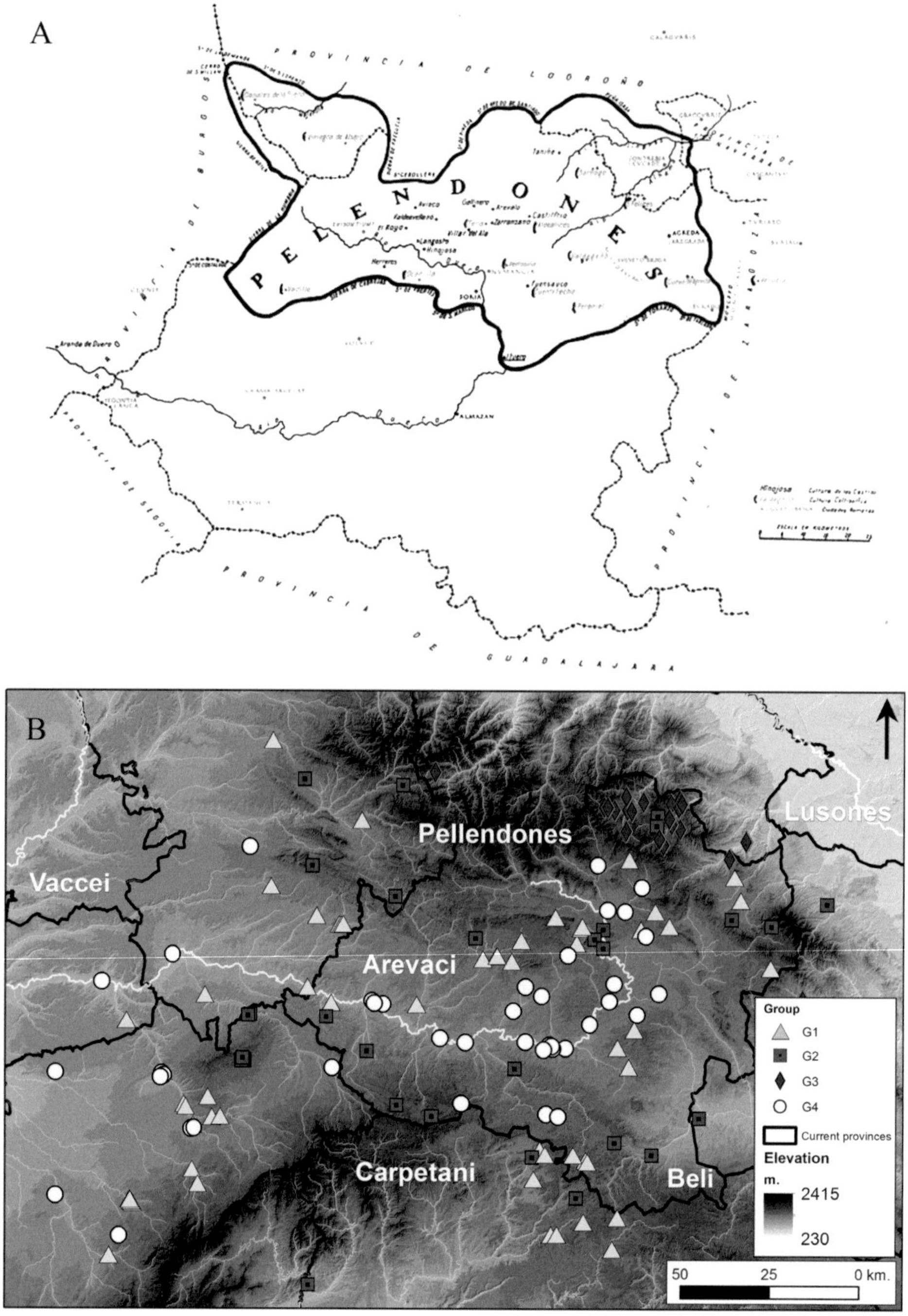

Figure 6: The ethnic groups: A. Taracena's proposal for the Pellendoni ethnic group (Taracena 1933, Figure 1). B. The general view of ethnic groups according our analysis.

in the Early Iron Age, although these two different strategies of inhabiting the landscapes grow stronger in the Late Iron Age.

The debate on ethnic identities has often formed part of the analysis of settlement distributions in Iron Age research in Iberia. In our region, Blas Taracena (1933) related the mountains with the Pellendoni ethnic group, while relating the valley with the militaristic Arevaci depicted in the references of the classical authors. However, if we understand ethnic identities as the materialisation of different ways of doing things (González-Ruibal 2003, 122-123), we could support Taracena's proposal not only through the variances in shape, size and relationship to specific landscapes detected in our analysis, but also through the different ways of construction of domestic contexts or attitudes to death with differential visibilities of cemeteries.

A change of context: the beginning of Roman settlement in the 1st century BC

The 1st century BCE was a complex period, full of changes and transformations and with sparse data in many cases. Rome conquered the area and imposed a new social model which was reflected in several aspects of people's life, such as the settlement pattern. Some sites from the previous period continued to be occupied, especially the cities which evolved around the government of the territories. In spite of the centralisation process conducted by the Romans, the cities maintain a strong indigenous character by mixing together local products and Roman imports, which were still scarce.

The new settlement pattern was characterised in the wider area of Celtiberia by the construction of new central places close the previous sites, except in the valley of the Upper Duero River (which could be related to the Arevaci), where Rome rebuilt these centres in the same places (Jimeno 2011, 263-266). This adds difficulties when it comes to identifying the material culture of this period, as we have seen in Numantia where the archaeological levels between the destruction of the Celtiberian city in the 133 BCE and the Sertorian War (75-72 BCE) have only been preserved in very specific areas of the site, near the defensive walls (Jimeno *et al.* 2012, in press).

The tendency to settle on hilltops was abandoned and the number of sites in landform class 10 decreases with Roman control, while new sites were established in relatively flat environments where plains (class 5) became the favourite landform, followed by upper slopes or mesas (class 7).

The PCA results are quite similar to those from the preceding periods. PC1 and PC2 comprise 80.28% of the explained variance. Landform classes 1, 6, 7 and 5 define the variability from PC1, whilst landform classes 3, 4 and 10 neatly define the variability for PC2.

The cases were clustered into 3 groups via K-means (G1, G2 and G3). From the analysis, we find a greater homogeneity in landforms compared with the previous periods without a marked polarisation of the landscapes. Three groups can be observed, in which G1 includes 14 sites with a clear predominance of plain landforms (class n. 5) in the valley bottoms. G3 brings together 12 sites strongly related to class 6 (open slopes). Finally, between the two, G2 is the larger group, with 26 sites, 46% of them continuing from the previous period.

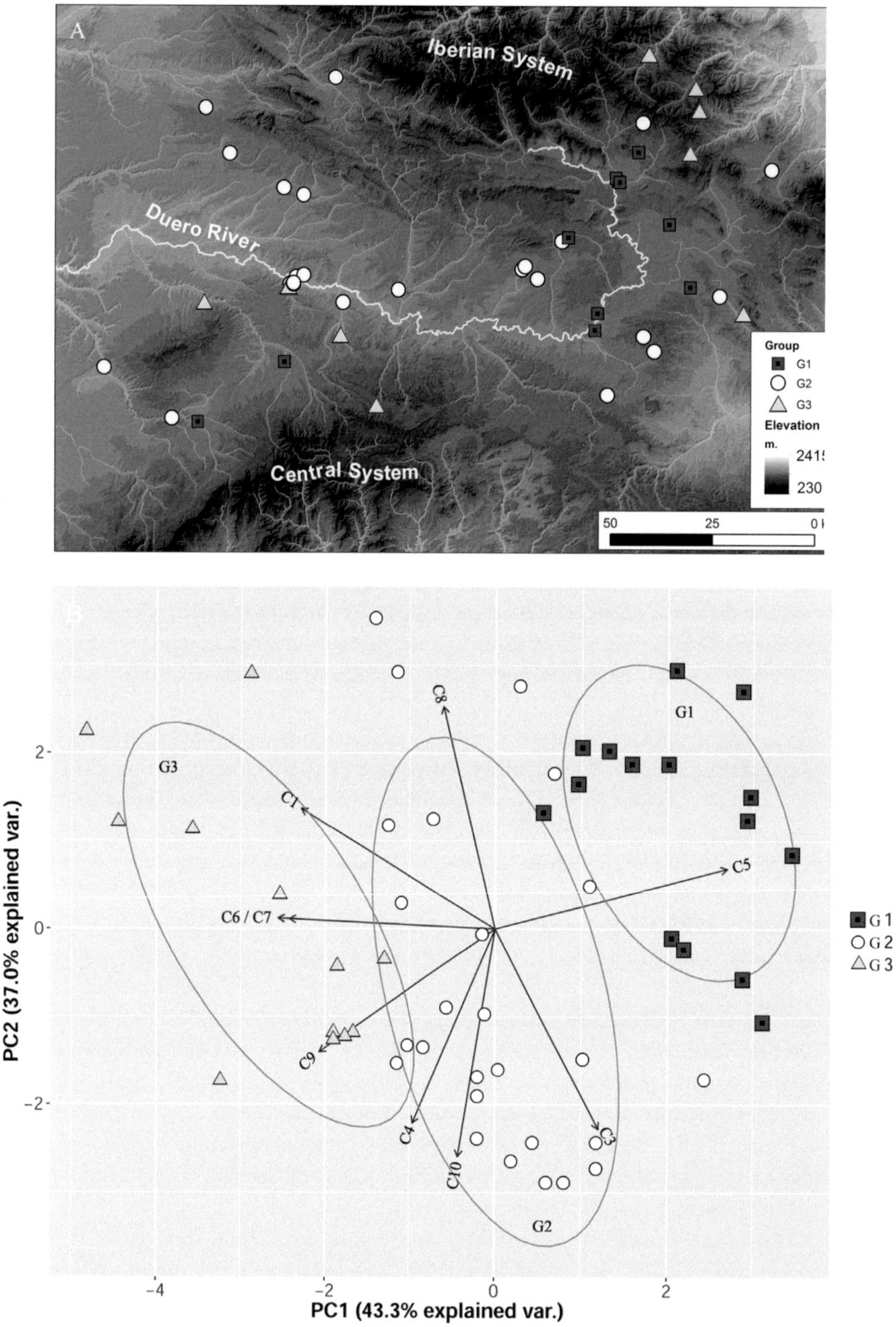

Figure 7: Groups in the Celtiberian-Roman period: A. Distribution map. B. Scatter plot representing the projection of the cases (n=52) for the two first components and the K-means groups.

The strong preference for flat open environments is illustrative of the clear Roman influence in the choice of landforms and how, throughout the region, they found similar locations to build their settlements as the result of various social, political and ideological changes. As far as this is concerned, Appian (*Hisp.* 16, 99) collects references to the changes of settlement location, such as in the city of Termes where Titus Didius forced the inhabitants to move their houses to the plain and live without defensive walls as a punishment for their insubordination.

Conclusions: beyond landforms

As we have suggested, the relationship between people and landscapes is bidirectional. People create landscapes as a reflection of the different aspects of their identities and the power relations which arise within the communities in which they live. At the same time, the materiality of the environments, their shapes, colours, potentialities, create people. Throughout this analysis we have seen how the material aspects of the landscapes remain, in spite of cultural changes.

The methodology we have chosen works consistently with large data sets (327 sites for the global analysis) and with high resolution DEMs as the best way to characterise the regional variability of the data. It must be highlighted that the method is robust and valid for different fields of study to assess diverse patterns of settlement and even for exploring trends in large datasets, such as that used in this paper.

The results have sought to provide a more objective landform classification than the traditional visual characterisation which has been used in this region until now and it has allowed us to relate areas which had previously been studied separately. Following the analysis, we are able to observe some similarities in settlement patterns and material culture, enabling us to explore the differences in terms of identities and landscape strategies. However, new efforts and methods need to be developed in order to understand the probable causes for variability in smaller regions, which is one of our future aims. Consequently, we can focus on the people's ideology as a whole, in which landscape is just one piece of an extremely complex puzzle.

References

Alfaro, E. 2005. *Castillejos y Villares. Modelos de poblamiento antiguo en el interior del Sistema Ibérico.* Soria: Soria Edita.

Barrio, J. 1991. *La segunda Edad del Hierro en Segovia.* Madrid: Universidad Autónoma de Madrid, PhD Thesis.

Borobio, M. J. 1985. *Campo de Gómara. Carta Arqueológica de Soria.* Soria: Publicaciones de la Excma. Diputación de Soria.

Fernández-Götz, M. and Krausse, D. 2013. Rethinking Early Iron Age urbanisation in Central Europe: the Heuneburg site and its archaeological environment. *Antiquity* 87, 473-487.

González-Ruibal, A. 2003. *La experiencia del otro. Una introducción a la etnoarqueología.* Madrid: Akal.

Heras, E. 2000. Aproximación a la evolución del poblamiento en el suroeste de la provincia de Soria durante la Edad del Hierro y la etapa del Alto Imperio, in: Baquedano, E. (coord.). *Soria arqueológica: a José Luis Argente.* Soria: Diputación Provincial de Soria, 205-238.

Jenness, J. 2006. Topographic Position Index (tpi_jen.avx) extension for ArcView 3.x, v. 1.2. Jenness Enterprises, accessed 1st July 2016 from http://www.jennessent.com/arcview/tpi.htm.

Jimeno, A. 2011. Las ciudades celtibéricas de la Meseta Oriental. *Complutum* 22/2, 223-276.

Jimeno, A. and Arlegui, M. 1995. El poblamiento en el Alto Duero, in: Burillo, F. (coord.). *Poblamiento Celtibérico. III Simposio sobre los Celtíberos.* Zaragoza: Instituto Fernando el Católico, 93-126.

Jimeno, A., Chaín, A., Quintero, S., Liceras, R. and Santos, A. 2012. Interpretación estratigráfica de Numancia y ordenación cronológica de sus cerámicas. *Complutum* 23/1, 203-218.

Jimeno, A., Liceras, R., Quintero, S. and Chaín, A. in press. Unraveling Numantia: Celtiberian and Roman settlement (Soria, North-Central Spain), in: *Archaeology in the River Duero valley (Spain and Portugal): from Paleolithic to Medieval Age. New Perspectives and advance in the investigation of the past.* Cambridge: Cambridge Publications.

Lorrio, A. 2005. *Los Celtiberos.* Madrid: Real Academia de la Historia.

Morales, F. 1995. *La Altiplanicie Soriana. Carta Arqueológica de Soria.* Soria: Publicaciones de la Excma. Diputación de Soria.

Olsen, B. 2007. Genealogías de la asimetría: por qué nos hemos olvidado de las cosas. *Complutum* 18, 287-291.

Pascual, A.C. 1991. *Zona Centro. Carta Arqueológica de Soria.* Soria: Publicaciones de la Excma. Diputación de Soria.

Pascual, P. and Pascual, H. 1984. *Carta Arqueológica de La Rioja. I.- El Cidacos.* Calahorra, Colección de Amigos de Calahorra.

Pike, R., 1988. The geometric signature: Quantifying landslide-terrain types from digital elevation models. *Mathematical Geology* 20, 491-511.

Revilla, M. L. 1985. *Tierra de Almazán. Carta Arqueológica de Soria.* Soria: Publicaciones de la Excma. Diputación de Soria.

Revilla, M. L. and Jimeno, A. 1986-1987. La dualidad de la cultura castreña en la provincia de Soria. *Zephyrus* 39-40, 87-101.

Romero, F. 1991. *Los castros de la Edad del Hierro en el Norte de la provincia de Soria.* Valladolid: Universidad de Valladolid.

Romero, F. and Lorrio, A. 2011. El origen del poblamiento celtibérico en el Alto Duero. *Complutum* 22/2, 95-127.

Sacristán, J.D. 2011. El urbanismo vacceo. *Complutum* 22/2, 185-222.

Sacristán, J.D. 2007. *La Edad del Hierro en la provincia de Burgos.* Burgos: Publicaciones de la Excma. Diputación de Burgos.

Sacristán, J.D. 1989. Vacíos Vacceos. *Arqueología espacial 13, 77-88.*

Taracena, B. 1954. Los pueblos celtibéricos, in: Menéndez Pidal, R. (coord.). *Historia de España I, (3), Los pueblos prerromanos.* Madrid, 195-299.

Taracena, B. 1941. *Carta Arqueológica de España, Soria.* Madrid: CSIC, Instituto Diego Velázquez.

Taracena, B. 1933. *Tribus celtibéricas: 'Pelendones'.* Guimaraes: Sociedad da Martins Sarmento.

Weiss, A. D. 2001. *Topographic position and landforms analysis.* Poster Presentation, ESRI User Conference, San Diego, CA.

Notes on contributors

Raquel Liceras-Garrido is an FPU pre-doctoral researcher at the Complutense University of Madrid, funded by the Spanish Ministry of Education, Culture and Sport, having completed her BA in History and an MA in Geographic Information Technologies in the same institution. She has also been a visiting research student at the Universities of Durham, Edinburgh and Bradford (UK). She has been a member of the Archaeological Research Team of the Numantia archaeological site since the beginning of her research career. Her PhD research is focused on the identities and power relationships of the communities of the Upper Duero River during the Iron Age, supervised by Dr. Alfredo Jimeno and Dr. Enrique Cerrillo-Cuenca.

Enrique Cerrillo-Cuenca is an archaeologist who has worked in several universities and research centres in Spain and Portugal. His last position (2009-2015) was a Ramon y Cajal Fellowship at the Spanish National Research Council. He currently belongs to the CHAIA research group based at the University of Évora (Portugal). He earned his PhD in 2003 with a thesis on Neolithic settlement in the inner basin of the Tajo River. He has carried out research in several fields of Prehistory, such as the Mesolithic, Neolithic, megalithic landscapes and rock art in South America. As far as the methodological scope of his research is concerned, his areas of expertise are LiDAR and aerial photography in archaeology, photogrammetry, supercomputing in archaeology and the application of spatial technologies.

Alfredo Jimeno-Martínez is a Senior Lecturer in Prehistory at the Complutense University in Madrid. He has carried out extensive research focused on the Iberian Meseta from the Bronze Age to the Romanisation period, which is reflected in some 20 books, 150 papers and numerous presentations at meetings and conferences. In the international arena, he has been the co-director of different projects on the Blue Nile River as a member of the Archaeological Mission of the Complutense University in Sudan (1990-1999) and the director of the CAI, the service for archaeometric research and analytical methods in archaeology at the Complutense University (2009-2014). Since 1994, he has been the Director of the Numantia archaeological site for the research, dissemination of scientific knowledge and promotion of cultural heritage.

The contribution of GIS to the analysis of the distribution of Roman caves between the Ebro River and the Pyrenees

Leticia Tobalina Pulido,[1] *Benoît Pace*[2] *and Alain Campo*[2]

Introduction

The subject of cave occupation during the Roman era has hardly ever been dealt with in detail by archaeologists. Until the 1980s, the lack of data and the scarcity of information concerning finds, along with the absence of specialised research on this theme, have shown both the lack of interest and the difficulties involved in such research.

Nevertheless, since that time, a degree of progress has been made and a clear increase in research on this topic can be noticed. New studies, although they can still be considered as marginal, have come about and have enabled researchers to reconsider the occupation of these types of sites during the Roman era (for France, Réchin and Dumontier 2013; Balmelle *et al.* 2001, 215-217; Raynaud 2001; for Spain, Fanjul Peraza 2011; Gutiérrez Cuenca *et al.* 2012; Utrilla *et al.* 2014; Rubio 2014). These new additions to this area of research show how important it is to reconsider the status of these sites, which clearly appear as "*un type d'établissement beaucoup plus fréquent qu'on ne le croit généralement, qui échappe aux réflexions conduites sur l'organisation des territoires de montagne ou de piémont*" (Dumontier and Réchin 2013, 97).

1 University of Navarre/ Université de Pau et des Pays de l'Adour leticiatobalina@gmail.com.
2 Université de Pau et des Pays de l'Adour Benoit.pace@yahoo.fr alan-64@hotmail.fr.

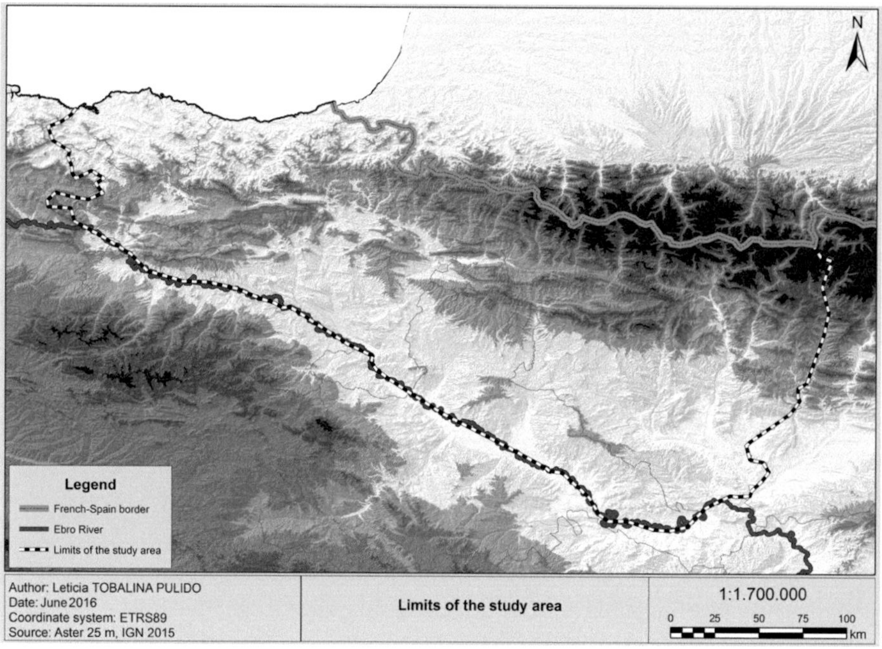

Figure 1. Geographic area of the research in this article.

In the light of this situation, the POEM[3] research project, an initiative of the ITEM laboratory belonging to the *Université de Pau et des Pays de l'Adour*, has enabled a study group to be set up in order to deal with this matter from a new perspective via the application of an appropriate analysis method. In keeping with the axis of "The seasonal flow of men and animals and the forms of space occupation which ensue" this research group has focused solely on the study of these types of frequencies.

Over the following pages, which will examine the case of the areas between the Ebro and the Western Pyrenees (Figure 1), briefly analysing the current state of research concerning this topic. We shall then proceed to carry out a review of the method we have applied (geodatabases and G.I.S.). Finally, we shall present the initial results of our analysis.

Cave occupation during the Roman era: a marginal subject

Studies concerning cave occupation during the Roman era from the perspective of spatial analysis are still scarce, as has been pointed out by Dumontier and Réchin: "Jusqu'à une époque récente, les niveaux d'époque romaine rencontrés lors des fouilles de cavités ont souvent été victimes d'un certain désintérêt" (Dumontier and Réchin, 2013, 97). This quote highlights one of the main difficulties encountered

3 The POEM research project (*«Mobilités et Échanges dans les Pyrénées Occidentales et leurs Piémonts»*, 2013-2016), directed by Fr. Réchin, is a diachronic and multidisciplinary research project. Its objective is to identify and characterise the different types of occupations between the Western Pyrenees and their foothills.

when dealing with this subject and stigmatises this study by indirectly implying that it is marginal. However, it is more likely that the lack of importance attributed to this theme is due to a somewhat biased interpretation of this phenomenon, that of an isolated population, especially during the socially unstable periods which were prominent in the Late Empire (Gutiérrez Cuenca *et al.* 2012). If we look into the main syntheses concerning the Roman population, with the exception of a few cases (*cf.* Esteban Delgado 1990), chapters on the topic of caves are very short, and are often associated with marginal occupations.

The origins of this research date back to the beginning of the 20th century, when archaeologists began to observe the presence of Roman artefacts, mainly from the Late Empire, in prehistoric caves. At that time, their interpretations had little to back them up. For example, De Saint-Périer stated: "*Peut-être la grotte fut-elle l'objet d'un culte à l'époque gallo-romaine, car nous avons retrouvé, dans les niveaux superficiels, de grands bronzes du Haut-Empire, qui semblent avoir été jetés de l'extérieur en manière d'offrande*" (De Saint-Périer 1935, 65). It was only in the 1960s that more reasoned interpretations of these phenomena were developed (Apellániz Castroviejo 1972; 1973; Février 1978).

Initial interpretations linked these occupations to the unstable periods of the Late Empire (for example the invasions which occurred at the end of the 3rd century and the beginning of the 5th century (Février 1978) and to a continuation of prehistoric ways of life and pastoralism (Apellániz Castroviejo 1973). In the 1980s, debates over understanding the presence of Roman artefacts intensified, with a dichotomy between those who believed in the continuation of pastoralism (López Rodríguez 1985, 146-150) and those who supported the idea of insecurity (Martinez Salcedo and Unzueta Portilla 1988, 61-64). The scarcity of excavations and/or of surveys conducted, which did not take into consideration the importance of stratigraphy, was incredibly detrimental to the proper development of new interpretations. What is more, this situation even validated such hypotheses in scientific publications for many years.

The most recent studies published deal mainly with regional syntheses. One of the most prominent French examples is, without a doubt, the study carried out by C. Raynaud (2001) (see also Réchin and Dumontier 2013; Ralite *et al.* 2012). This research, focused on the southeast region of France, brought forward new and relevant issues concerning the general study of caves. In the Murcia region of Spain, L. López-Mondéjar (2009) linked this phenomenon to Iberian sepulchral practices. In Asturias, A. Fanjul Peraza (2011) conducted an exhaustive study which included caves from other historical periods. The novelty brought forth from his work was the inclusion of mythology as a source, which enabled a better understating and provided new interpretations concerning certain caves which belonged to his area of research. According to this author, caves are "*como pequeños establecimientos agropecuarios de producción mixta, en aquellos lugares que no se encuentran en zonas de montaña*" (Fanjul Peraza 2011, 105). E. Gutiérrez Cuenca *et al.* (2012) contributed largely to advances made in research concerning the occupation of caves during the Late Antiquity in Cantabria, offering new alternatives to traditional interpretations, particularly concerning the *Arlanpe* cave: "*evidencias de un comportamiento ritual, con un componente mágico-religioso*"

(Gutiérrez Cuenca *et al.* 2012, 247). In the case of Guadalajara, E. Gamo Pazos (2013) linked Roman reoccupation of caves with their presence during the Bronze Age and those of mountainous sites, a process which falls within the territorial restructuration which began in the 3rd century (Gamo Pazos 2013, 215). Also worthy of mention is M. Rubio's (2014) work on the "*Subbética Cordobesa*", which is a case study of the caves and highlights that none of the caves studied documented "*una ocupación en sentido estricto, sino más bien una presencia ocasional y un uso discontinuo*" based on collected material and most of all on the limited quantities of Roman artefacts found (Rubio 2014, 217).

The contribution of GIS to the analysis of Roman caves

The POEM project: A presentation of the common geodatabase

The use of G.I.S. has considerably facilitated the study of noteworthy corpora of data for nearly two decades (Rodier 2011, 9; Wheatley and Gillings 2002; Mayoral Herrera *et al.* 2011). As, in the words of X. Rodier, "*L'information archéologique étant spatiale par nature*" (Rodier 2011, 9), it seemed particularly adequate to use this tool in order to carry out new research on the occupation of Pyrenean caves during Antiquity. Our approach, which arose from an exploratory approach (data collection, statistical analyses, the creation of templates), provides us with a more systematic and global outlook on these occupations. Thus, the former enables us to deal precisely with the main aspects of which they are made in order to measure their invariants or the specifics which pertain to the studied area, and to compare them to other regions. Here, we shall present solely an overview of our work due to the fact that the recording and analysis of the data is still in progress.

Ever since it was set up in 2013, the POEM project has relied greatly on the use of G.I.S. and on the creation of common databases, which provide the different axes which are addressed (axis n°1: "*la circulation saisonnière des hommes et des bêtes et les formes d'occupation de l'espace qui leur sont associées*"; axis n°2: "*les échanges de biens transpyrénéens ou parallèles à la chaîne*"; axis n°3: "*les conceptions partagées des espaces funéraires et des sanctuaires*"). In order to proficiently process this data, several databases (the purpose of which is to centralise the information pertaining to the identification and characterisation of specific sites) have been created. Initially, the model was created for a research project concerning the "*campements de piémonts pyrénéens*" (Pace *et al.* in press) based on significant field data, the originality of which arose both from a typology of the encountered structures and the problem of identification which ensued.[4] This model, therefore, seemed useful to us for our research on the Roman occupation of caves, as it met the same standards in defining these original sites for the studied period.

4 Excavation reports describing its occupations have often been difficult to use within a synthesis or a general database because of the heterogeneity of typologies established to define these sites. Recording information has required the creation of a more homogeneous typology and of a more flexible recording system.

Structure and methods for georeferencing data

Created in ArcGIS software, the physical model of the geodatabase consists of the link between four main tables, to which several secondary tables are added (Figure 2). The latter are intended to describe and make an inventory of the finds. The main tables result from three properties which characterise any archaeological information (Rodier 2011, 42), in other words, the following criteria: location (where?), dating (when?), and identification (what?).

In particular, these three properties are translated into our geodatabase as follows: an initial point shapefile dedicated to spatial information (named "*sites*") enables us to recover all the basic information regarding the site (coordinates (System 1989 UTM 30T/31T), historiography, bibliography, summary of findings, etc.); a table intended for temporal information ("*occupations*"), the purpose of which is to inform us of the number of different occupations and of their respective chronologies; and finally two other tables meant to answer the "what" question: In other words, the characterisation of anthropogenic structures encountered on-site ("*structures*") and the uncovered artefacts ("*mobilier*").

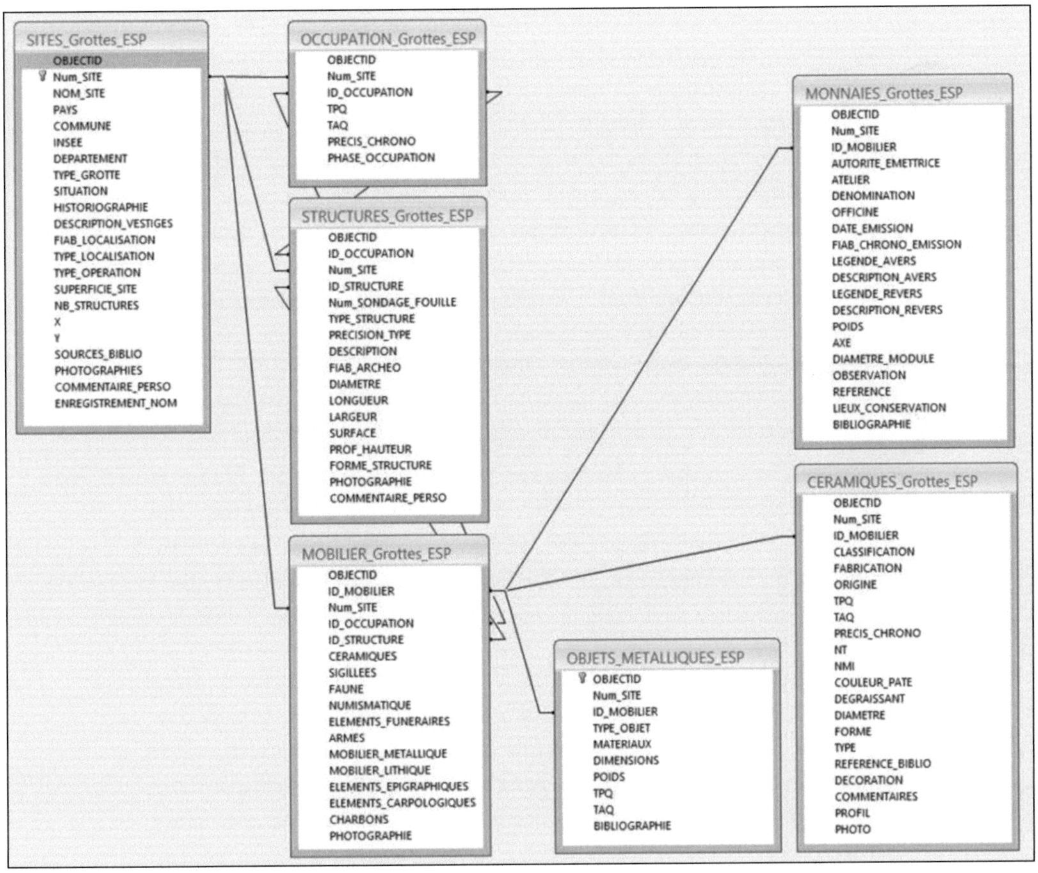

Figure 2. Structure of the geodatabase.

The variations in the types of finds in the caves, as well as their respective specificities, have also led us to create three secondary tables in order to classify these artefacts according to their own typology. Each category of discovered artefact, therefore, possesses its own table, its own features, giving us the possibility to be more precise as far as the saved information is concerned and to reduce the number of fields initially intended for the "*mobilier*" table, the purpose of which is to centralise information regarding the whole of these elements according to their nature and site. A link has therefore been created between the main "*mobilier*" table and each of these secondary tables.[5] Thus, the following tables can be found: one dedicated to referencing currencies, one for metallic artefacts and one for ceramics.

As a result of the general scales used for our research (corresponding to the area ranging from the Pyrenees and the Ebro), only the "*sites*" table is georeferenced. The multiplication of points meant to precisely geolocalise the various elements which compose a site would not have been relevant because of the scale employed. Geolocalisation was then made at the site level (via the centroid of discoveries), which implies the indirect georeferencing of the "*structures*", "*mobilier*" and "*occupations*" tables, the latter being linked (1 to N type) to the "*sites*" shapefile.

Finally, it should be remembered that this geodatabase was used by all the researchers involved in this project. Its creation came about thanks to a common effort on the part of each of the members. Data collection was an individual endeavour with the data later being shared with the other project members, meaning that the findings may be processed autonomously by each member of the research team (either individually or in small groups). Each member thus takes charge of the table(s) pertaining to his/her specialty(ies), thereby increasing the precision of the processed information. Therefore, the method achieves its aim thanks to the very structure of the database; each of the linked tables can be taken individually and completed the same way, at various speeds.

As has been mentioned above, our area of research (the Western Pyrenees and the Ebro River) extends across the Autonomous Communities of the Basque Country and Navarra, the province of Huesca and part of the province of Zaragoza (Aragón). The Ebro forms the northwest/southeast border of our study area and the Noguera Ribagorzana River marks the eastern limit (it also marks the approximate limit between the western and central Pyrenees).

Spatial Analysis

Although this technique is still quite rare in the field of research we have chosen, archaeologists have gained interest in the tools used by geographers in order to analyse the occupation of space more specifically, regardless of the period. G.I.S. are an integral part of this range of tools, yet their usefulness is limited since "*desde la arqueología no se pueden entender ni enunciar los fenómenos espaciales y temporales como sucede en otras esferas de conocimiento, ya que el objeto de estudio de los arqueólogo/as es una fracción material de un conjunto extenso de acciones/procesos realizados en el pasado*" (Castillejo 2013, 134). Many projects have involved the use

5 Integrating a table dedicated to referencing animal remains will be a future endeavour.

of G.I.S. in order to model anthropogenic phenomena (González Insua 2013; Fiz 2013; Castillejo 2013; De Soto and Carreras 2009). It should be noted that these results require a great amount of attention in order to draw correct conclusions. In our area of study, the only study to have been published concerning G.I.S. analyses is that of P. Utrilla *et al.* (2014). In this article, the authors conduct "*un análisis espacial GIS con el que poder caracterizar su localización geográfica, prestando especial atención al análisis de visibilidad*" (Utrilla *et al.* 2014, 695). They search for an explanatory model based on the chronology of the artefacts and the location of the cave. However, the G.I.S. analyses which were conducted were not developed in depth and only pertained to visibility from the entrance of each cave.

The study of caves during the Roman Era presents particularities which make it difficult to model them with G.I.S., since "deben ser las particularidades de cada contexto las que permitan definir posibles interpretaciones. Para elaborar propuestas sólidas se deben tener en cuenta un número importante de variables" (Gutiérrez Cuenca *et al.* 2012, 242), for example, both the external and internal specificities of each cave, mainly location, measurements, positioning, density of findings, context, etc. (Gutiérrez Cuenca *et al.* 2012, 242).

Due to the many methodological difficulties involved, this research is an initial approach to the geographic distribution of caves containing Roman remains.[6] Spatial analyses have been conducted in ArcGIS 10.1, with the ASTER 25 m DTM (Digital Terrain Model) and the Spanish national vector maps BCN 200. The archaeological data are part of the POEM database, supplemented by L. Tobalina and A. Campo (Tobalina Pulido *et al.* 2015). The analyses we have undertaken are as follows: a general breakdown of the caves with an analysis of density (Kernel analysis) and altitude; a link between the caves and the main Roman cities (5 and 10 km buffers); and finally, a general analysis of the different locations of caves containing coins from the Roman era.

Cave distribution analysis

In order to picture the geographic distribution of caves, we conducted a Kernel density analysis (Figure 3), a statistical density method, as opposed to a parametric one, which uses "*como base de cálculo la distribución puntual en el espacio de los asentamientos sin observar en principio condicionantes geográficos*" (Fiz 2013, 95). This enabled us to make a more precise observation concerning the concentration of sites in each zone of our area of research. As a result, three characteristics seem to stand out in terms of geographic distribution. Firstly, it is important to point out that the vast majority of the caves considered (78.46%) are located in the present-day Autonomous Community of the Basque Country. In this region, a high density of caves can be found which do not have "*paralelos, por el momento, ni en Cantabria ni en Asturias, provincias* [vecinas] *con un importante número de cuevas y cuya romanización tiene muchos rasgos en común con la de los espacios costeros de Bizkaia y Gipuzkoa*" (Gutiérrez Cuenca *et al.* 2012, 240). This high concentration

6 We presented the complete G.I.S. study during a «POEM» colloquium in April 2017 at the *Université de Pau et des Pays de l'Adour.* The publication should be released by the end of 2018 and will present the results of our research.

of caves is most likely a response to the archaeological research carried out in this region since the end of the 19th century (Apellániz Castroviejo 1973; 1972; Barandiarán 1967). Moreover, three levels of concentration can be noted: one towards the Atlantic coast, a second one inland towards the *Basque Mountains* and a third one in the *Arcena* and *Cantabria* sierras, south of the province of *Alava*. This distribution seems to respond to both geographic and geological factors, with a high concentration in karstic areas, which supports C. Raynaud's research (2001, 452). This distribution creates an important contrast between the areas which have been explored and studied most in the Basque Country and those which are less studied, as is the case of Navarra, Zaragoza and Huesca.

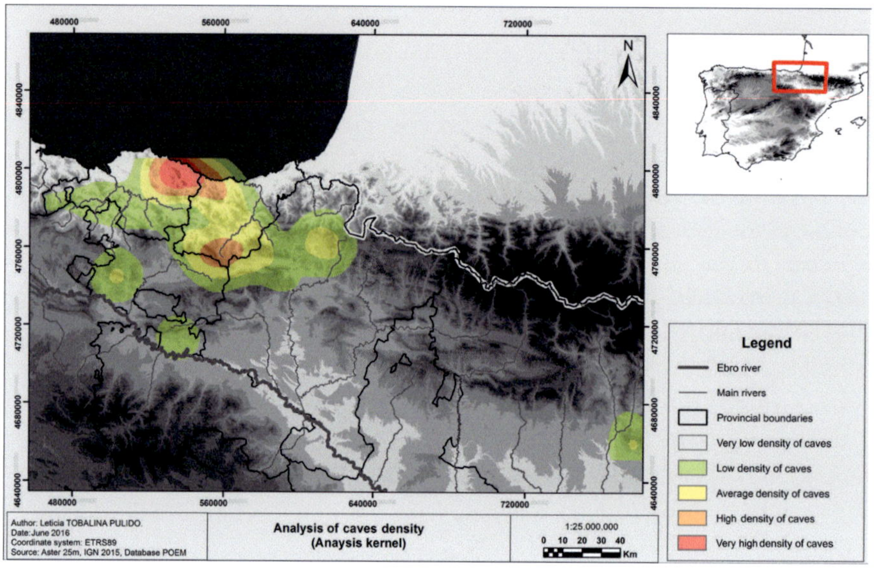

Figure 3. Analysis of density of caves (Kernel Analysis).

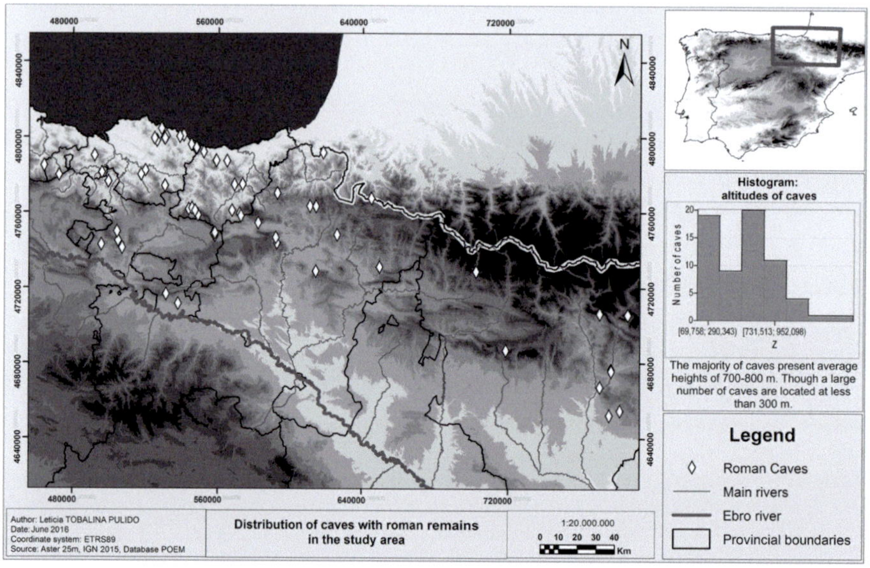

Figure 4. Distribution of caves with Roman remains in the study area. A histogram of altitudes (inset).

In order to establish the altitude of each cave, we created an altitude histogram based on the Aster DEM elevation. This has enabled us to calculate and find the most common altitude for these types of sites (Figure 4).

Proximity analyses: rivers and Roman cities

Two types of connections have been established: on the one hand, the proximity between caves and waterways and on the other, the proximity between caves and the main Roman towns. Firstly, our aim was to determine the existence of a link between the caves and the watercourses in order to confirm if certain minimal conditions (such as proximity to water springs, accessibility, etc.) qualify certain caves as settlements. Then, a spatial analysis was carried out based on the proximity of these sites to waterways (up to 250 and 500 m) (Figure 5). In order to do this, we used the hydrography layer of the BCN 200 map of Spain. As opposed to certain traditional hypotheses, which potentially qualify caves as settlements (Apellániz Castroviejo 1972; Gil Zubillaga 1997), the proximity of the caves to waterways is not significant since the majority of these caves are located less than 1,000 metres from a watering hole. This does not invalidate older theories but shows how restrictive and constraining living in caves was in terms of self-sufficiency. This is even more remarkable if we consider the difficulty involved in bringing water to these caves, which are located on steep slopes and are often difficult to reach. It would, therefore, be necessary to check if any alternative springs existed nearby.

As far as the link between the caves and major Roman towns (Figure 6) is concerned, the criteria which determine the extent of the area of influence of the towns in the surrounding environment are variable due to the inherent differences between towns (depending on their size and status). This is why G. Adams (2006) based his studies on an area spanning 5 km around Roman towns in order to determine the suburban zone of cities, such as Pompei and Stabiae, a distance which "*prendrait à peine plus de deux heures de route aller-retour, soit un déplacement aisément réalisable en une demi-journée*" (Pace 2015, 247-248). Others prefer larger extensions, for example 10 km starting from the Aurelian wall or 50 km from Rome (Buzón Alarcón 2011, 16).

Given that the towns in our area are either small or medium-sized, we chose to establish areas of influence spanning between 5 to 10 km. In this way, we noted certain links between caves and Roman towns, particularly in the case of *Forua* and *Labitolosa*. In both cases, a large concentration of caves is found around the cities in a buffer of 5 km, a phenomenon which had previously been pointed out by E. Gutiérrez Cuenca *et al.* (2012, 240). Though a very small number of caves (12.08 %) seem to be linked to the area closer to a town, we can suppose that they mostly belong to the city's suburban territory.

Yet what can be said about their function(s)? Were they places of worship? Shelters? Hoards? Places to make cheese? Some authors have tried to systematise the function of the caves during the Roman era by focusing on their specificities. In the 1990s, in a study based on Roman caves in Britain, K. Branigan and M. J. Dearne (1992) created a synthesis establishing criteria to classify the potential forms of occupation. In Spain, I. Aguilera (1996) established four types of uses: continuous occupation involving settled economic activity; settlements for

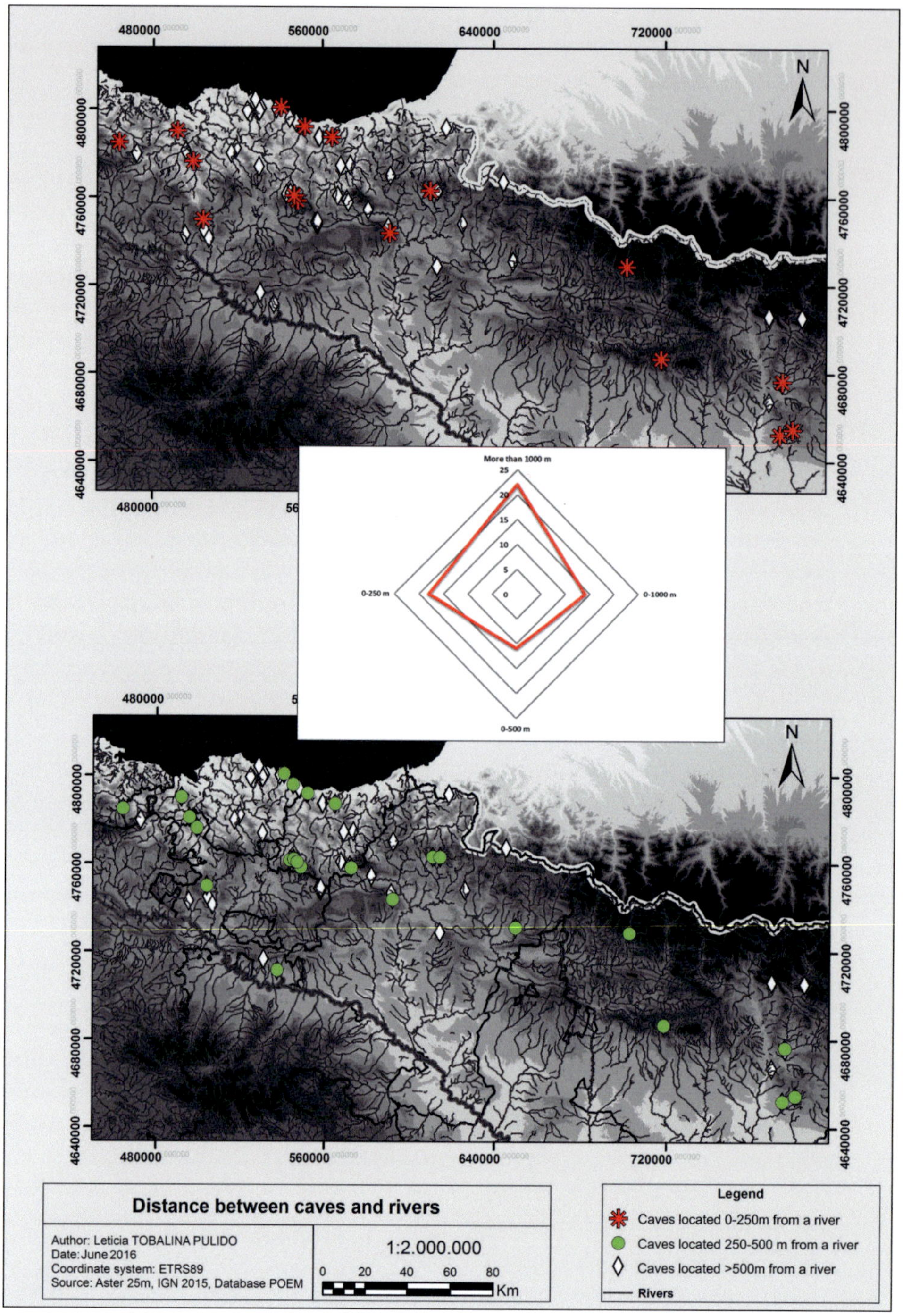

Figure 5. Distance between caves and rivers.

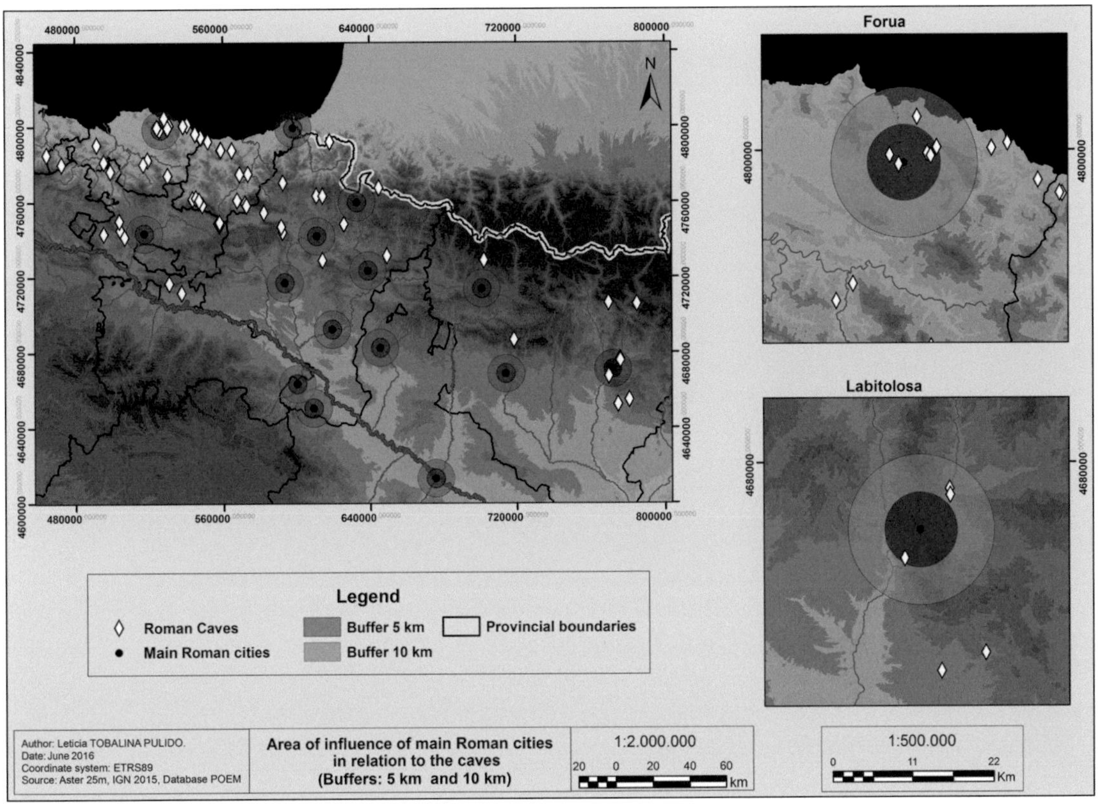

Figure 6. Area of influence of main Roman towns in relation to the caves.

economic purposes (pastoralism); hideouts during unstable periods and, finally, places of asceticism or hermitage. J. A. Quirós and B. Bengoetxea (2010) used and modified K. Branigan and M. J. Deane's chart, but maintained the criteria based on types of artefacts and the size and form of the caves.[7] Despite these attempts, systematising this phenomenon has proved to be an extremely difficult task as we are not able to find a single classification model which applies to each case (Gutiérrez Cuenca *et al.* 2012, 243).

Analysing the geographic distribution of caves containing Roman era coins

The study of Roman coins unearthed in these caves has enabled us to produce reliable dating indices and, more significantly, to highlight research inadequacies. There is a substantial presence of coins and they have been found in almost one third of the cases (29.23%). However, the number of coins unearthed is not a substantial indicator due to factors such as the loss of data over time, the disappearance of artefacts and incomplete excavations.

7 M. Rubio suggests adding the artefact chronology to this criteria because "*determinados procesos históricos pueden verse reflejados en éstas*" (Rubio 2014, 218).

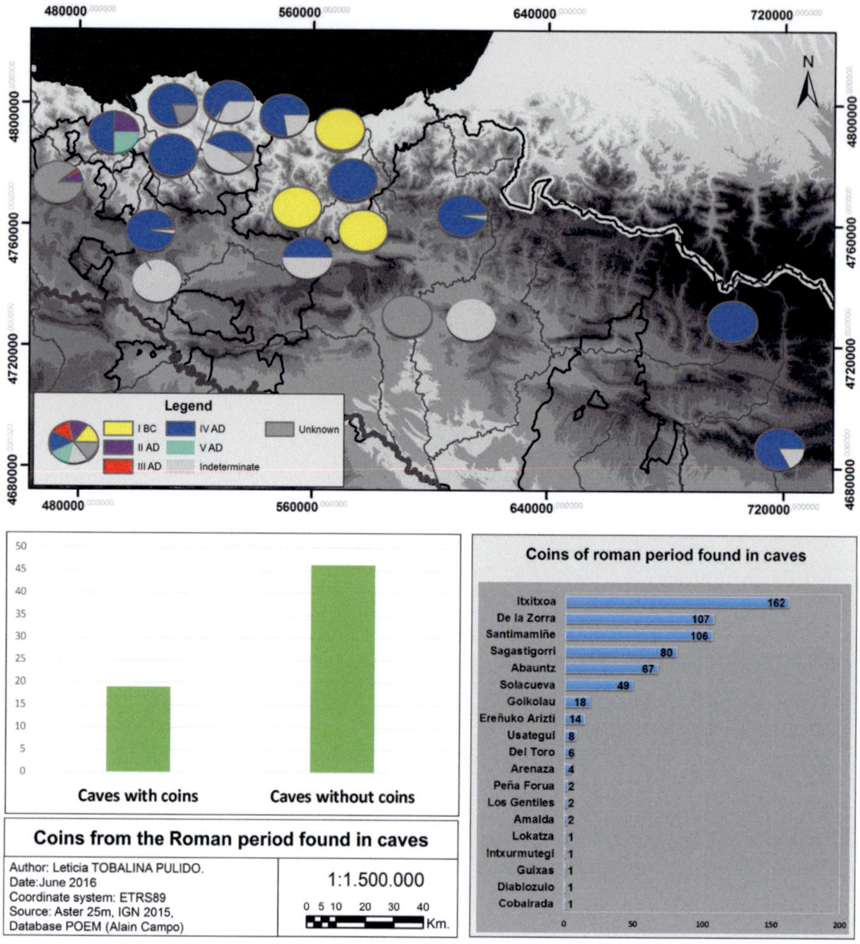

Figure 7. Coins from the Roman period found in caves, I.

The majority of coins which have been discovered can be dated to the second half of the 4th century A.D. This has often been interpreted as a consequence of the barbarian invasions of the beginning of the 5th century A.D. (Utrilla *et al.* 2014; Tudanca Casero 1997, 397-398). This theory has been widely developed without taking into consideration contradictory elements, such as the dating of monetary artefact (Tobalina Pulido *et al.* 2016). In keeping with this, bronze coins minted after 408 A.D. are scarce, although it must be pointed out that some have been identified in the caves we have studied. Therefore, how can the presence of a coin dated after 408 A.D. in the *Abauntz* cave be explained if it is, in fact, a cache resulting from the barbarian invasions?

Doubts over caves inhabited during the 3rd century A.D. should also be noted. Aside from the *La Zorra* cave, which is an exception, we have noticed that coins dating back to the 3rd century A.D. have been mixed with later mints. This finding confirms I. Pereira's (1974, 242) previous theory of a filtering of currencies during the 3rd century A.D. on the rare occasions on which they crossed the Pyrenees.

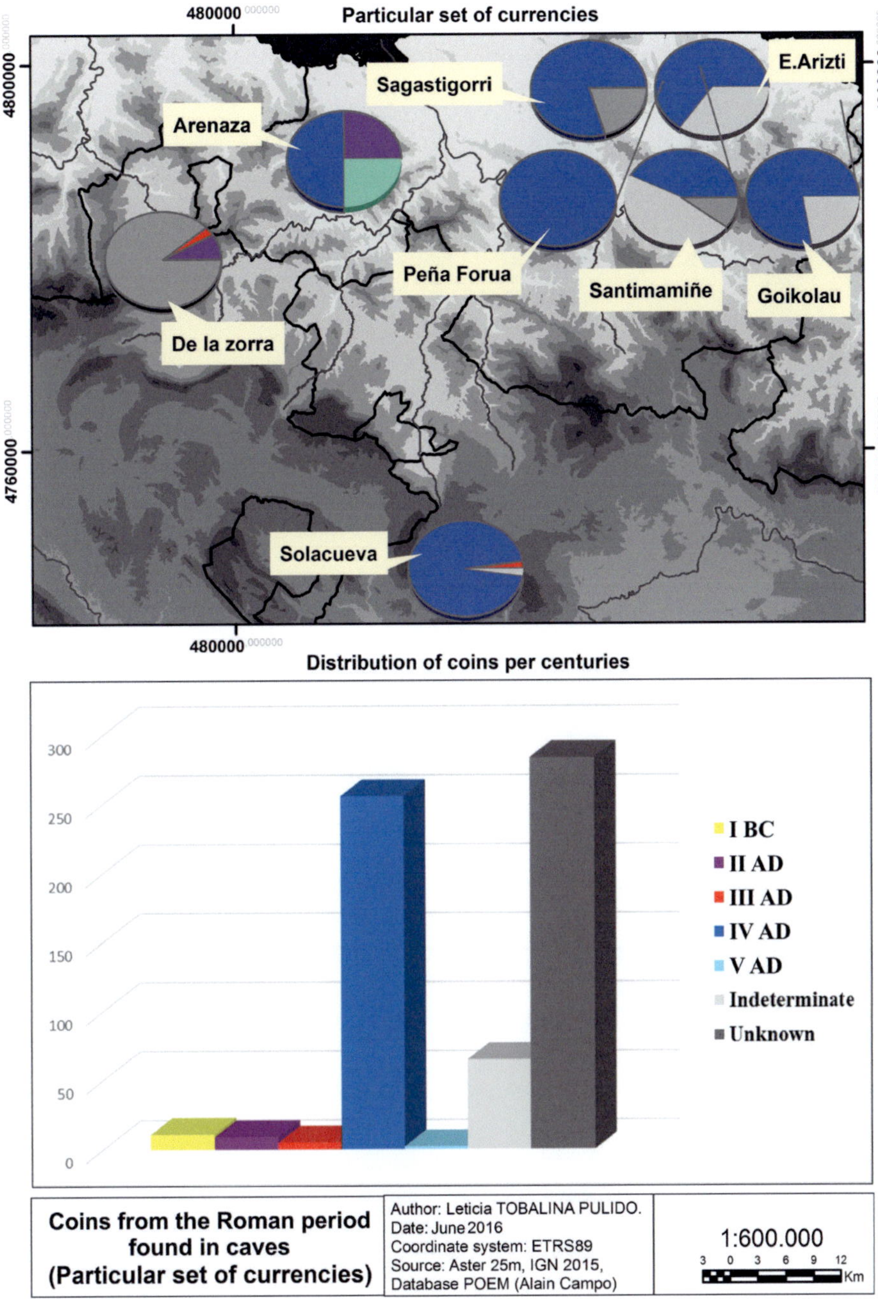

Figure 8. Coins from the Roman period found in caves, II.

Thus, the theory according to which caves served as hideouts during the barbarian invasions of the 3rd century A.D. does not seem convincing in our area of research.

If the strong presence of Roman artefacts dating from the 4th century A.D. is considered as a remnant of hideouts due to unstable periods, why is it that no coins from the 3rd century A.D. are present in the caves we have studied? Why are the monetary artefacts of such poor quality in these sites, which are considered to be caches, if people would logically have brought their most prized possessions with them?

Thus, it seems absolutely necessary to put these interpretations into perspective and to study in greater detail the finds from each of these caves. This method is essential to the development of new archaeological hypotheses concerning the occupation of these caves during Antiquity.

Conclusion

We have carried out an initial synoptic analysis concerning the occupation of caves during the Roman era. This has led us to formalise various questions in order to gain a better understanding of this subject. The caves we have studied, which were all formed in karstic areas, suffer from an analysis of spatial distribution biased by the lack of research conducted in the regions studied (Huesca and Navarra are the best examples). Let us hope that this area of research will develop in these locations in order to clarify our judgement. Moreover, the results brought up by our different analyses have not been clear enough to consider these caves as actual elements belonging to the territorial domain of the neighbouring cities. Though some cases seem to fit to this theory (as it is the case for *Forua* and *Labitolosa*), it seems likely to us that some others are rather linked to extra-urban or peripheral territories, even if the data we currently have at our disposal cannot confirm this thesis.

The caves are most often located on two altitudinal levels (between 70 and 300 m and between 600 and 800 m), a factor which can considerably change their function, and, in general, seem to be located 1 km away from waterways. This is a factor which can offer a different perspective to their potential function.

A final analysis has enabled us to study another type of finds within the caves: coins. These have been documented in about one third of the caves we studied, and are mostly from the middle of the 4th and beginning of the 5th centuries A.D. Given the scarcity of coins from the 2nd and 3rd centuries A.D., we can, once again, cast doubt on the function of these caves as settlements or hideouts, triggered by the "barbarian invasions" of the 3rd and 5th centuries A.D.

Finally, we have planned new explorations in order to further review and clarify the occupation of several caves during the Roman era. This endeavour will be undertaken in 2016 in the region of Navarra, in collaboration with a team of prehistoric archaeologists and diggers. The results will be published in the POEM project's final publication, in which we shall conduct a study gathering data from both sides of the Pyrenees.

Acknowledgements

We would like to thank Camille Leclère (Université de Pau et des Pays de l'Adour, Camille_leclere@hotmail.fr) for translating the article (French – English). P. Tobalina for his help with some small adaptations and E. Carlsson-Brandt and F. Fanjul for their review of the English text. We would also like to thank V. Mayoral, C. Parcero and P. Fábrega for the opportunity they have given us to present our research in this book.

References

Adams, G. 2006. *The Suburban Villas of Campania and their Social Function*, BAR International Series 1542, Oxford: British Archaeological Reports.

Aguilera Aragón, I. 1996. La ocupación tardorromana de la cueva del Moro. *Bolskan* 13, 133-137.

Apellániz Castroviejo, J. M. 1972. La romanización del País Vasco en los yacimientos en cueva. *Estudios de Deusto* 20, 305-310.

Apellániz Castroviejo, J. M. 1973. Corpus de materiales de las culturas prehistóricas con cerámica de la población de cavernas del Pais Vasco meridional. *Munibe. Antropología-Arqueología*, 366/Suplemento.

Barandiarán, J. M. 1967. Excavaciones en el Montico de Charratu y en Saracho. *Estudios de Arqueología Alavesa* 2, 7-20.

Braningan, K. and Dearne, M. J. (Eds.). 1992. *Romano-British cavemen: cave use in Roman Britain*. Oxford: Oxbow.

Buzón Alarcón, M. 2011. Reflexiones acerca del suburbio en la ciudad romana. *Romvla* 10, 7-42.

Castillejo, A. M. 2013. Percepción y cuantificación de la variabilidad arqueológica a escala intra-site: una revisión de los formatos de datos y sus implicaciones analíticos. *Revista Internacional de ciencia y tecnología de la información geográfica* 13/1, 131-153.

De Saint-Périer, R. 1935. Quelques œuvres d'art de la grotte d'Isturitz. *Bulletin de La Société Préhistorique de France* 32/1, 64-77.

De Soto, P., and Carreras, C. 2009. La movilidad en época romana en Hispania: aplicaciones de análisis de redes SIG para el estudio diacrónico de las infraestructuras de transporte. *HABIS* 40, 303-324.

Dumontier, P. and Réchin, F. 2013. Une grotte pyrénéenne occupée au début de l'époque romaine: le site d'Apons à Sarrance (Pyrénées – Atlantiques), *Aquitania* 29/ Supplément, 97-143.

Esteban Delgado, M. 1990. *El País Vasco atlántico en época romana*. San Sebastián: Universidad de Deusto.

Fanjul Peraza, A. 2011. Las últimas cuevas. Observaciones en torno a la ocupación de las cuevas astur leonesas. *Arqueología y territorio medieval* 18, 91-116.

Février, P. A. 1978. Problèmes de l'habitat du Midi méditerranéen à la fin de l'Antiquité et dans le haut Moyen Âge. *Jahrbuch Des Römisch-Germanischen Zentralmuseums Mainz* 25, 208-247.

Fiz, I. 2013. Métodos estadísticos y funciones SIG: una propuesta de modelado del poblamiento en el ager Tarraconensis. *Archivo Español de Arqueología* 86, 91-112.

Gamo Pazos, E. 2013. Cuevas y alturas: reocupación de hábitats prerromanos en el Bajo Imperio en la provincia de Guadalajara, in: Álvarez Jiménez, D., Sanz Serrano, R. and Hernández de la fuente, D. (eds.). *El espejismo del bárbaro: ciudadanos y extranjeros al final de la Antigüedad.* Madrid: Biblioteca Potestas, 213-240.

Gil Zubillaga, L. 1997. Hábitat tardorromano en cuevas de la Rioja Alavesa: los casos de Peña Parda y Los Husos I (Laguardia, Álava). *Isturitz* 8, 137-149.

González Insua, F. 2013. Aproximación a la distribución espacial de sitios arqueológicos de la Prehistoria Reciente en la Ría de Arousa (Pontevedra). *Gallaecia* 32, 129-167.

Gutiérrez Cuenca, E., Hierro Gárate, J.Á., Ríos Garaizar, J., Gárate Maidagan, D., Gómez Olivencia, A., and Arceredillo Alonso, D. 2012. El uso de la cueva de Arlanpe (Bizkaia) en época tardorromana. *Archivo Español de Arqueología* 85, 229-251.

López Rodríguez, J. R. 1985. *Terra Sigillata hispánica tardía decorada a molde de la Península Ibérica,* Valladolid: Secretariado de Publicaciones.

Martínez Salcedo, A. and Unzueta Portilla, M. 1988. *Estudio del material romano de la cueva de Peña Forua.* Deusto: Universidad de Deusto.

Mayoral Herrera *et al.* 2011. *Tecnologías de información geográfica y análisis arqueológico del territorio.* Madrid: CSIC.

Pace, B., Béague, N., Chopin, J.F., and Le Couédic, M. (in press). Les « *campements de piémonts pyrénéens* » de la Protohistoire à l'Antiquité: apports des systèmes d'informations géographiques (S.I.G.) et perspectives spatiales, in: *Pays pyrénéens et environnement, Fédération historique de Midi-Pyrénées, 150 ans de la Société Ramond, Bagnères-de-Bigorre (12-14 juin 2015).*

Pereira, I., Bost, J.P. and Hiernard, J. 1974. *Les Monnaies, in Fouilles de Conimbriga.* Paris: Diffusion E. De Boccard.

Quirós Castillo, J.A. and Bengoetxea Rementería, B. 2010. *Arqueología III. Arqueología Medieval y Posmedieval.* Madrid: UNED.

Raynaud, C. 2001. L'occupation des grottes en Gaule méditerranéenne à la fin de l'Antiquité, in: *Les campagnes de la Gaule à la fin de l'antiquité. Actes du colloque Montpellier, 11-14 mars 1998.* Antibes: APDCA, 419-471.

Rodier, X. 2011. *Information spatiale et archéologie.* Paris: Errance.

Rubio, M. 2014. Vestigios de ocupación romana en cuevas naturales de la Subbética Cordobesa. Nuevas hipótesis Interpretativas. *Antiqvitas* 26, 205-225.

Tobalina Pulido, L., Campo, A., Duménil, V. and Pace, B. 2016. Historiografía, metodología y problemática en el estudio de la frecuentación de las cuevas naturales en época romana entre el Ebro y el Garona. *Antesteria* 5, 197-206.

Tobalina Pulido, L., Duménil, V., and Campo, A. 2015. Fréquentations des grottes durant l'époque romaine: le cas de la Navarre. *Cuadernos de Arqueología de La Universidad de Navarra* 23, 123-161.

Tudanca Casero, J. M. 1997. *Evolución socioeconómica del Alto y Medio Valle del Ebro en época Bajoimperial romana.* Logroño: Instituto de Estudios Riojanos.

Utrilla, P., Laborda, R. and San Sebastián, M. 2014. La reocupación de cuevas prehistóricas del Prepirineo oscense en época romana, in Duplá Ansuategui, A., Escribano Paño, M. V., Sancho Rocher, L. and Villacampa Rubio, M. A. (eds.). *Miscelánea de estudios en homenaje a Guillermo Fatás Cabeza*. Zaragoza: Instituto Fernando el Católico, 693-704.

Wheatley, D., and Gillings, M. 2002. *Spatial technology and archaeology. The archaeological application of GIS*. New York.

Notes on contributors

Leticia Tobalina Pulido: PhD student in archaeology (University of Navarre/ *Université de Pau et Pays de l'Adour*) supervised by F. Réchin, J. Andreu and J. Sánchez Velasco. She is a graduate in History and Art History and has a Master's degree in archaeology. She works on the research of land structuring during the Late Antiquity period between the Ebro and the Pyrenees and, more particularly, specialises in GIS analysis. She is an associate researcher of the POEM project, has published several articles in specialised publications and has participated in several colloquiums.

Benoît Pace: PhD student in archaeology (*Université de Pau et Pays de l'Adour*) supervised by F. Réchin. He is an associate researcher of the POEM project and works on the organisation and use of spaces in Southern Aquitaine during the Gallo-Roman period, specialising more particularly on GIS analysis and databases. He has published several articles in specialised publications and has participated in several colloquiums.

Alain Campo: PhD in archaeology (*Université de Pau et Pays de l'Adour, UPV, supervised by F. Réchin et J.J. Larrea.*). He is an associate researcher of the POEM project specialised in Roman numismatics. He has studied several monetary treasures and has published several articles in specialised publications and participated in several national and international colloquiums.

The potential of the Geographic Information Techniques for the analysis of the morphology and settlement patterns of the Roman military sites of early imperial era in Iberia

José Manuel Costa-García[1]

Roman military archaeology in Iberia

Just a dozen Roman military sites were known to the North of the river Douro by the end of the 1980's (Morillo Cerdán 1991). Although the development of urban archaeology and the improvement of field survey techniques began to change this scenario during the next decade (Morillo Cerdán 2002b), archaeological research was by then mainly focused on the study of Roman permanent settlements, and an important part of the academic world did not pay enough attention to (if not refused to accept) the discovery of the first temporary sites linked with the Cantabrian Wars (29-19 BC) (Peralta Labrador 2002b). Despite these early objections, this little revolution was on at the beginning of the 21st century: not only the number of sites were more than doubled, but also the archaeological evidence begun to diversify. Several marching camps, small fortifications and siege scenarios dating from the Augustan era were found in the northern mountains (Camino Mayor *et al.* 2007; Peralta Labrador 2006). This situation naturally motivated the gradual abandonment of the static visions based on the intensive study of the permanent settlements.

The progressive and increasing use of remote sensing techniques (such as historical and modern aerial photography, satellite imagery, aerial LiDAR or photogrammetry) and geographic information systems (GIS) has boosted this

---

1 Departamento de Historia, Universidade de Santiago de Compostela. Praza da Universidade, s/n 15703 Santiago de Compostela (Spain) josemanuel.costa@usc.es.

process in the last years.[2] Furthermore, the development of specific methodologies for the detection and analysis of these temporary sites is now a reality (Costa-García *et al.* 2016; Menéndez Blanco *et al.* 2013). As a result, the number of Roman military sites documented in northern Iberia exceeded the hundred by the end of 2015, and about 90% of them could be defined as temporary (Camino Mayor *et al.* 2015). This impetus also contributed to overflow the traditional areas where the Roman military presence was studied. The discoveries are no longer limited to the Spanish Northern Plateau or to the Cantabrian and Asturian mountainous regions: the Galician and northern Portuguese territories start to shown their potential (Costa-García *et al.* 2015; Costa-García *et al.* 2016; Orejas *et al.* 2015).

The existence of fresh archaeological data should encourage the development of new analytical perspectives. However, Roman military archaeology in Iberia lacks a tradition of solid, synthetic studies. Although we can find some catalogues and conference proceedings whose objective was to put together all the archaeological data available (Camino Mayor *et al.* 2015; Morillo Cerdán 2002a; 2006; Morillo Cerdán and Aurrecoechea 2006; Morillo Cerdán *et al.* 2009; Serna Gancedo *et al.* 2010), it is difficult to detect a real methodological uniformity in them. To a certain extent, this situation could reveal the youth of Roman military archaeology as an autonomous discipline in Iberia, but it looks like the amending of this historical neglect is not a research priority at the present time. The efforts on documenting and cataloguing the recent discoveries have perhaps relegated this kind of reflexive approaches to a second plane once more. In many aspects, the most basic interpretative archaeological work still needs to be done, and this situation naturally slows down the development of wider studies.

Maybe this is one of the reasons why the analyses focused on the way in which the theoretical principles of the *metatio castrorum* were locally adapted (morphological studies, locational analysis...) are practically unknown in Iberia (Costa-García 2011; 2013; González Fernández and Vidal Encinas 2005). Of course, the main aim of this paper is not to fill all those gaps, but to explore how GIS could help to do it. It should be noted that the reliability of the following analyses might suffer because of the existence of a heterogeneous data base: not only our knowledge of these sites is uneven due to the reasons stated above, but they also show a diversified

2 In recent times, the Spanish government has developed a strong policy of open access to the geographic data. The PNOA program (National Plan of Aerial Orthophotography) of the IGN (National Geographic Institute) has released the majority of the historical and modern aerial photography via WMS services or direct download (http://pnoa.ign.es/). The raw LiDAR data of the several flights covering the entire national territory (2009-2014) is also accessible to the common user (http://pnoa.ign.es/coberturalidar). This situation has not only provided a powerful tool to the archaeologist but also contributed to the democratisation of the archaeological research as a whole. Unfortunately, the situation in Portugal does not allow us to be that optimistic, since the majority of the aerial photography must be requested to the Geographic Institute of the Portuguese Army (https://www.igeoe.pt/) and the LiDAR coverage is limited to a narrow strip of the coastal area. Only the iGEO portal (http://www.igeo.pt/) offers recent orthophotography through a WMS service.

typology and an irregular distribution. From a total number of 120 individuals,[3] 86 sites were temporary camps, 21 could be described as *castella* or small outposts, 8 were "permanent" fortress or forts, 3 have been defined as defensive lines, and only 2 of them could be classified as towers. This implies that some typologies are going to be necessarily underrepresented in the following analyses. In order to explore the potential of the Geographic Information Techniques to answer some questions related with the Roman art of establishing a fortification, these analyses have been grouped under three main topics: morphology, defensive solutions and locational pattern.

Morphological diversity

A close reading of the ancient theoretical treatises on *metatio castrorum* reveals that the election of a morphological solution was a matter of secondary importance when building a marching camp. Indeed, the final decision was determined by two main rules: to guarantee enough inner space for comfortably pitching the tents; and to extend the outer perimeter of the fortification only inasmuch as the garrison could ensure its defence (Veg. *Mil.* 1.22-23, 3.8). However, these very same treaties state that the square or rectangular modules were preferred by the Roman army for practical reasons, since the organisation and distribution of the inner space was quite easier this way (Polyb. 6.32-33; Ps.-Hyg. 21; Joseph. *BJ* 3.5.1-4).

In the last years the joint use of aerial photography, LiDAR data and GIS has allowed us to review and re-catalogue many Roman military sites in northern Iberia. Thanks to these tools it was possible to document the perimeter defences more precisely, so the reconstruction of the original layouts and modules is easier now (Figure 1). Further analyses on these data seem to confirm the suggestion of the classical treatises: the majority (29.2%) of the non-permanent Roman military sites located to the North of the river Douro showed a rectangular plan (Figure 2). In fact, this percentage could be increased with the addition of the quadrangular exemplars (3.8%), and the irregular (6.6%) or partially destroyed sites (12.3%) which seem to follow an original square/rectangular pattern. Even trapezoidal figures (6.6%) could be considered a deviation of this general model.

The oval-circular fortifications follow in number (12.3%). This shape was indeed the best suited for the two theoretical rules given above (Richardson 2004), so it is not surprising to find that this solution was the preferred for small *castella* or fortlets (28.6%): the space needed to be optimised due to the limited number of soldiers garrisoned here. On the contrary, other polygonal solutions, such as triangular (1.9%) or pentagonal (1%) figures, were in general far less representative.

Finally, several cases coexist under the label of irregular sites (15.1%). The irregularity of some of them could be the result of the expansion and/or reoccupation of an existing enclosure, as the impressive exemplars of Cildá (Peralta

3 For this study were only selected those sites in which the existence of archaeological structures related with the Roman army was rigorously attested. The different phases of each site were counted separately as individuals. Due to the strict space restrictions of this paper, the citations regarding these sites have been limited to the minimum, but they can be found in the most recent monographic studies on the subject (Camino Mayor *et al.* 2015; Costa García 2013; Morillo Cerdán 2014).

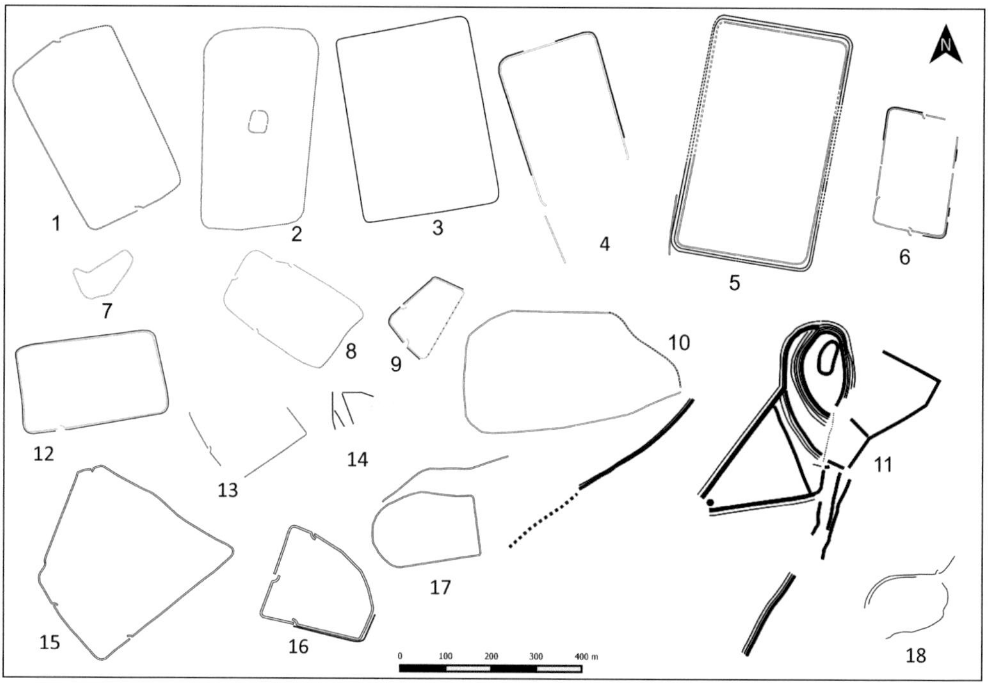

Figure 1: A sample of the morphological diversity: A Penaparda (1), El Pico el Outeiro (2), Villalazán (3), Monte dos Trollos (4), Los Llanos (5), Valdemeda (6), A Recacha (7), El Chao de Carrubeiro (8), Moyapán (9), A Pedra Dereta (10), Monte Currieḷḷos (11), Ganda das Xarras (12), El Chao (13), Picu Viyao (14), El Xuegu la Bola (15), A Cortiña dos Mouros (16), Ḷḷagüezos (17), and La Garita (18).

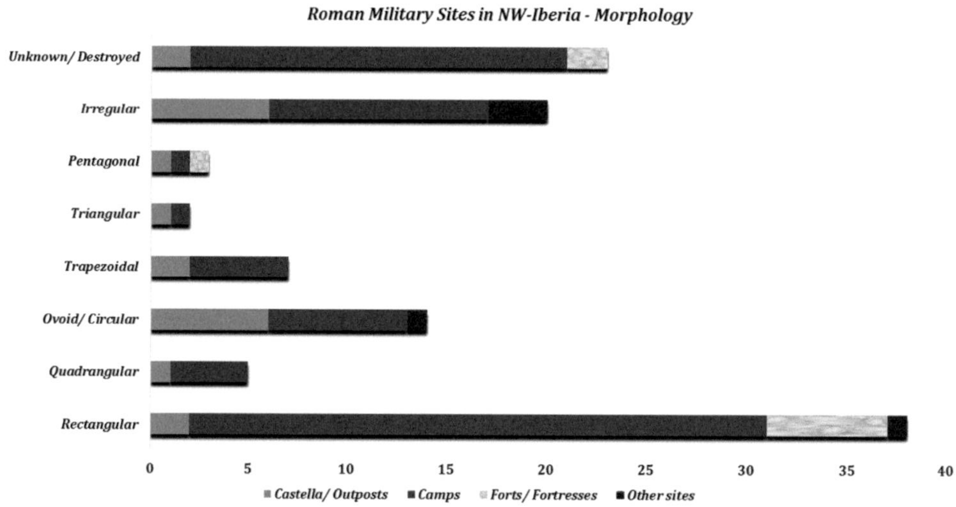

Figure 2: Roman military sites in Iberia: typological and morphological classification.

Labrador 2002a) and Monte Currieḷḷos (Camino Mayor *et al.* 2005) proved years ago. Other settlements were established on uneven locations to control a route or a natural passage, so the regular modules were completely altered if they were not dismissed (Peralta Labrador 2011). The final shape of the defensive lines created to slow the advance of the enemy across the mountain ranges was also very dependent on the local topography, reflecting the fidelity to a very similar strategic principle. Lastly, many little Roman fortifications were also founded over Late Iron Age settlements, and the reuse of some sections of the old defences by the Roman army implied the creation of irregular defensive perimeters (Fernández Vega and Bolado del Castillo 2011; Peralta Labrador 1999; Torres Martínez *et al.* 2011).

The discovery in recent times of more and more irregular sites led us to a last qualitative observation: are these numbers truly representative of a historical reality, or either are they reflecting our tendency to easily detect some (canonical) shapes over other less obvious manifestations of the Roman military presence? Although a strict distribution pattern cannot be detected (Figure 3), data clearly show that the local topography usually determined the election of the morphological solution when establishing a military settlement. That is why a more heterogeneous group can be found in the mountainous areas than in the plains, and it could also explain why the forced adaptation and redrawing of the theoretical modules (that is, the "irregularity factor") was more frequent in the former zones.

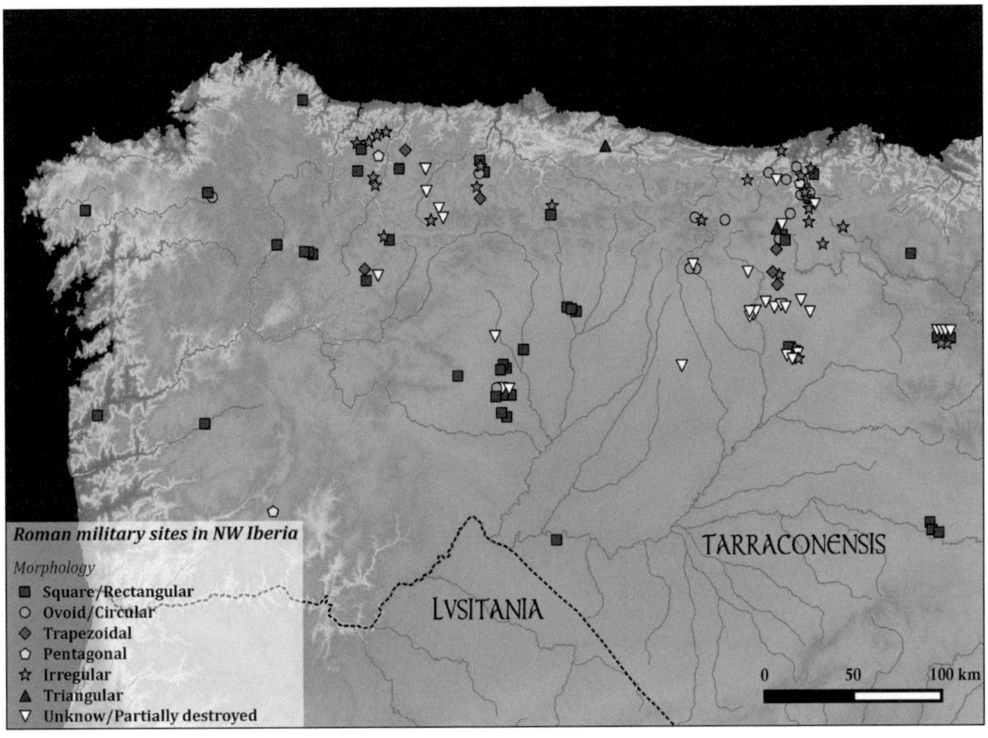

Figure 3: Morphological distribution of the Roman military sites in Iberia north of the river Douro.

The sites of A Penaparda, El Pico el Outeiro and A Pedra Dereta (Figures 2 and 4) can help to explain this situation. These three marching camps are located over a traditional mountain route which connects the Galician eastern area and the western Asturian territories (Menéndez Blanco *et al.* 2014). They show a similar extension (about 10 ha), but very different shapes (from an almost rectangular layout to an irregular one). Even if we cannot reject a diachronic interpretation for these enclosures, it rather seems that they belonged to a single unit which built its camps adapted to the local topography during its advance across the enemy territory. On the contrary, the impressive concentration of Roman camps close to Herramélluri (La Rioja), reveals how easy was to build up rectilinear defensive perimeters in plain areas (Didierjean *et al.* 2014).

Regarding the permanent settlements, in some cases the actual plan of the sites is not known in detail due to the singularities of urban archaeology, as it happens in Herrera de Pisuerga (Palencia) and Astorga (León) (González Fernández 1997; Illarregui Gómez 2002). The regular playing-card layout seems to have been the rule here (47%), but some slight deviations due to the local topography or to the particular evolution of the fortifications were also attested in León (Morillo Cerdán and García Marcos 2003) and Rosinos de Vidriales 2 (Zamora) (Carretero Vaquero *et al.* 1999). Other precise observations can be stressed in those sites explored at a larger extent. For instance, the size (ca. 2.6 ha), module (6:5) and internal distribution of the forts of Bande and A Cidadela (Galicia) are quite similar, revealing that these fortifications were probably raised following the same plan, if not by the same unit (Costa-García 2011).

Defensive solutions

The use of remote sensing techniques has allowed us to better document the defences of many sites. The combined use of this information and the evidence coming from traditional surface surveys and/or archaeological excavations, provides an interesting amount of contrasted data. After reviewing all these pieces of evidence, it could be stated that the Roman military sites of the early imperial times in Iberia do not generally show complex defensive systems (Figures 3 and 4). For instance, the temporary fortifications with more than a single ditch are quite unusual and these are mainly located in the Spanish Northern Plateau. Considering that it was in the mountainous zones of Asturias and Cantabria where the Augustan campaigns took place, this is a surprising fact. It has been noted that the hardness of the soils often hindered the digging of a proper ditch in these mountainous regions, so the camps and small outposts were frequently defended by nothing more than a rudimentary rampart made of earth and stones (Serna Gancedo *et al.* 2010). Regarding other defensive features, the gates protected by *clauiculae* have been documented in almost one quarter (24%) of the marching camps and small fortifications related with he Asturian-Cantabrian wars: the simplest variation of inner *clauicula* was by far the preferred solution. On the contrary, the use of *titula* has been barely attested in those sites so far.

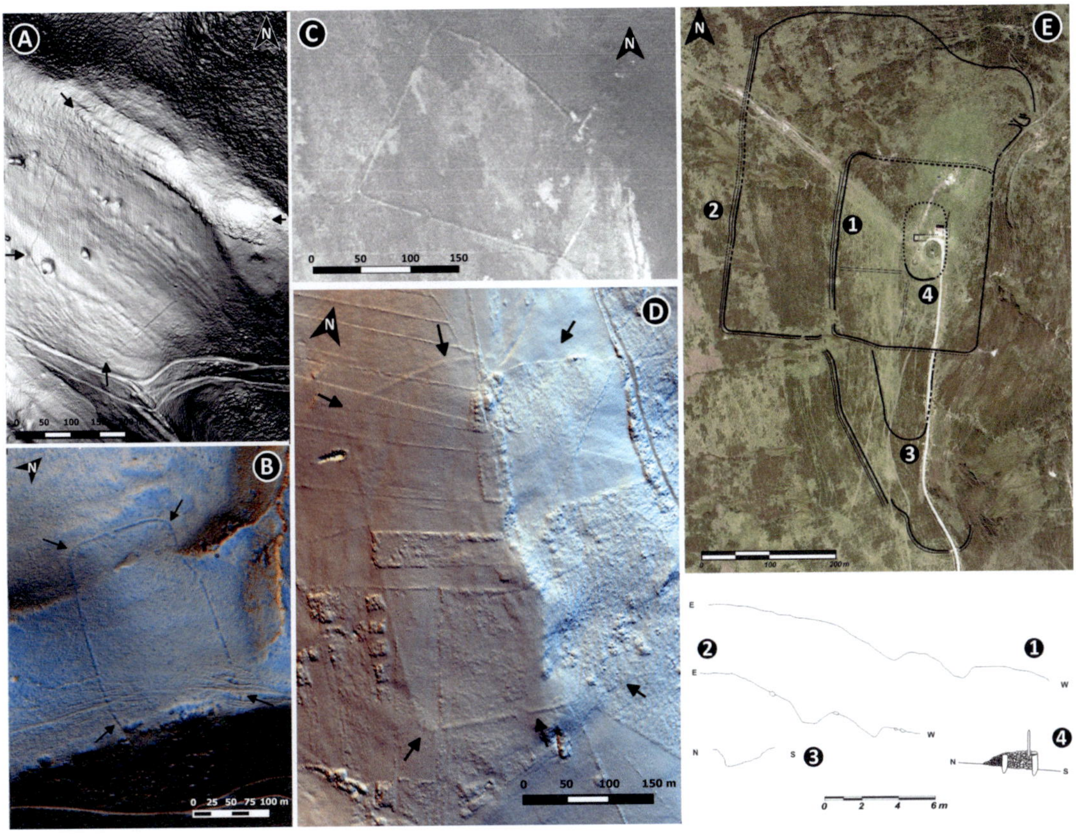

Figure 4: The potential of aerial photography and the aerial LiDAR for the study of Roman military sites. Some morphological and defensive solutions detected at El Xuegu La Bola (Asturias) (A), El Chao de Carrubeiro (Asturias) (B), A Cortiña dos Mouros (Galicia-León) (C), A Penaparda (Galicia) (D), and Cildá (Cantabria) (D).

For these very reasons, the complexity of certain fortifications in the mountainous areas should be taken into account when studying the episodes of the Cantabrian-Asturian wars. The famous sites of Monte Currieḷḷos (Asturias) or Cildá (Cantabria) are impressive examples of the art of castrametation in northern Iberia (Camino Mayor *et al.* 2005; Peralta Labrador 2002a) (Figures 2 and 4). However, the multiple defensive perimeters, the terraced platforms or the use of *bracchia* are not only found in those massive fortifications. An increasing degree of morphological complexity can be perceived among the sites discovered thanks to the use of remote sensing techniques in the last years. For example, the annex of the camp of Monte dos Trollos (Galicia) (Costa-García *et al.* 2016) resembles that of La Poza 1 (Cantabria) (Cepeda Ocampo 2006); Picu Viyao and perhaps A Pedra Dereta (Asturias) (González Álvarez *et al.* 2011; Menéndez Blanco *et al.* 2014) show triangular *bracchia* similar to those of Monte Currieḷḷos; Cueiru, Ḷḷagüezos, Llaurienzo (Asturias), Cildad or La Cabaña (Cantabria) (Costa-García *et al.* 2016; Hierro Gárate *et al.* 2015; Martín Hernández 2015) also present multiple defensive lines or enclosures marking out different camping areas. Similarly, the excavations at Monte Bernorio and Santa Marina (Cantabria) revealed the reoccupation of Late Iron Age hillforts after their conquest, the reinforcement of their old defences, and the construction of new, stone walls (Fernández Vega and Bolado del Castillo 2011; Torres Martínez *et al.* 2011). This behaviour had been previously detected in La Espina del Gallego (Cantabria) at the start of the 20th century (Peralta Labrador 1999).

However, our knowledge about the defences of the Julio-Claudian permanent military bases is very limited. The data mainly comes from achaeological interventions developed during the decades of 1990 and 2000. After that, the lack of economic resources for the systematic excavation of sites placed in rural areas, or either the difficulties inherent to urban archaeology have slowed down the research. *Fossae duplices* and earth ramparts were detected in Astorga (González Fernández 1997) and Rosinos de Vidriales (Carretero Vaquero and Romero Carnicero 1996). The defensive system of the legionary camp of León is better known (Morillo Cerdán 2010): the Augustan defensive system comprised a box rampart and at least one ditch. It was replaced in Tiberian times by a new *agger* built with turf bricks (*caespites*). The oldest archaeological contexts of Herrera de Pisuerga have not been properly clarified (Illarregui Gómez 2002; Morillo Cerdán *et al.* 2006).

From the Flavian era onwards masonry walls began to be the norm. It has been attested the petrification of the old earth ramparts at the legionary fortress of León (Morillo Cerdán and García Marcos 2003) and the auxiliary fort of Rosinos de Vidriales (Carretero Vaquero *et al.* 1999). On the contrary, the forts of Bande (Rodríguez Colmenero and Ferrer Sierra 2006) and A Cidadela (Caamaño Gesto 1997) were built in stone *ex novo*. Unfortunately, no comparative analyses of these evidences have been developed yet, and only the slight morphological variations showed by towers and gates have been used as dating elements. Some architectural reforms and alterations have been also detected in many of these structures, but they were never studied as a whole. The use of remote sensing techniques such as photogrammetry or laser scanning could help to fill this gap by following the specific methodologies of archaeology of architecture (Mañana-Borrazás *et al.* 2002, 2016).

The majority of the sites in use at the beginning of the 2nd century AD show only one ditch surrounding the walls, and usually not very wide or deep. Considering that the only military emergency in Hispania since the end of the Asturian-Cantabrian Wars was the brief uprising of Galba in 68 AD (Suet. *Galba* 8-10), this is not surprising. Unfortunately, the building of the Tetrarchy walls of León destroyed the remains of the external defences of the only legionary fortress documented in this period (Morillo Cerdán 2010).

Settlement patterns

The ancient writers who dealt with the *castra metatio* stressed the importance of selecting a proper location when establishing a camp (Polyb., 6.32-38; Ps.-Hyg. 57; Veg. *Mil.* 1.22, 3.8). Even if some degree of ambiguity can be detected in those treatises, two main issues were commonly observed: the place needed to be "safe", and the presence of certain natural resources in the proximities was preferable.

Still, these strict rules could usually be taken as a simple but wise advice. Like many other aspects of the castrametation, the general ideas of a safe and well supplied location relied more on the tight balance of some variables than on the strict observation of each of them. For instance, a location that is strong by nature does not necessarily imply a great altitude, but it should not be overlooked from a higher ground. Similarly, water was an essential resource, but its proximity could be judged as a positive or a negative aspect. The close presence of a water spring or a river guarding the flank of the settlement was always welcome, but the locations prone to flooding should be utterly avoided.

To a great extent, those ideas recorded by the ancient writers could be synthesised as variables to be included in a GIS-based location analysis. Issues such as relative height, local relief, slope, nearby presence of water courses or optimal visibility are easily collectable and quantifiable. Unfortunately, other elements cannot be reconstructed *a priori* due to the lack of reliable information about the ancient environment.

Of course these issues are not unsurmountable, but they demand supplementary archaeological work. The distribution of forested areas and pasture lands or the location of water courses and natural springs have changed since ancient times, but new palaeoenvironmental analyses could provide very useful information. Likewise, the exact position of the enemy and the direction of the march could be also inferred from the distribution of the gates in a camp (Ps.-Hyg. 56; Veg. *Mil.* 1.23), so a more accurate knowledge of the defensive perimeters is mandatory. In short, many difficulties could be overcome through specific, micro-level analyses of these sites following the most basic principles of landscape archaeology. Unfortunately, archaeological research on the Roman military in Spain has showed little interest in these approaches up to date.

Considering these limitations and the strict spatial boundaries of this work, only a selective sample of 30 sites (25% of the total individuals) has been analysed through GIS software.[4] Regarding the variables taken into consideration, they were grouped under two main categories in order to define the locational and visual control patterns of the sites.

The locational study (Parcero Oubiña and Fábrega Álvarez 2006; Stančič *et al.* 2000) comprises four subordinated analyses:

a. The classification of the position of the settlement bearing in mind some generic situations recorded by the ancient manuscripts, such as the morphology of the land (i. e. plain, hill, mountain, valley…), the location on a hilltop or a slope, and the presence of higher ground in the surroundings.
b. The calculation of the relative height of the site -topographic prominence. Five buffers based on Roman measures were defined in order to analyse this factor from close to far ranges: ¼, ½, 1, 3 and 6 Roman miles (ca. 370,

4 Both ArcGIS 10.3 and QGIS 2.10 software were used for these calculations. The 5m LiDAR-based digital elevation models (DEM) were granted by the Spanish National Plan of Aerial Orthophotography (PNOA). More precise (1m) digital terrain models (DTM) were also built after the interpolation of raw LiDAR data of the PNOA using LAStools (Repidlasso GmbH) and SAGA GIS software.

740, 1500, 4500 and 9000 m). Relative Height was calculated as follows: RH=(SH-MA)/STD, being RC the relative height, SH the height of the site, MA the mean altitude of the analysis area, and STD the standard deviation of the values of the mentioned area. A positive result implies a prominent position of the site, while a negative one indicates that it is under the mean height values.

c. The study of the local gradient. This issue is important to define two aspects: the natural strength of the position and the water runoff (a basic logistic observance). It can also help to define the orientation of the site, since the most elevated point was commonly reserved to place the tent of the commander and the front door (the one which heads to the enemy) should be located in the farthest wall (Ps.-Hyg. 56; Polyb. 6.32). The gradient was measured from the epicentre of the site following four main axes, and the results are expressed in percentage values. G = ((H1-H2)/D)*100, being H1 the height at the epicentre of the site, H2 the height at the end of every axis, and D the horizontal distance between H1 and H2.
d. The proximity of water streams and to what extent they help to reinforce the defensive position.

A second GIS-based approach included both visibility and visual prominence analyses (Llobera 2001; 2003; Wheatley 1995; Zamora-Merchán 2013), in order to understand not only the way in which the sites effectively controlled the surrounding areas but also if they occupied a visual prominent place within those landscapes. Five visual range areas were defined following the Roman-mile buffers previously mentioned. These analyses were executed through the Viewshed algorithm of ArcMap 10.3. For the visibility calculations the OFFSETA was set in 2.8 m (the sight height of a man -1.6 m- plus the height of a regular agger -1.2 m-),

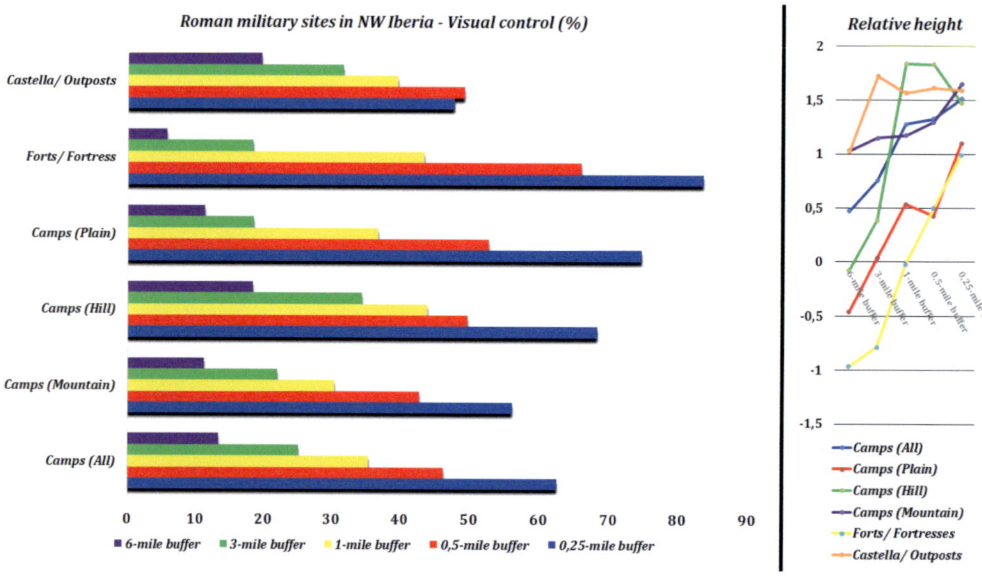

Figure 5: Roman military sites in Iberia: visual control of the surroundings and relative height.

the radius limited to 10000 m and the Z Factor to 1.7 m (the height of a person). For the "visibilization" calculus the OFFSETA was set in 1.7 m and the OFFSETB in 1.2 m (the agger only).

The results show that, to a great extent, the marching camps follow the theoretical principles of "safe location". Their relative heights are over the mean altitude values of the surrounding areas and that usually guarantees a quite good visual control, at least in the closest ranges (below 1 mile) (Figure 5). Nevertheless, specific patterns can be also detected after grouping the sites in accordance to their generic location.

When compared with the overall, the relative height of marching camps in mountainous areas is far above the average (over +1), but this does not imply a better visual control due to the complex geography of these regions.[5] Indeed, they could be defined as highly selective: the control of key positions such as mountain passages, mountain ranges, or the very accesses to the camp were prioritised over the complete seizure of the surroundings. The attempts of blocking mountain routes or immediately accessing water springs were issues commonly observed. In short, the final location of the sites was strongly dependent on strategic factors. As one could expect, steep slopes (5-27%) are also frequent, and there is a marked trend to establish the rear part of the camps on the higher terrain, so front gates are usually located on a lower ground. Altogether, these issues are probably reflecting that those settlements belonged to war scenarios.

The mountainous *castella* and lookout posts should be analysed separately. They were placed on the most elevated but less accessible locations, so they were frequently isolated.[6] The enhancing of visual control in the long distance[7] was made at the cost of a lower control over the short distance, except over key positions and natural routes through mountain ranges and valleys. Those objectives were probably of certain importance during war and post-war eras.

The Roman camps established in flat lands show less prominent locations, but their visual control is among the largest at the short distance,[8] where river fords, natural routes and passages are commonly found. A strategic use of watercourses is also attested, and the locations near river confluences granted the protection of more than 50% of the defensive perimeter. Thus, these positions derived from a strongly planned defensive behaviour and a strategic exploitation of the surrounding landscape. Gentle slopes (<1%) allowed a smooth water runoff and the articulation of regular "playing-card" layouts.

Finally, the camps located on hills usually occupied prominent positions over plain or valley areas, which implies a more extended and uniform visual control.[9] Once again the overlooking of key positions such as natural passages or river fords is frequently recorded. Likewise, meanders and river confluences were commonly selected when placing these sites. The values of terrain gradient were very dependent on the morphology of the specific hill (from <1% to ca. 15%) but these

5 Percentage values from closest to farthest range: 56.1%, 42.8%, 30.4%, 22% and 11.3%.

6 Relative height values are over +1 and they even reach +3.

7 Percentage values from closest to farthest range: 47.7%, 49.2%, 39.6%, 31.5% and 19.6%.

8 Percentage values from closest to farthest range: 74.7%, 52.7%, 36.8%, 18.5% and 11.3%.

9 Percentage values from closest to farthest range: 68.3%, 49.7%, 44%, 34.4% and 18.4%.

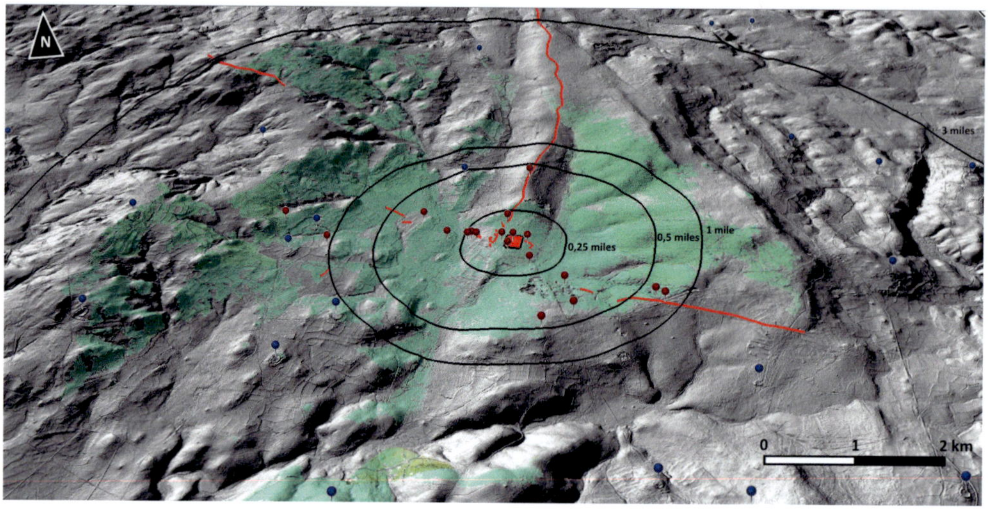

Figure 6: The fort of A Cidadela. Visual control in relation with the Roman (red) and Late Iron Age (blue) sites.

only occasionally conditioned the layout of the enclosures to a great extent. The placing of the rear part of the camp on the upper area of the hill was a tendency.

If we look at the so-called permanent sites we will find out that they were usually positioned not in "safe locations", but in places suitable for daily life. Plains or hills with soft slopes were preferred when establishing a fort or fortress, even if they were overlooked by higher grounds in the nearby, as happened in Rosinos de Vidriales, Bande or A Cidadela.[10] However, these locations granted a large visual control over the immediate surroundings as well as a notable prominence in the short and medium distance. For instance, the fort of A Cidadela visually controlled more than 50% of the land within a 1 mile distance and the visual prominence of the site was over 80% within the same area.

The preference for watersheds and river confluences can almost be considered a rule in these cases. It might be reflecting an attempt of delimiting a sort of military exclusion zone, since the presence of structures other than those of the camp is very limited in these areas. However, the archaeological landscapes showed a great complexity beyond those natural boundaries (Figure 6), so the question here should be if the logistic principles were stronger than the strategic ones regarding the location of those military sites. But, once again, this kind of diachronic analyses demand the existence of solid studies sensitive to the methodologies of landscape archaeology.

Discussion

Although the gradual introduction of new techniques and methodologies has increased the number of discoveries in the last decades, the archaeological study of the Roman military presence in Spain is still anchored in descriptive approaches. The aim of this paper was to break this inertia by showing that useful information

10 Relative height values usually became negative from less than a mile onwards.

can be extracted from the combined study of these sites, even applying the simplest statistical and locational analyses. Moreover, one way to overcome the heterogeneity and incompleteness of the archaeological data is indeed by asking transversal questions. It is not possible to gather certain datasets if we skip the bigger picture.

Even if the conclusions of this study are preliminary, they do not lack any significance. Precisely in this very moment, when the accumulation of data breaks the traditional uniformity of the archaeological record linked with the Roman Army, the definition of morphological, defensive, locational and behaviour patterns becomes fundamental for a correct characterisation of the Roman military presence in Iberia. Furthermore, this impetus should lead us to the building of predictive models and to the re-examination of the old archaeological catalogues in the search of new data. In this regard, many fortified settlements have been catalogued in the north-western area of Iberia under the label "castro" (hillfort), and even if the majority of them actually are Early or Late Iron Age hillforts, many others could belong to very different historical periods, from the Bronce Age to the Medieval Era. Sites such as A Recacha (Asturias-Galicia) (Menéndez Blanco *et al.* 2011), A Cortiña dos Mouros (León-Galicia) (Vidal Encinas 2015) (Figure 4) or O Cornado (Galicia) (Gago Mariño and Fernández Malde 2015), among others, were initially classified as Late Iron Age "castros". We know now that they are Roman military settlements. The reoccupation of native settlements pointed above could be another example of the "invisibility" of the Roman military presence in the archaeological record.

The use of GIS-based analyses for the study of Roman military sites in Iberia is far from being ended. A detailed examination of the gradients could help us to determine to what extent the inner space of these fortifications could be used as camping ground, and to carry out some demographic estimations on that basis (Zorn 1994; Porčić 2011). The discrimination of visibilities could also give us some interesting hints about which were the primary surveillance objectives from the perimeter defences or what area of the fortification was actually controlled from the commander's tent (Wheatley 2004; Bevan and Lake 2013). More complex analyses of accessibility and defensibility of the sites could be also implemented (Llobera *et al.* 2011; Parcero Oubiña 2013). However, these are only little pieces of a bigger process: in order to really understand the impact of the Roman army over the territory, it is necessary to analyse the fortifications within their archaeological landscapes through comprehensive, diachronic, and multidisciplinary studies.

Acknowledgements

I would like to thank César Parcero, Victorino Mayoral and Pastor Fábrega for inviting me to collaborate in this volume and for their advice throughout the courses on GIS and remote sensing techniques. I must also remember the other members of the Romanarmy.eu project (J. Fonte, A. Menéndez Blanco, D. González Álvarez, R. Blanco-Rotea, V. Álvarez Martínez, M. Gago), whose daily work to a great extent contributed to gather the data catalogue here analysed.

References

Bevan, A. and Lake, M. (eds.) 2013. *Computational Approaches to Archaeological Spaces*, Walnut Creek: Left Coast Press.

Caamaño Gesto, J. M. 1997. Sondeos arqueológicos en la muralla del campamento romano de Cidadela. *Gallaecia* 16, 265-284.

Camino Mayor, J., Estrada García, R. and Viniegra Pacheco, Y. 2005. El campamento romano de la vía Carisa y la conquista de la Asturia Trasmontana, in: Fernéndez Ochoa, C. and García Díaz, P. (eds.). *III Coloquio Internacional de Arqueología de Gijón: Unidad y diversidad en el Arco Atlántico en época romana*. Oxford: Archaeopress, 65-75.

Camino Mayor, J., Peralta Labrador, E. and Torres Martínez, J. F. (eds.). *Las Guerras Astur-Cántabras*. Gijón: KRK Ediciones.

Camino Mayor, J., Viniegra Pacheco, Y., Estrada García, R., Ramos Oliver, F. and Jiménez Moyano, F. 2007. El campamento y la vía de la Carisa. Reflexiones arqueológicas y militares, in: J. Fernández-Tresguerres (ed.). *Astures y romanos: nuevas perspectivas*. Oviedo: Real Instituto de Estudios Asturianos, 61-94.

Carretero Vaquero, S. and Romero Carnicero, M. V. 1996. *Los Campamentos Romanos de Petavonium (Rosinos de Vidriales, Zamora)*. Zamora: Fundación Rei Alfonso Henriques.

Carretero Vaquero, S., Romero Carnicero, M. V. and Martínez García, A. B. 1999. Las estructuras defensivas del campamento del Ala II Flavia en Petavonium (Rosinos de Vidriales, Zamora), in: De Balbín Behrmann, R. and Bueno Ramírez, P. (eds.). *II Congreso de Arqueología Peninsular: Zamora, del 24 al 27 de septiembre de 1996*. Madrid: Universidad de Alcalá – Fundación Rei Afonso Henriques, Vol. IV, 183-194.

Cepeda Ocampo, J. J. 2006. Los campamentos romanos de La Poza (Cantabria), in: Morillo Cerdán, A. (ed.). *Arqueología militar romana en Hispania II: Producción y abastecimiento en el ámbito militar*. León: Servicio de Publicaciones de la Universidad de León, 683-690.

Costa García, J. M. 2011. La castrametación romana en el Noroeste Peninsular: Algunos apuntes para su estudio. *Férvedes* 7, 215-223.

Costa García, J. M. 2013. *Arqueología de los asentamientos militares romanos en la Hispania altoimperial (27 a. C. – ca. 280 d. C.)*. Santiago de Compostela: USC – PhD Dissertation.

Costa García, J.M., Blanco Rotea, R., Gago Mariño, M. and Fonte, J. 2015. Novedades sobre la presencia del ejército romano en el occidente galaico, in: Camino Mayor, J., Peralta Labrador, E. and Torres Martínez, J.F. (eds.). *Las Guerras Astur-Cántabras*. Gijón: KRK ediciones, 285-289.

Costa García, J. M., Fonte, J., Menéndez Blanco, A., González Álvarez, D., Gago Mariño, M.,

Blanco-Rotea, R., Álvarez Martínez, V. 2016. Roman military settlements in the Northwest of the Iberian Peninsula. The contribution of historical and modern aerial photography, satellite imagery and airborne LiDAR. *AARGnews* 52: 43-51.

Didierjean, F., Morillo Cerdán, Á. and Petit-Aupert, C. 2014. Traces des guerres, traces de paix armée: l'apport de quatre campagnes de prospection aérienne dans le nord de l'Espagne, in: F. Cadiou and M. N. Caballero (eds.). *La guerre et ses traces. Conflits et sociétés en Hispanie à l'époque de la conquête romaine (IIIe-Ier s. a.C.)*. Bordeaux: Ausonius, 149-179.

Fernández Vega, P. Á. and Bolado del Castillo, R. 2011. El recinto campamental romano de Santa Marina (Valdeolea, Cantabria): Un posible escenario de las guerras cántabras. Resultados preliminares de la campaña de 2009. *Munibe* 62, 303-339.

Gago Mariño, M. and Fernández Malde, A. 2015. Un posible recinto campamental romano en O Cornado (Negreira, Galicia). *Nailos* 2, 229-251.

González Álvarez, D., Álvarez Martínez, V., Jiménez Chaparro, J. I., Menéndez Blanco, A., Colloto Montero, J. 2011. ¿Un nuevo establecimiento militar romano en la Asturia Transmontana?. El Picu Viyao (Piloña, Asturias). *Férvedes* 7, 225-234.

González Fernández, M. L. 1997. *La fortificación campamental de Asturica Augusta*. Astorga: Ayuntamiento de Astorga.

González Fernández, M. L. and Vidal Encinas, J. M. 2005. Recientes hallazgos sobre el campamento de la Legio VII Gemina en León: la situación de los Principia y la configuración de los Latera Praetorii. *Boletín del Seminario de Arte y Arqueología. Sección Arqueología* 71, 161-184.

Hierro Gárate, J. Á., Gutiérrez Cuenca, E. and Bolado del Castillo, R. 2015. Avances en la identificación de nuevos escenarios del Bellum Cantabricum, in: Camino Mayor, J., Peralta Labrador, E. and Torres Martínez, J. F. (eds.). *Las Guerras Astur-Cántabras*. Gijón: KRK Ediciones, 197-205.

Illarregui Gómez, E. 2002. Acerca de los campamentos altoimperiales de Herrera de Pisuerga y su entorno, in: Morillo Cerdán, Á. (ed.). *Arqueología Militar Romana en Hispania*. Madrid: CSIC – Ed. Polifemo, 155-166.

Llobera, M. 2001. Building Past Landscape Perception With GIS: Understanding Topographic Prominence. *Journal of Archaeological Science* 28(9), 1005-1014.

Llobera, M. 2003. Extending GIS-Based Visual Analysis: the Concept of Visualscapes. *International Journal of Geographical Information Science* 17, 25-48.

Llobera, M., Parcero Oubiña, C. and Fábrega Álvarez, P. 2011. Order in movement: a GIS approach to accessibility. *Journal of Archaeological Science* 38(4), 843-851.

Mañana-Borrazás, P., Blanco-Rotea, R., Sánchez-Pardo, J. C. 2016. Fast 3D recording techniques: a low-cost method for the documentation and analyisis of scattered architectural elements as part of the EMCHAHE project, in: Kamermans, H., de Neef, W., Piccoli, C., Posluschny, A. G. and Scopigno, R. (eds.). *The three dimensions of Archaeology*, Oxford: Archaeopress, 99-110.

Martín Hernández, E. 2015. El Mouro. Castrametación en la vía de la Mesa (Belmonte de Miranda/Grao, Asturias), in: Camino Mayor, J., Peralta Labrador, E. and Torres Martínez, J. F. (eds.). *Las Guerras Astur-Cántabras*. Gijón: KRK Ediciones, 239-247.

Menéndez Blanco, A., González Alvarez, D., Álvarez Martínez, V. and Jiménez Chaparro, J. I. 2011. Nuevas evidencias de la presencia militar romana en el extremo occidental de la Cordillera Cantábrica. *Gallaecia* 30: 145-165.

Menéndez Blanco, A., González Alvarez, D., Álvarez Martínez, V. and Jiménez Chaparro, J. I. 2013. Propuestas de prospección de bajo coste para la detección de campamentos romanos de campaña. El área occidental de la Cordillera Cantábrica como caso de estudio. *Munibe* 64, 175-197.

Menéndez Blanco, A., González Alvarez, D., Álvarez Martínez, V. and Jiménez Chaparro, J. I. 2014. Campamentos romanos de campaña en el Occidente de Asturias. *Excavaciones Arqueológicas en Asturias* 2007-2010: 245-251.

Morillo Cerdán, Á. 1991. Fortificaciones campamentales de época romana en España. *Archivo Español de Arqueología* 64, 135-190.

Morillo Cerdán, Á. (ed.) 2002a. *Arqueología Militar Romana en Hispania*. Madrid: CSIC – Ed. Polifemo.

Morillo Cerdán, Á. 2002b. Conquista y estrategia: el ejército romano durante el periodo augusteo y julio-claudio en la región septentrional de la Península Ibérica, in: Morillo Cerdán, Á. (ed.). *Arqueología Militar Romana en Hispania*. Madrid: CSIC – Ed. Polifemo, 67-93.

Morillo Cerdán, Á. (ed.) 2006. *Arqueología militar romana en Hispania II: Producción y abastecimiento en el ámbito militar*. León: Servicio de Publicaciones de la Universidad de León.

Morillo Cerdán, Á. 2010. Sistemas defensivos en los campamentos romanos de León, in: *V Congreso de Obras Públicas Romanas. Las Técnicas y las construcciones en la Ingeniería Romana*. Córdoba: 463-477.

Morillo Cerdán, Á. 2014. Arqueología Militar romana en Hispania: balance de dos décadas de investigación (2000-2014), in: Martínez Ruiz, E. and Cantera Montenegro, J. (eds.). *Perspectivas y novedades de la Historia Militar: una aproximación global*. Madrid: Ministerio de Defensa, Vol. 1, 25-58.

Morillo Cerdán, Á. and Aurrecoechea, J. (eds.) 2006. *The Roman Army in Hispania*. León: Univesidad de León.

Morillo Cerdán, Á. and García Marcos, V. 2003. Legio VII Gemina and its Flavian fortress at León. *Journal of Roman Archaeology* 16, 275-286.

Morillo Cerdán, Á., Hanel, N. and Martín Hernández, E. (eds.) 2009. *Limes XX. Estudios sobre la frontera romana*. Madrid: CSIC – Ed. Polifemo.

Morillo Cerdán, Á., Pérez González, C. and Illarregui Gómez, E. 2006. Herrera de Pisuerga (Palencia). Introducción histórica y arqueológica. Los asentamientos militares, in: García-Bellido, M. P. (ed.). 2006. Los campamentos romanos en Hispania (27 a.C.-192 d.C). El abastecimiento de moneda. Madrid: CSIC – Instituto Histórico Hoffmeyer – Instituto de Historia – Polifemo, Vol. 1, 305-323.

Parcero Oubiña, C. 2013. Midiendo decisiones locacionales: Una aproximación a la evaluación de la defensibilidad efectiva de sitios arqueológicos fortificados. *Comechingonia* 17, 57-82.

Parcero Oubiña, C. and Fábrega Álvarez, P. 2006. Diseño metodológico para el análisis locacional de asentamientos a través de un SIG de base Raster, in: Grau Mira, I. (ed.). *La aplicación de los SIG en la arqueología del paisaje*. Alicante: Universidad de Alicante, 69-91.

Peralta Labrador, E. 1999. El asedio romano del Castro de la Espina del Gallego (Cantabria) y el problema de Aracelium. *Complutum* 10, 195-212.

Peralta Labrador, E. 2002a. Los campamentos de las Guerras Cántabras de Iguña, Toranzo y Buelna (Cantabria), in: Morillo Cerdán, Á. (ed.). *Arqueología Militar Romana en Hispania*. Madrid: CSIC – Ed. Polifemo, 327-338.

Peralta Labrador, E. 2002b. Los campamentos romanos de campaña (castra aestiva): evidencias científicas y carencias académicas. *Nivel Cero* 10, 49-87.

Peralta Labrador, E. 2006. La revisión de las guerras cántabras: novedades arqueológicas en el norte de Castilla, in: Morillo Cerdán, A. (ed.). *Arqueología militar romana en Hispania II: Producción y abastecimiento en el ámbito militar*. León: Servicio de Publicaciones de la Universidad de León, 523-547.

Peralta Labrador, E. 2011. Campamentos romanos en Cantabria. *Castillos de España* 161-162-163, 23-26.

Porčić, M. 2011. An exercise in archaeological demography: estimating the population size of Late Neolithic settlements in the Central Balkans. *Documenta Praehistorica* 38, 323-332.

Richardson, A. 2004. *Theoretical Aspects of Roman Camp and Fort Design*. Oxford: Archaeopress.

Rodríguez Colmenero, A. and Ferrer Sierra, S. (eds.) 2006. *Excavaciones arqueológicas en Aquis Querquennis. Actuaciones en el campamento romano (1975-2005)*. Lugo: Unión Fenosa – Grupo Arqueolóxico Larouco – Universidade de Santiago de Compostela – Fundación Aquae Querquennae-Vía Nova.

Serna Gancedo, M. L., Martínez Velasco, A. and Fernández Acebo, V. (eds.) 2010. *Castros y castra en Cantabria: fortificaciones desde los orígenes de la Edad del Hierro a las guerras con Roma. Catálogo, revisión y puesta al día*. Santander: Acanto.

Stančič, Z., Veljanovski, T. and Podobnikar, T. 2000. Understanding Roman settlement patterns through multivariate statistics and predictive modelling, in: Lock, G. (ed.). *Beyond the map. Archaeology and Spatianl Technologies*. Amsterdam: IOS Press, 146-157.

Torres Martínez, J. F., Serna Gancedo, A. and Domínguez Solera, S. D. 2011. El ataque y destrucción del oppidum de Monte Bernorio (Villarén, Palencia) y el establecimiento del castellum romano. *Habis* 42, 127-149.

Vidal Encinas, J. M. 2015. La aportación de las infraestructuras de datos espaciales al conocimiento de nuevos sitios castreños en la provincia de León. *Férvedes* 8, 25-34.

Wheatley, D. 1995. Cumulative viewshed analysis: a GIS-based method for investigating intervisibility, and its archaeological application, in: Lock, G. and Stančič, Z. (eds.). *Archaeology and GIS: A European Perspective*. London: Routledge, 171-185.

Wheatley, D. 2004. Making space for an archaeology of place. *Internet Archaeology*, 15

Zamora-Merchán, M. 2013. Análisis territorial en arqueología: percepción visual y accesibilidad del entorno. *Comechingonia* 17, 83-106.
Zorn, J. R. 1994. Estimating the Population Size of Ancient Settlements: Methods, Problems, Solutions, and a Case Study. *Bulletin of the American Schools of Oriental Research* 295 (295), 31-48.

Notes on contributor

José Manuel Costa-García has a PhD in Archaeology (2013) and BA in History (2006) with a honours degree from the University of Santiago de Compostela. He obtained an Advanced Studies Diploma in Archaeology from the same university (2008). He also got a FPU predoctoral grant (2008-12) and a postdoctoral grant of the Galician Regional Government (2016-2019). Active member of the archaeological research team in the Roman fort of A Cidadela (Sobrado dos Monxes, A Coruña) (2007-2010 and 2016). His specialization is the study of Roman military sites combining specific methodologies coming from archaeology, ancient history, epigraphy and new technologies.

Rethinking Tafí: a political approach to the landscape of a Southern Andean Formative community

Jordi A. López Lillo[1]

Introduction: an aboriginal repolitization

In 1977, the French ethnographer Pierre Clastres (1994, 166) stated that a primitive society can be defined as *a multiplicity of undivided communities which obey the same centrifugal logic.* It was the response of an Americanist to a widespread problem in social interpretation dating back, precisely, to the European discovery of America: the mixture, within the framework of liberal cosmogonies (such as the thinking of Locke or Adam Smith), of the *logical fiction* of a so-called "state of nature" with an alleged *historical event*; a mixture which in turn leads to the "discovery" of Prehistory under the imagination of savages as "primitives" (Gosden 1999). As Todorov puts it, from the Enlightenment onwards, the conceptualisation of Humanity as a universal category conceals a strong by-product by which "the existence of a human substance actually 'other' is rejected, beyond a simply imperfect state of 'ourselves'" (Todorov 1998, 157). In fact, the thread starting with the 19th century's social evolutionism, pervaded with the idea of "progress", promoting the common archaeological assumption of a "band - tribe - chiefdom - State" typological sequence is obviously enough to require no further comment.

Nevertheless, it is important to note the teleological bias incorporated into this kind of hypothesis, especially inasmuch as it runs the risk of leading us to forget its instrumental character (being a heuristic tool in a usually negative process of identification) and leads us towards a naturalisation of our own endocultural assessments without any serious attempt to cross-check. For example, this is what led to the *primitive barter* scenario in order to explain trading behaviour in modern societies, which is stubbornly upheld, even nowadays, by some authors despite the fact that a "savage polity" weaving its social tissue by such a trade

---

1 Centro de Investigaciones "María Saleme de Burnichon". Universidad Nacional de Córdoba. Argentine Republic jordilopezlillo@gmail.com.

has never been found (Humphrey and Hugh-Jones 1992; Graeber 2011). On the other hand, it is also what underlies the *primitive communism* altar on which real ethnographic data are sacrificed. Both constructs share, and therefore inadvertently secure, an economy-oriented point of view which assumes politics only as a social epiphenomenon, even if the ethnographic record has actually started to point towards quite the opposite (Godelier, 1999; 2014). Both take part in a polarised argument between individualistic and holistic approaches with the same result: an evident inability to acknowledge even the possibility of different behaviours bearing different cultural targets, and therefore, bring forth an "Anthropology of operative logics", which seems to be the only way of comprehending the *non-linear reality* revealed by the acceleration of contextual criticism since about 1970.

In this context, the Clastrean point of view might be privileged in re-evaluating social and political systems within the different cultural scenarios unearthed by Archaeology and also, as a matter of fact, by in re-evaluating own culture. By now, the detection of some precise institutions, which willingly prevent the formation of the State by promoting centrifugal logic over the centripetal,in canse of systemic over-stress, produced a major change in a field in which politics could be hardly imagined without *domination* in a Weberian sense (except for a few noteworthy exceptions, *e. g.* the "ordered anarchy" of Evans-Pritchard, and some of his followers' statements about primitive law and government (Gluckman 1973; Hoebel 2006; Mair 1962).

The aim of this contribution is to join the re-evaluation of the archaeological record and, thus, of the historical processes it reflects through the new light provided by this response in addition to exploring its potentialities. In order to do so, I will turn to one of the best known case-studies in the Formative period of North-Western Argentina, the Tafí tradition, and, more specifically, to the analysis of its landscape.

As will be seen, if, finally, some achievement emerges from this effort and Tafí history is shown in a more parsimonious way than before, it will come neither from a qualitative novelty in our knowledge of its material features, nor from a ground-breaking technological input in its analysis but rather, from a pronounced theoretical rearmament. It is not by chance that the communication which gave birth to this paper was originally entitled "A classical GIS implementation for a novel interpretation of the Formative landscape in Southern Andes".

The Tafí tradition, in traditional Archaeology

Located in the Tucumán Province, nearly 120 Km from its capital at San Miguel, the Tafí Valley corresponds to the *keshua* biotope or "temperate high valleys" 2000 to 2500 MASL in this case, between the *yunga* forests and the chaco-pampean plateau to the east, and the *puna*, to the west. Together with its position on the main road through the Cumbres Calchaquíes towards the higher Yocavil Valley, the prairie vegetation typical at this altitude has evidently played a major role in the archaeological knowledge of the region, dating back to a work as famous as Ambrossetti's notes on the so-called *menhirs* in 1897.

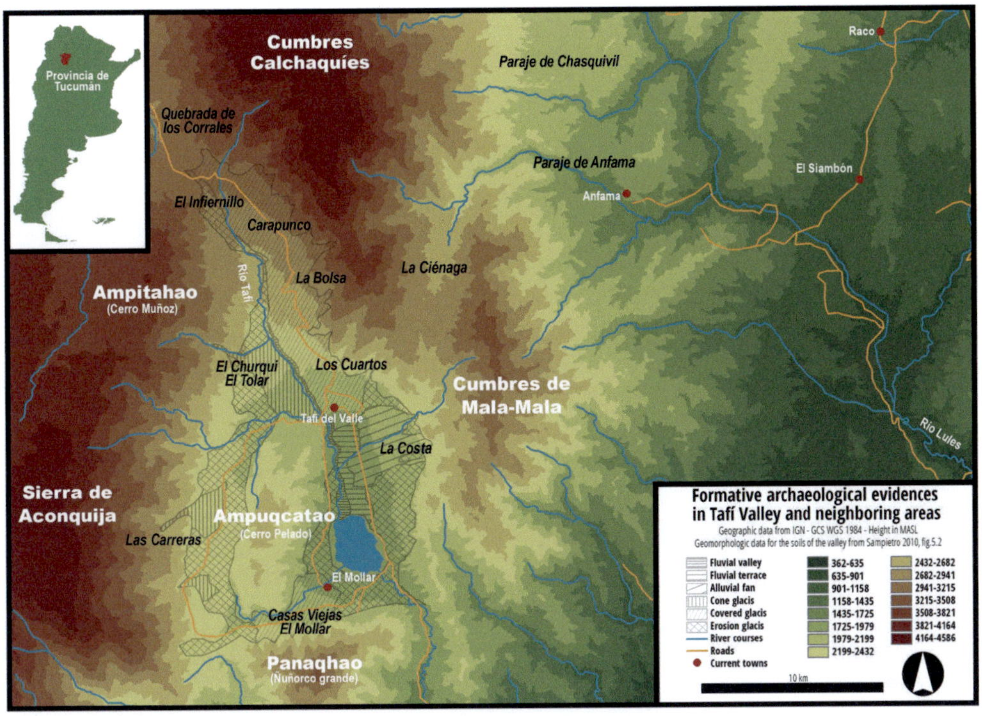

Figure 1: Formative evidence in the Tafí Valley and neighbouring areas (Tucumán Province); note that most of the Yunga forest, to the east, is still unknown to Archaeology due to environmental difficulties.

Furthermore, without any doubt, the more prominent elements in the record are the hundreds of widely scattered, highly surface-visible, circular stone structures dating from the Formative period, from *c.* 250 BC to 850 AD, in which the first agropastoralist settlements appeared in the Tafí valley. From Casas Viejas on the southern edge; all around the Ampuqcatao hill, noticeably denser at sites such as Las Carreras and El Churqui on the hillsides of Aconquija or La Costa and Los Cuartos, towards the Mala-Mala hills; to the north, at La Bolsa, Carapunco and El Infiernillo: the tear-shaped valley is practically covered with this kind of structure (figure 1). Most of them have been interpreted as house clusters and closely-related productive sites. In fact, the "Tafí Household Pattern" has become a common descriptor in regional archaeological analysis (Di Lullo 2012; Oliszewski *et al.* 2013; Álvarez Larraín and Lanzelotti 2013), emphasising its homogeneity both in a more extensive area than the valley and over almost a millennium. In turn, it reveals an unavoidable problem for any archaeological interpretation since this homogeneity, together with the paucity of data resulting from test pits and open-area excavations, generates a delicate "synchronic apriorism" by which all visible structures are supposed to have been occupied simultaneously at a cultural peak. However, the absolute dates available confirm an extremely long-term occupation for these houses (*e.g.*, between cal. 130-260 and 650-780 AD for La Bolsa sector 1, Unit 14 case (Salazar 2010).

This Formative household pattern consists of circular or sub-circular semi-subterranean rooms, ranging in size between 2 and 20 m^2, connected to a larger unroofed courtyard or patio. As can be seen in the aforementioned Unit 14 (figure 2), virtually the only fully excavated Tafí household (*cf.* Salazar *et al.* 2007; Sampietro 2010), a single entrance connected the patios to the outside areas. These clusters vary in size and number of attached rooms but in all cases the structures maintain the same spatial organisation. The entrance and pathways illustrate greater fluidity within the house than between the indoor and outdoor spheres and, to a certain extent, the areas of activity detected indicate a centripetal organisation. In this sense, the common presence of burial sites within the patios is remarkable, especially since stratigraphic sequences such as those unearthed at Unit 14 demonstrate that it was the result of complex depositional processes. This makes it quite evident that the corpses of ancestors were not socially dead but were continuously taken out from their tombs, shown, worshipped, fed, and buried again. What is more, this archaeological bundle indicates that they were part of daily activities, and the material configuration of the burial site generated a permanent interaction with the living.

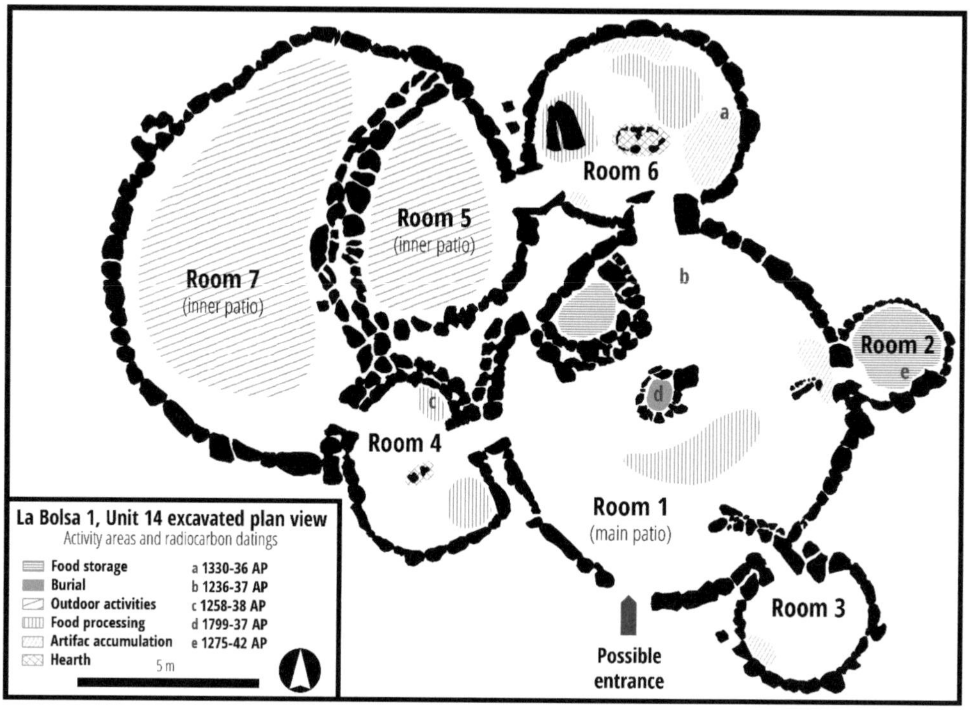

Figure 2: Planimetric plan view of Unit 14, La Bolsa 1 site, indicating areas of activity identified by Salazar (2010); note that not all attached rooms had identical functions, at least during the last occupancy prior to the abandonment of the structure. All AMS radiocarbon analyses were made from charcoals on the occupation floor, except for the dating of the burial site.

Indeed, this characteristic has consistently been bound together with the cultural interpretation of the other prominent element of the Tafí archaeological record: the carved monoliths, confusingly called *menhirs*. It is highly likely that these stones are to be identified as *wanka* or lithified ancestors (Duviols 1979; García Azcárate 1998). According to this widespread Andean belief, after death some people turn into stone with a specific ancestral status and are used by the living as strong landmarks of identity, territoriality and fertility. Unfortunately, despite these monoliths being recognised as early as the 19th century, relatively little information was recorded about their original location. Given that more than a hundred of them were relocated en masse to La Angostura during the nationalistic military process of the 1970s, and then to an archaeological park at El Mollar, any possibility of further spatial analysis is virtually impossible, although it is known that between thirteen and sixteen pieces were recovered around the Casas Viejas Mound and that others were found within household contexts.

Indeed, the Casas Viejas Mound is an outstanding entity and, of course, its discovery has traditionally influenced the entire social interpretation of this archaeological culture. Also located at El Mollar, within the south-eastern quarter of the valley, the mound was partially excavated by González and Núñez Regueiro (1962) and was interpreted as a ceremonial centre. It is mostly made up of consumption debris and, leaving aside the aforementioned monoliths, some inhumations were also found, presumably from a slightly later time with respect to the dating of its construction. Given this characterisation, according to current social theories for Andean prehistory (Tartusi and Núñez Regueiro 1993; Tarragó 1999), the presence of such a "central spot" has been used to hold, to some extent, the hypothesis of an emerging ritual chieftainship over Tafí communities. But only *to some extent* because the evident homogeneity both of pottery and architectural features throughout the chronological sequence, in turn, has prevented the furthering of the model, as was the case for other nearby regions such as Ambato, in the province of Catamarca (*cf.* Cruz 2007; Laguens 2014; Gordillo 2007; Zuccarelli 2012). Bearing this in mind, Núñez Reigueiro's first proposals in the 1970s consisted of a neo-evolutionistic explanation with a two-phase development named after the sites of La Angostura and Carapunco and relying mostly on the presence or absence of certain ceramic typologies. Under this framework, the early dates for the Casas Viejas Mound placed its ceremonial use in the La Angostura phase but, above all, it emphasised a *cultural stasis* for Tafí communities throughout the First Millennium AD.

The 1980s witnessed the appearance of a processual approach which reverted the former by proposing a hypothetical process of increasing social complexity based on settlement patterns. Unfortunately, however, it was based entirely on surface data, which was virtually the only option available. Again, Berberián and Nielsen (1988) distinguished two models, with chronological implications. In the first one, known as Tafí I, based on the Carapunco site pattern which is strongly centred on the household and is, therefore, widely scattered, agricultural strategy is extensive and there is little investment in infrastructure. Given that the polynucleated domestic groups seem to be quite clearly self-sufficient, the model proposes a slow vegetative growth which is reflected archaeologically in the progressive aggregation of rooms and other unroofed areas in domestic compounds around a main patio. During a later stage of evolution, known as Tafí II, based on the La Bolsa site pattern, this growth led to a concentrated settlement pattern and the emergence of true villages. Nevertheless, this higher level of social integration, which supposedly developed after *c.* 500 AD, also lacks any archaeologically recognisable hierarchies. Therefore, the intra-social segregation had to be restrained in some way.

The crux of the question is that, as is the case with other materialistic paradigms, Processualism pays little or no attention to problems such as the social and cultural mechanisms securing this restraint and therefore reproduces the depoliticisation of non-hierarchical societies. As a result, these sorts of interpretations are quite compatible with the apriorism by which all social order tends towards a rationalistic centralisation, as materialised at the Casas Viejas Mound and its presumed ritual chiefdom. However, the specific cultural grounds of any reason, the character and the logic woven behind a given centre, are seldom truly explored.

In addition, this objection is fully coherent with the conclusions drawn from one of the most important systematic surveys carried out, not by chance, at the La Bolsa and Carapunco sites by an archaeological team from the Universidad Nacional de Córdoba (UNC), once directed by Berberián and Nielsen, whose main authors point out that (Franco Salvi *et al.* 2014, 318):

"The narratives which deal with the social process experienced by the First Millennium inhabitants of the Tafí Valley have some elements in common. Basically, a significant rupture between modes of social organisation, cultural patterns and ways of producing is expected [...]. But the observation of this rupture in the material record is, at least, blurry. The evidence presented here (pottery, architecture, radiocarbon dating, etc.) leads us to think that such a rupture has little empirical basis".[2]

This is not to mention the fact that the first amendment to the "increasingly centred polity" hypothesis arises more quickly as the centre begins to be decentralised, such as when this survey found another mound at La Bolsa, in the northern half of the Tafí Valley.

Revisiting the Tafí "pathways"

One of the major projects of the UNC team over the course of the last few years has been the full topographical recording of surface evidence in wide areas of the valley. The aim is not only to document the current archaeological remains, but also to provide the necessary basis for carrying out a comprehensive spatial analysis and for planning further, more in-depth, interventions. Last but not least, the survey of archaeological remains constitutes one of the cornerstones of the collaboration with native communities, fostering mutual knowledge, comprehension and the preservation of heritage within their territory.

Given these aims, a single topographical form was designed. Among other descriptions and numerical identifiers, it comprises fields on "shape" (circular, linear or rectangular structures, stone pile, canal or fixed grinding element), a preliminary "functional interpretation" (patio, room, isolated hut, corral, platform

2 All translations from Spanish by the author

Shape
> Polygon
>> Area in square meters
>> Chrono-typology
0. No data 1. Formative room 2. Formative patio 3. Formative isolated 4. Formative productive 5. Historic 6. Second millennium prehispanic 7. Formative house cluster (addition of [shape/polygon/chrono-typology/1+2]) 8. Stone pile
> Point
>> House cluster (centroid of [shape/polygon/chrono-typology/2])
>>> Magnitude index (number of attached [shape/polygon/chrono-typology/1])
>> IPS (centroid of [shape/polygon/chrono-typology/4])
>>> Magnitude index (mean of [shape/area in square meters])
>>> Arrangement in a cluster (centroid of [shape/polygon/chrono-typology/3])
0. No data 1. No 2. Yes
>> Isolated hut
>> Grinding element
>>> Type
0. No data 1. Vertical pressure 2. Horizontal pressure
>>> Arrangement in a cluster
0. No data 1. No 2. Yes
>Polyline
Degree of intervention
0. No data 1. Mapped from aerial or satellite imagery 2. Topographied 3. Surface gathering 4. Pit test 5. Excavated
Topographic series
0. Out of series 1. La Bolsa 1 2. La Bolsa 2 3. La Bolsa 3 4. Carapunco 1 5. Carapunco 2 6. Carapunco 3
Dating
0. No 1. By diagnostic artifact (only applicable when [Degree of intervention/3] or further) 2. By radiocarbon
Notes (text field)

Table 1: Organisational diagram for Tafí GIS, itemizing the categories employed in this study as well as some bases for future developments. With this in mind, the conversion from polygons into points for analytical purposes should be noted, especially in the case of house clusters. Also worthy of note is its potential as an archaeological chart.

field (*andén*), enclosed field (*canchón*) or indeterminate) and a preliminary "chronological adscription" (Formative, Regional Development, Inca or Historical periods). In this way, it was grounded on the proven chrono-typology designed by Berberián and Nielsen, and adapted by Salazar (2010) and Franco Salvi (2012). However, the new form introduces some slight differences: it is centred on architectural features regardless of their integration into larger built compounds, as a characteristic facilitating subsequent GIS processing. Indeed, despite this objective necessarily being postponed for the time being, these kinds of exercises may, in the near future, support the development of a more comprehensive archaeological chart. Hence, the emphasis placed on including some remains beyond those belonging to the Formative. A quick look at the organisational GIS diagram (table 1) will confirm this, whilst showing at a glance the internal structure of our data.

As mentioned above, the survey is limited to the northern edge of the valley, specifically to the hillsides of La Bolsa and Carapunco, the archaeological landscapes of which lay the basis for the cultural-ecological perspective on the history of the Tafí communities. Surface gathering of sherds and test pits carried out alongside the topographical survey do not reveal any significant chronological differences between the two sites (Franco Salvi *et al.* 2014). Therefore, whilst waiting for deeper open-area interventions, the "synchronic apriorism" still remains unavoidable. What is more, it seems to be the only meaningful approach at the present time. Therefore, the analyses presented here were carried out with the data from the two sites jointly, scattered over an area of almost 750 hectares. This sample (figure 3) comprises 155 house clusters (HC) with more than 550 rooms, 303 productive closed-structures among corrals and *canchones* (in fact, most of them probably served both functions alternatively so, for analytical purposes, were grouped as indeterminate productive structures (IPS)) and 71 isolated huts (IH), the function of which remains unclear. In addition, there are some constructed remains which are not currently included in the study, such as stone piles resulting from the *andenería* outfitting, fixed grinding elements and, above all, the *andenería* itself, which is not always well-defined on the surface. Finally, the La Bolsa Mound was also included: the *circumstances of its centrality* are, precisely, what we are trying to define.

Given that the household was quite obviously fundamental in Tafí communities, the study of the latter as a whole necessarily has to begin with the former. In this way, house clusters are reflected in the GIS taking the centroid of its main patio. A basic way of comparing them, given the state of our knowledge, is by a magnitude index equivalent to the number of roofed rooms attached to make up the cluster. In addition, it is important not to forget the fact that this index does not imply any direct demographic statement. In the case of La Bolsa 1, Unit 14, Salazar (2010; *cf.* Sampietro and Vattuone 2005) found that only two rooms contained a hearth, while another two were inner patios for these rooms, accessible exclusively through each room. Another was possibly a sort of storeroom. And the last one had been altered too much to be able to state anything about the activities carried out in it. Certainly, this picture belongs to the last occupation of the building before its abandonment around the end of 8th century and shows decisively the kind of problem we are dealing with.

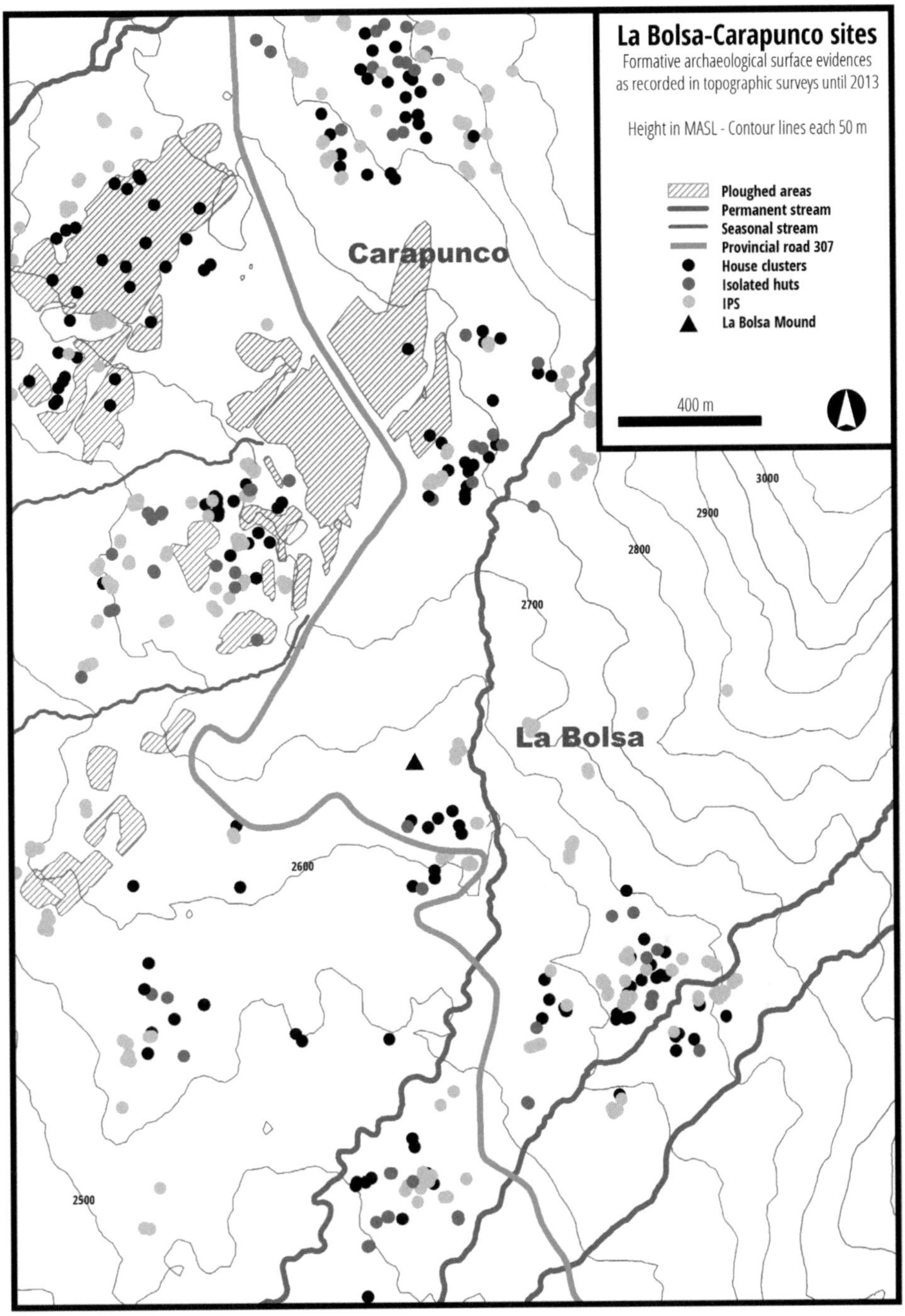

Figure 3: Formative evidence on the hillsides of Carapunco and La Bolsa in the northeast of the Tafí Valley. Despite the fact that other types of structures are partially visible on the surface and, hence, were recorded topographically, analyses presented in this study revolve around house clusters, indeterminate productive structures (IPS) and isolated huts, and their spatial relation with the La Bolsa Mound.

Magnitude index	Number of house clusters	%	Number of attached rooms	%
1	18	12%	18	3%
2	23	15%	46	8%
3	44	28%	132	25%
4	24	15%	96	17%
5	24	15%	120	22%
6	15	10%	90	16%
7	6	4%	42	8%
8	1	1%	8	1%
Total	**155**	**100%**	**552**	**100%**

Table 2: Some figures on the magnitude indexes of the house clusters sampled, and the allocation of attached rooms among them.

Most HCs, specifically 58% of the whole sample (table 2), have between 3 and 5 attached rooms. Less than a third of clusters have less than 3 rooms with only 5% of the HCs presenting 7 or more rooms. Their distribution on the map stresses the gaps between major units which, with the exception of the couple formed by the only eight-roomed house cluster and an adjacent seven-roomed one at Carapunco 1, to the northeast of the road, are located at an average distance of 400 to 600 m. The same is true for the nucleations they form, as can clearly be seen in a density analysis. If the directional distribution of settlements is taken into account (as shown in a standard deviation ellipse for HC points, weighted by its magnitude indexes), an orographic pattern is confirmed (figures 4 and 5). Above all, the Formative population appears to have been spread according to the parallel boundaries of the southern foothills of the Cumbres Calchaquies, to the east, and the upper Tafí river, to the west. There are, furthermore, no "central" inhabited structures or clusters, but rather a wide clear space between the La Bolsa Mound and the median centre of the sample.

In accordance with the relationship of this settlement pattern with the mound, the Tafí landscape could be characterised as centrifugal, due to the fact that this singular structure (and, therefore, the political centre it may represent) is outlined more as a common space than as a hierarchical magnetic pole. In fact, its location close to the southern rim of a slight depression both dissociates the spot from the nearer inhabited area, at La Bolsa 2, around 100 m to the south, and negatively determines its viewshed (figure 6). None of the northern settlements can be seen from the top of the mound and, despite some scattered HCs also being visible to the southwest, the only noteworthy nucleation is the distant sector of La Bolsa 1. Therefore, this architectural *unicum* can definitely not be considered as a strategic control entity. For the same reason, the mound itself is not a highly perceived spot in the landscape. Indeed, it should be remembered that it was not archaeological noticed until the UNC's topographical surveys. However, the huge field where it is located was, perhaps, a little more prominent. Thus, turning to the cumulative viewshed derived from the addition of the visible areas from each HC (figure 7), the adjacent zone to the mound presents an index which is moderately higher than that of many of the inhabited nucleations. However, in no way does the position of the mound stand out, nor it is comparable with the omnipresence of the Cumbres Calchaquíes. Furthermore, this analysis reveals a limited and relatively homogeneous intervisibility among house clusters.

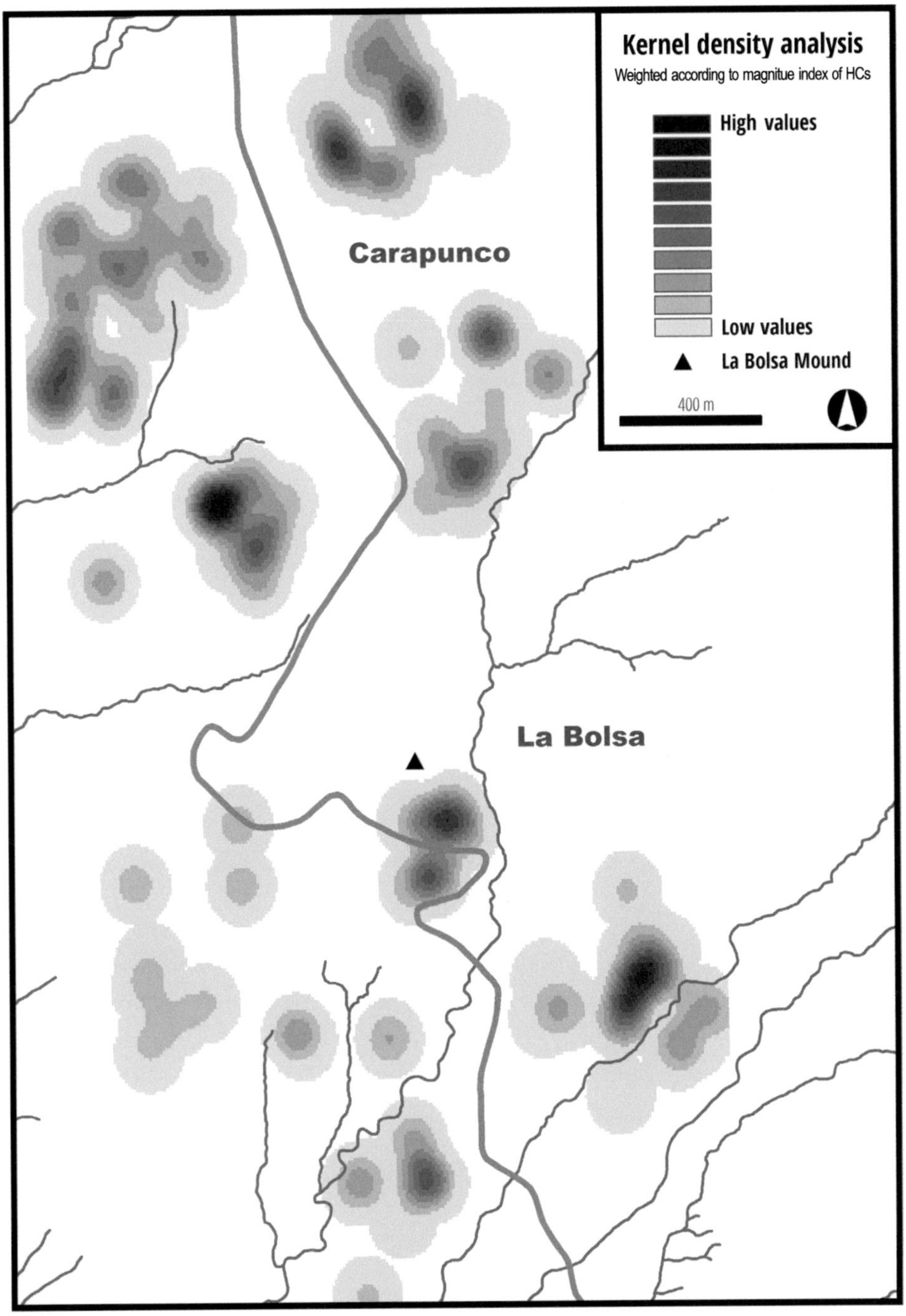

Figure 4: Kernel density analysis, weighted according to magnitude index of HCs.

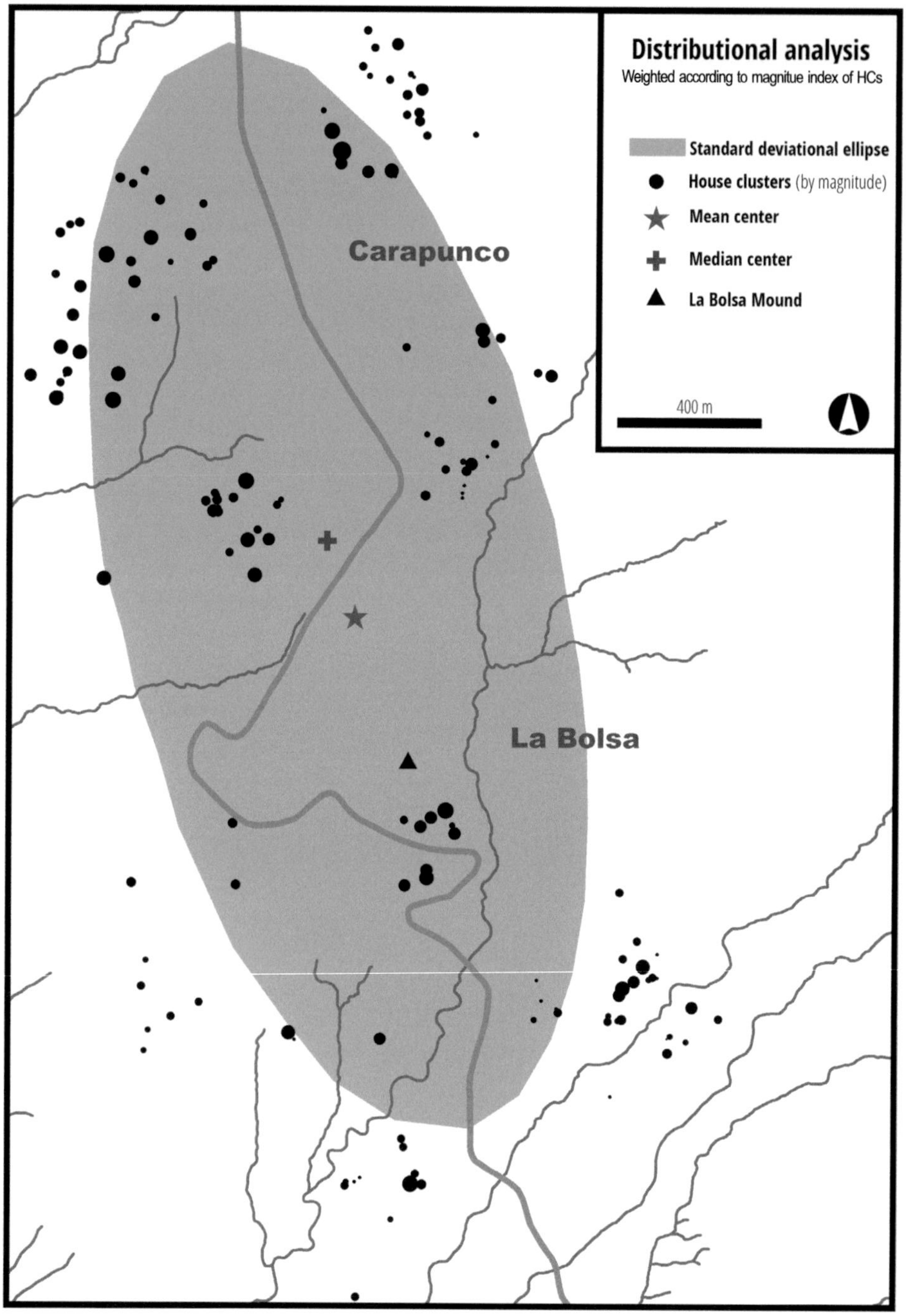

Figure 5: Distributional analysis, weighted according to magnitude index of HCs. It should be noted that the size of the points on the map also reflects the number of attached rooms composing each house cluster. It should also be noted that the shape of the standard deviational ellipse and, especially, the mean and median centres may vary if more data are added in further surveys from beyond the study area. In any case, current results seem to establish quite a definite logic. Therefore, no significant changes in meaning are expected despite the more or less mathematically centred appearance of the mound in the future.

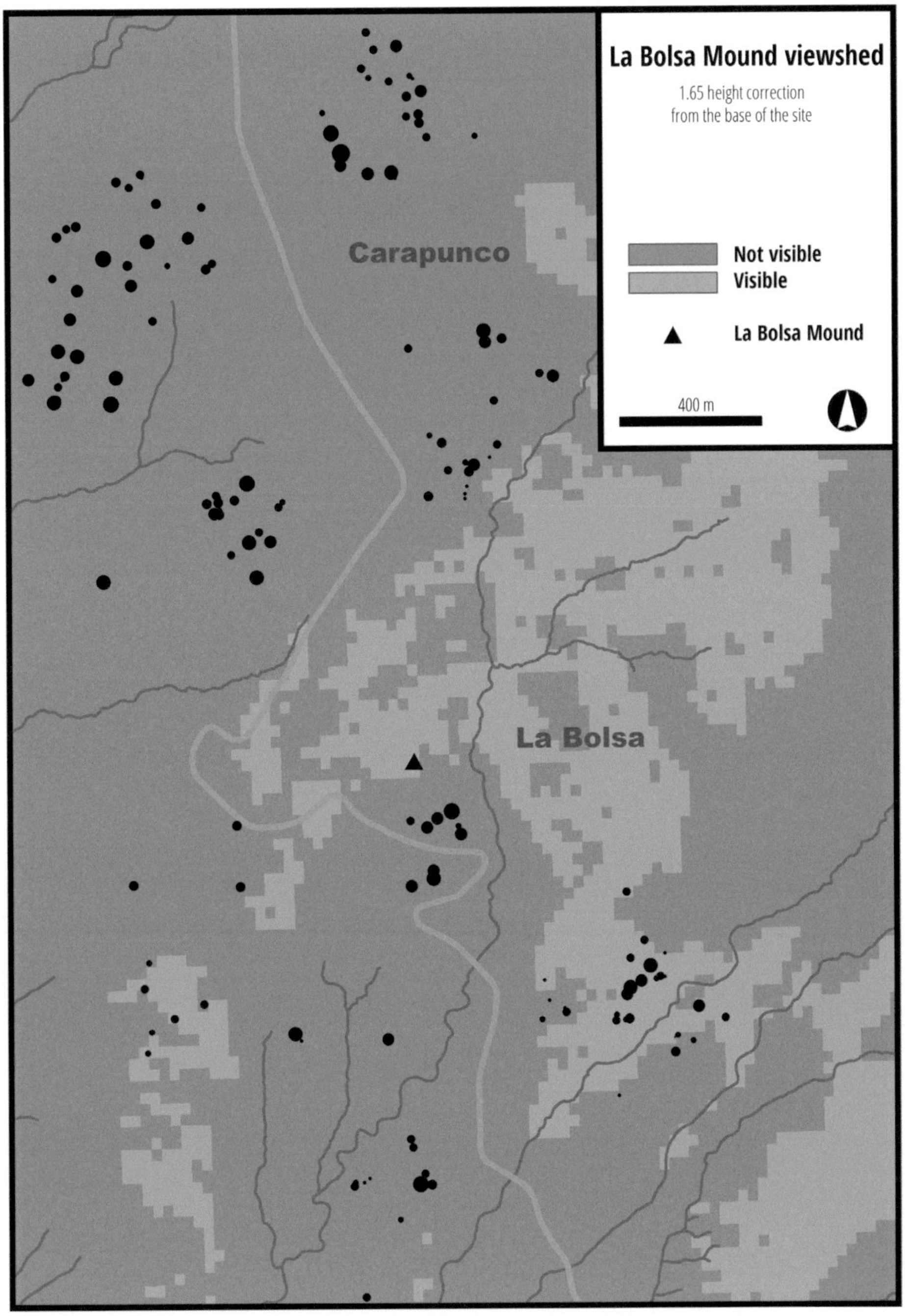

Figure 6: La Bolsa mound viewshed.

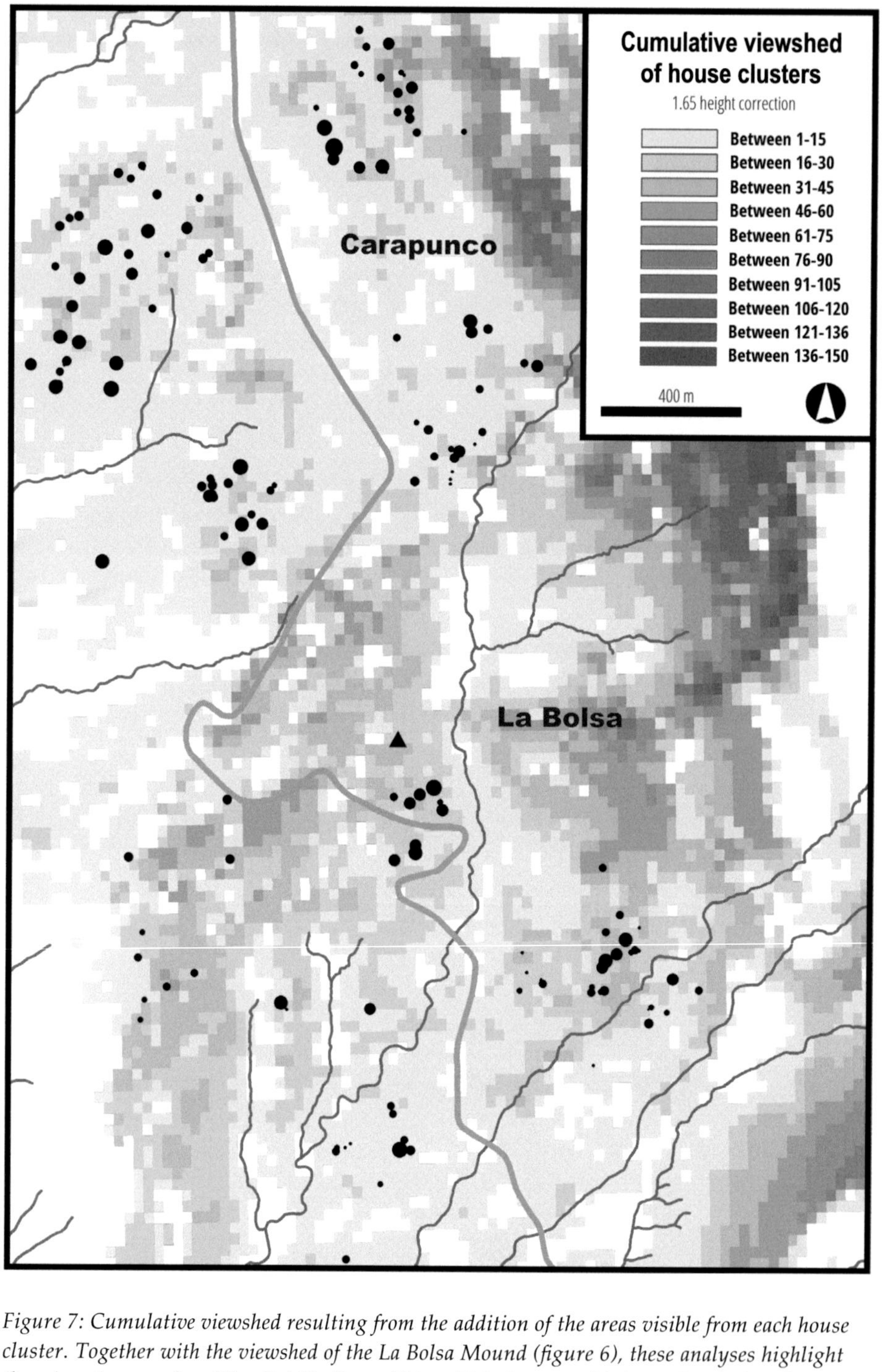

Figure 7: Cumulative viewshed resulting from the addition of the areas visible from each house cluster. Together with the viewshed of the La Bolsa Mound (figure 6), these analyses highlight the micro-topography of the slight hollow within which the mound is located and, therefore, its central but not dominant position in the archaeological landscape.

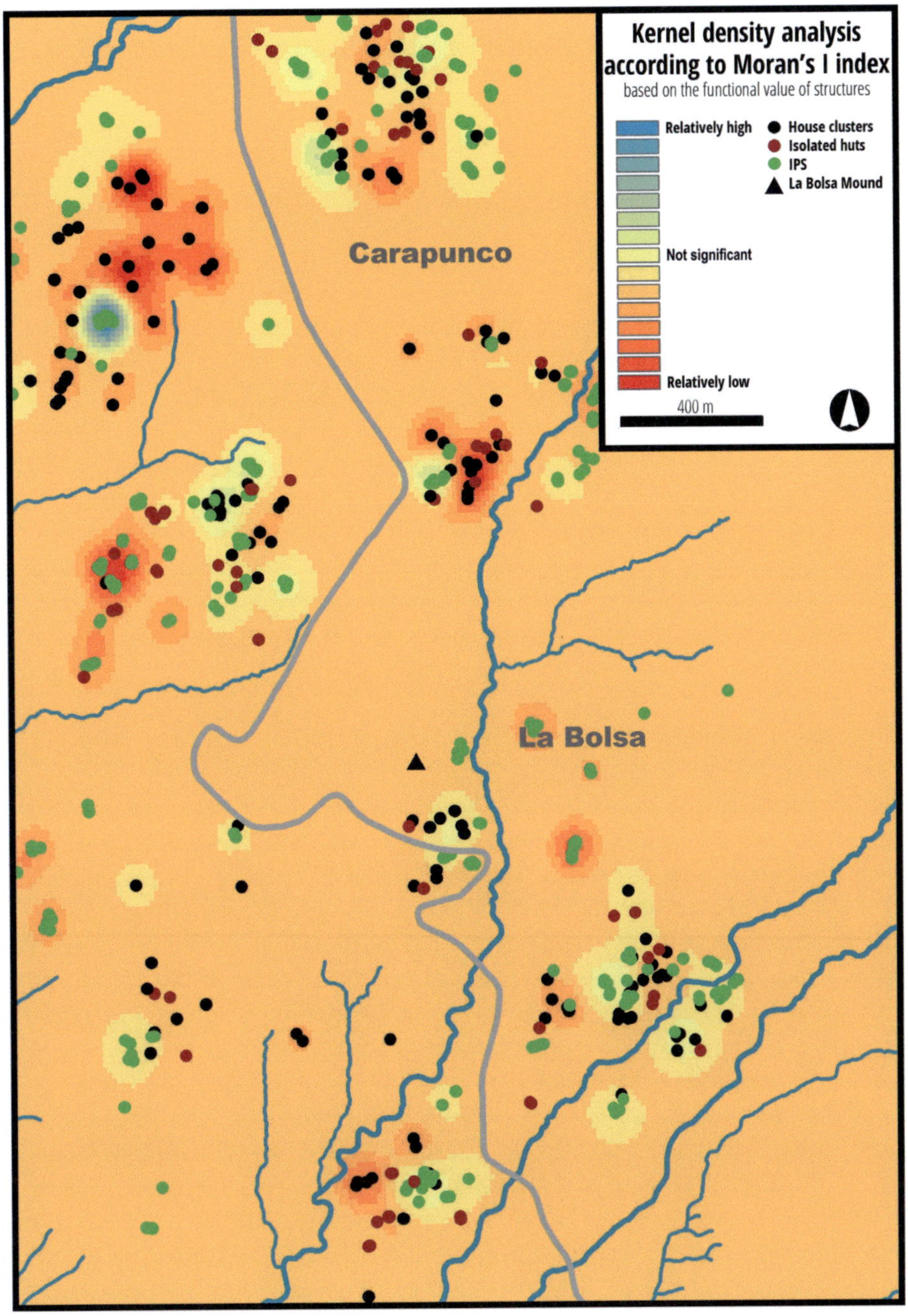

Figure 8: Kernel density analysis weighted by Moran's I index based on functional adscription; This can be compared with figure 3 in order to crosscheck the sharpest edges shown at Carapunco 2 with contemporary ploughed areas.

Some further remarks stand out from a Moran's I analysis regarding the functional adscription of the whole sample, also including IPSs and IHs (figure 8). As is known, the goal of this index is to set hot and cold points where a cluster of entities presents significant statistical variations in the spatial distribution of a given trait; the processing of the results by means of a density diagram shows those edges quite clearly. In fact, due to their relative softness, inhabited nucleations appear as more or less fuzzy spatial entities with permeable borders and even a low centripetal tendency towards themselves. It is certain that there is an exception at Carapunco 2 (to the northwest of the road) but if we take into account the fact that it is a ploughed area today, we can convincingly explain the lack of small units between the bigger and massive HCs, which are harder to remove by ploughing.

Finally, faced with the absence of a functional discretionality or any other maximal structuring focal points, the Formative landscape of Tafí (or, at least, the northern edge of the valley) seems to be something else besides "centrifugal": it is "continuous". In a cultural environment such as this, each household arises as the main central spot for the practices of its inhabitants, and probably also their self-representations: their *primary social identifiers*. Or at the least, it arises as the best defined and, therefore, the strongest one despite the *feasible possibility of some kind of middle-range grouping* which may be coherent with the nucleation instance, to some extent and degree, which, for sure, is still undefinable.

Discussion: Centrifugal landscape, continuous landscape and savage polities

Having reached this point, it is time to resume the thread started in the introduction by turning towards the ethnographic record of the neighbouring Amazonian lowlands and beyond. If we take for granted the fact that all human groups live, and have always lived, as a distinctive trait of cultural-semiotical gregarious animals, according to structuring-structured institutions, the government of stateless societies has traditionally been a major concern in both Anthropology and Prehistory. It is quite notable that the polysemic sense of "government" may imply here a biased interpretation. It refers to *the ruling faction* of a society, *the powerful*, although it also refers to the *system of political power* and this does not necessarily entail the existence of such kinds of factions. This is where Clastres comes in. In *Society against the State* he wondered about the inner logic of the so-called "savage society"; and the beginning of the answer could not be more suggestive: "power is exactly what these societies intended it to be. And as this power is -to put it schematically- nothing, the group thereby reveals its radical rejection of authority, an utter negation of power" (Clastres 1989, 43). Perhaps it is a somewhat naive starting point, but it had been dangerously forgotten until then.

First of all, his approach meant a radical development of Sahlins' intuitions regarding the under-productiveness of the Domestic Mode of Production, and hence the dismissal of every hypothesis relying primarily upon economic causality (Clastres 1994, 105-118; Sahlins 1972). Indeed, even the chieftainship models suggested in "Poor man, rich man, big man, chief" were put under quarantine by its own author due to the reasonably expected interferences in the indigenous political behaviour once they had been colonised by Europeans (Sahlins 1963,

291n13). Indeed, later Melanesia ethnographers and historians have stressed other sources of "power" as the backbone of politics, especially those exhibiting a given agent in a more or less specific *transcendence towards a sphere of significance beyond the cultural idea for "humanity"* (*cf.* Needham 1976; Dureau 2000; Lindstrom 1984; 1990; Görlich 1999), with certain economic practises being its epiphenomena, particularly those related with an *out-of-human-logic scale of consumption* (*cf.* Bataille 1988). All of this is entirely coherent with Clastrean statements on Amazonian chieftaincies, namely that their political sphere is arranged in the knowledge that "power", in other words, the hazy social attribution referred to in our culture as "power" is essentially a manifestation of natural, non-human agency, the same that gives birth to the universe and, therefore, to the laws within which the community-of-the-humans lives.

"Hence, far from giving us the lacklustre image of an inability to resolve the question of political power, these societies astonish us by the subtlety with which they have posed and settled the question. They had a very early premonition that power's transcendence conceals a mortal risk for the group, that the principle of an authority which is external and the creator of its own legality is a challenge to culture itself. [So] on discovering the great affinity of power and nature, as the twofold limitation of the domain of culture, Indian societies were able to create a means for neutralizing the virulence of political authority". (Clastres 1989, 44-45)

In his eyes, those means consist mainly of maintaining the chief completely overwhelmed, crushed in his social intercourse within the community, while outside it, a fierce war is carried on against the "other", according to the aforementioned centrifugal logic.

Maybe, if Clastres had not passed away prematurely, he would have deepened his theoretical formulation in order to resolve some voids and imbalances, which have been fairly pointed out. For example, Criado (2014) points out that, though inspiring, Clastres's toolkit is still somewhat vague for any direct analytical application. In my opinion, this is completely true, and is rooted in a poor systematisation of the dynamics woven between "power" and "authority", two quite different phenomena or, again, in the distinction between "dominancy" as put by Ethnology, and "domination" as a structured, coercive mutation in the intersection of authority and violence. What is more, perhaps driven by the "typological thought" denounced by Nielsen (1995), Clastrean statements have, too frequently, been taken as holistic social types instead of logics weaving the social tissue. Notwithstanding this, there is a lesson to be learned from Clastres's anarchist Anthropology: having paid attention to what savage societies tell us by themselves, there is a kind of political behaviour which cannot be properly termed; it is *neither prestate, nor stateless, but counterstate politics.*

However, what can this tell us about Tafí?

From my point of view, it is quite evident that a human group living under such demographic pressure simply cannot do without some kind of "rule of law". In fact, the community-of-the-humans is, above all, a moral communion found in mutual self-identification. This is why Gluckman (1973, 230-231) clarifies Evans-Pritchard's Nuer paradox by stating that *if anarchy is ordered, it is because there is law.* This is a opoint easier to express in languages which make a distinction herein, as does Latin, between *ius-lex* or the like. However, this does not entail the cultural recognition of any social agency invested with the power to create and manipulate this law. Recently, it has been recalled, after Nielsen, that "domestic architecture is one of the most effective means of spreading messages regarding identity, so it is actively handled in the negotiation of power" (Oliszewski *et al.* 2013, 42-43). Despite the still undefined use of "power" (which can probably be better explained in a wider sense, as political intercourse), for Oliszewski's team the more or less homogeneous materiality throughout the Tafí Valley and its adjacent areas may imply some sense of community beyond the local nucleations. This hardly means that all Tafí-pattern house clusters and settlements conformed to a single polity, but it gives a clue on the area, where we can expect similar kinds of political behaviour.

In fact, evidence recorded at La Bolsa-Carapunco fits in well with a Clastrean scenario. There is a proliferation of multiple self-centred, well-defined focal points, quite similar to one another, in which daily life took place. At the same time, these places acted as identifiers as is suggested by the management of burials within their patios. Consequently, house clusters seem to perform, at the same time, *the minimal and the main political role*, shaping a more or less continuous landscape in which no sharp-edges are noticeable. This in spite of the strong possibility of the existence of some instance of intermediate social grouping. If so, however, it seems to be beyond doubt that "village" and "segment" would be excessively heavy labels, at least until further archaeological investigation sheds more light on the precise character behind this more or less vague centripetal dynamic. However, at the same time, it is important to bear in mind that the actual extent of the statism of such social institutions is far from our all too frequently sclerotic views. Especially since, even in such classic cases as *The Nuer*, what is emphasised by a clear-eyed ethnographic observation is the *strategic processes of identification within a set of shared cognitive values*. Hence, "political relations are relative and dynamic. They are best stated as tendencies to conform to certain values in certain situations and the value is determined by the structural relationships of the persons who compose the situation" (Evans-Pritchard 1940, 137; *cf.* Karp and Maynard 1983).

In one way or another, it is here that the La Bolsa Mound, the uniqueness of which is self-evident, arises in a very different central role. As, without a doubt, was the case of Casas Viejas, the mound and the fields around it appear better defined as a scenario in which the collective ties are dramatized, bonding a political community which reproduces a counterstate logic while also securing the integration of a growing population; that is to say, one which enlarges, beyond Clastres's case studies, the self-identifying group within which some kind of human law is recognised and arbitrary violence is discouraged, albeit not at the cost of increasing domination. Hence the restraint noted by Berberián and Nielsen. Even though the La Bolsa Mound was also an arena in which internal tensions are shown, negotiated, or simply fought, they would have run on behalf of dominancy within a framework of fluid power and non-coercive forms of authority. This, in turn, gives back the full adaptive sense to Tafí polities from a point of view which is coherent with the objectives of the inhabitants of this landscape, rather than with a historical teleology towards the State. All things considered, perhaps it is time to start analysing the emergence of the State as an accident, as the blockage of the political fluidity during some kind of environmental cataclysm; and the so-called failed processes as the strictly normal aftermath for the cultural community-of-the-humans.

Acknowledgements

Expressing my gratitude to the teachers and organisers of the CSIC's courses on Geographic Information Technologies applied to Archaeology is far from being a formality but a genuine acknowledgement of their work and dedication. I am also indebted to John Gooseman and Julián Salazar for their inestimable remarks on a previous draft, though responsibility for the opinions and mistakes contained herein is entirely my own. Finally, this research was partially supported by a Santander Universities grant (Becas Iberoamérica-Jóvenes Profesores e Investigadores 2013), which allowed me to travel to Argentina in order to spend several months working with the archaeological team of the UNC. My final appreciation is for all of them: ¡va muro!

References

Álvarez Larraín, A. and Lanzelotti, S.L. 2013. Habitar y cultivar en el este del valle de Yocavil, in: Gordillo, I. and Vaquer, J.M. (eds.). *La espcialidad en arqueología: Enfoques, métodos y aplicación.* Quito: Abya Yala, 151-190

Bataille, G. 1988. The accursed share: An essay on general economy. New York: Zone Books

Berberián, E. and Nielsen, A.E. 1988. *Sistemas de asentamiento prehispánicos en el valle de Tafí.* Córdoba: Comechingonia

Clastres, P. 1989. *Society against the State: Essays in Political Anthropology.* New York: Zone Books

Clastres, P. 1994. *Archaeology of violence.* New York: Semiotext(e)

Criado Boado, F. 2014. Clastres: Ayer, hoy, siempre, in: Campagno, M. (ed.). *Pierre Clastres y las sociedades antiguas.* Buenos Aires: Miño y Dávila, 37-64

Cruz, P. 2007. Hombres complejos y señores simples: Reflexiones en torno a los modelos de organización social desde la arqueología del valle de Ambato (Catamarca), in: Nielsen, A.E., Rivolta, M.C., Seldes, V., Vázquez, M.M. and Mercolli, P. (comps). *Procesos sociales prehispánicos en el sur andino: La vivienda, la comunidad y el territorio.* Córdoba: Brujas, 99-122

Di Lullo, E. 2012. La casa y el campo en la Quebrada de Los Corrales (El Infiernillo, Tucumán): Reflexiones sobre la espacialidad en el primer milenio d.C. *Comechingonia* 16/1, 85-104

Dureau, C. 2000. Skulls, *mana* and causality. *The Journal of the Polyneasian Society* 109/1, 71-97

Duviols, P. 1979. Un symbolisme de l'occupation, de l'aménagement et de l'explotation de l'espace: Le monolithe *huanca* et sa fonction dans les Andes préhispaniques. *L'Homme* 19/2, 7-31

Evans-Pritchard, E.E. 1940. *The Nuer: A description of the modes of livelihood and political institutions of a Nilotic people.* Oxford: Clarendon Press

Franco Salvi, V. 2012. *Estructuración social y producción agrícola prehispánica durante el primer milenio d.C. en el valle de Tafí (Tucumán, Argentina)*. PhD Thesis. Córdoba: Universidad Nacional de Córdoba

Franco Salvi, V., Salazar, J. and Berberián, E. 2014. Paisajes persistentes, temporalidades múltiples y dispersión aldeana en el valle de Tafí (Provincia de Tucumán, Argentina). *Intersecciones en Antropología* 15, 307-322

García Azcárate, J. 1998. Monolitos-*huancas*: Un intento de explicación de las piedras de Tafí (República Argentina). *Chungara* 28/1-2, 159-174

Gluckman, M. 1973. *The judicial process among the Barotse of Northern Rhodesia (Zambia)*. Manchester: Manchester University Press

Godelier, M. 1999. *The enigma of the gift*. Chicago: University of Chicago Press

Godelier, M. 2014. *En el fundamento de las sociedades humanas: Lo que nos enseña la antropología*. Buenos Aires: Amorrortu

González, A.R. and Núñez Regueiro, V. 1962. Preliminary report on archaeological research in Tafí del Valle, N.W. Argentina, in: *Akten des 34 International Amerikanistenkongresses*. Vienna: Berger, 485-496

Gordillo, I. 2007. Detrás de las paredes: Arquitectura y espacios domésticos en el área de La Rinconada (Ambato, Catamarca), in: Nielsen, A.E., Rivolta, M.C., Seldes, V., Vázquez, M.M. and Mercolli, P. (comps). *Procesos sociales prehispánicos en el sur andino: La vivienda, la comunidad y el territorio*. Córdoba: Brujas, 65-98

Görlich, J. 1999. The transformation of violence in the colonial encounter: Intercultural discourses and practices in Papua New Guinea. *Ethnology* 38/2, 151-162

Gosden, C. 1999. *Anthropology and Archaeology: A changing relationship*. London: Routledge

Graeber, D. 2011: *Debt: The first 5000 years*. Brooklyn: Melville House

Hoebel, E.A. 2006: *The law of primitive man: A study in comparative legal dynamics*. Cambridge: Harvard University Press

Humphrey, C. and Hugh-Jones, S. 1992 (eds.): *Barter, exchange and value: An anthropological approach*. Cambridge: Cambridge University Press

Karp, I. and Maynard, K. 1983. Reading *The Nuer*. *Current Anthropology* 24/4, 481-503

Laguens, A. 2014. Cosas, personas y espacio social en el estudio de la desigualdad social: La trama de las relaciones en una sociedad diferenciada en la región andina de Argentina (ss. VI a X d.C.). *ArkeoGazte* 4, 127-146

Lindstrom, L. 1984. Doctor, lawyer, wise man, priest: Big-men and knowledge in Melanesia. *Man* 19/2, 291-309

Lindstrom, L. 1990. Big men as ancestors: Inspiration and copyrights on Tanna (Vanuatu). *Ethnology* 29/4, 313-326

Mair, L. 1962. *Primitive government*. Baltimore: Penguin Books

Needham, R. 1976. Skulls and causality. *Man* 11/1, 71-88

Nielsen, A.E. 1995. El pensamiento tipológico como obstáculo para la arqueología de los procesos de evolución en sociedades sin Estado. *Comechingonia* 8, 21-46

Oliszewski, N., Caria, M.A. and Martínez, J.G. 2013. Aportes a la arqueología del noroeste de Argentina: El caso de la Quebrada de Los Corrales (El Infiernillo, Tucumán). *MATS* 1, 26-56

Sahlins, M. 1963. Poor man, rich man, big man, chief: Political types in Melanesia and Polynesia. *Comparative Studies in Society and History* 5, 285-303

Sahlins, M. 1972. *Stone Age economics*. Chicago: Aldine-Atherton

Salazar, J. 2010. *Reproducción social doméstica y asentamientos residenciales entre el 200 y el 800 d.C. en el valle de Tafí, Provincia de Tucumán*. PhD Thesis. Córdoba: Universidad Nacional de Córdoba

Salazar, J., Franco Salvo, V., Berberián, E. and Clavero, S. 2007. Contextos domésticos del valle de Tafí, Tucumán, Argentina (200-1000 AD). *Werken* 10, 25-48

Sampietro, M.M. 2010. *Espacio, ambiente y los incios de la agricultura indígena en el Noroeste argentino: Un enfoque geoarqueológico*. Madrid: JAS

Sampietro, M.M. and Vattuone, M.A. 2005. Reconstruction of activity areas at a Formative household in Northwest Argentina. *Geoarchaeology* 20/4, 337-354

Tarragó, M.N. 1999. El Formativo y el surgimiento de la complejidad social, in Ledergerber-Crespo, P. (ed.). *Formativo sudamericano: Una revaluación*. Quito: Abya Yala, 302-313

Tartusi, M. and Núñez Regueiro, V. 1993. *Los centros ceremoniales del NOA*. Publicaciones del Instituto de Arqueología 5. San Miguel: Universidad Nacional de Tucumán

Todorov, T. 1998. *La conquista de América: El problema del otro*. Mexico: Siglo XX

Zuccarelli, V. 2012. *Arqueología de los paisajes agrarios surandinos: Aplicación de los SIG en el análisis de la problemática agraria en Catamarca Oriental, Argentina*. Saarbrücken: Editorial Académica Española.

Notes on contributor

Jordi A. López Lillo has an MA in Archaeology from the University of Alicante (UA), where he is currently working on his PhD on social theory focusing on anthropological reflections about the practical logics behind the historical emergence of the State. During this time, he has participated in research projects in Spain and Argentina, as well as co-directing the international exploratory workshop "Ζῷον ἐλεύθερον: Antropología política de las organizaciones sociales no estatales" in 2012. Currently, he is working as a researcher at the National University of Córdoba (UNC).

Landscapes of War. GIS applications in the study of the Spanish Civil War

Manuel Antonio Franco Fernández[1]

and Pedro Rodríguez Simón[2]

The Spanish Civil War of 1936 – 1939 has been a major field of historical research, producing a great number of studies, publications, and debates over the past few decades. And yet, the predominant approach to historical analysis has traditionally based itself almost exclusively on methods favouring written or graphic archival evidence. Oral accounts have only been considered useful resources for research over the past few years.

By comparison to this historiographical tradition, the study of material remains through archaeological methods has received less attention from academic researchers. While the archaeology of the contemporary past in general - particularly conflict archaeology - has been developing significantly in America and Europe since the late 1980s (Scott *et al.* 1989) and especially in the early 2000s (Buchli and Lucas 2001; Espenshade *et al.* 2002; Saunders 2001; 2002; 2007; Schofield 2005) in Spain we have had to wait well into the first decade of the twenty-first for this theoretical and methodological framework to be applied directly to the case of the Spanish Civil War (González-Ruibal 2007; 2016). Eventually, conflict archaeology has slowly made its way in our country and acquired more and more public and academic standing, although, as some authors have pointed out (Hernández 2007, 5; Rueda *et al.* 2015, 193-194), the study of the civil war has been hampered by the traumatic character of its content, and the weight of a tradition handed down by the Franco dictatorship. Both of these factors have caused rejection towards historical analysis, and approaches to the conflict as an uncomfortable topic, presented as a mere succession of military feats and battles.

---

1 antonio.franco.fernandez@gmail.com.
2 prsarqueologia@gmail.com.

Despite all the problems and ethical limitations (Moshenska 2008; González-Ruibal *et al.* 2014) conflict archaeology has managed to develop over the past few years. From a material perspective it has provided valuable data, relevant not only from a military point of view, but also for their social, political and anthropological implications.

The work we present here falls perfectly within this scope, and follows a research line born with a clear objective in mind: to complement and to contrast the information available from documents and oral accounts in order to document day-to-day micro-events and microhistories on the frontline and the rearguard. The intention of this project is to resituate historically those subaltern actors which lie neglected by military history, behind a fragmentary and humble archaeological record, but which is nonetheless capable of debunking myths and rediscovering hidden pasts. GIS tools provide a very useful resource for spatial analysis in war contexts, as they enable highly relevant data-gathering and the drawing of important conclusions. To illustrate this, in the first part of the present article we shall introduce the project which has been the backdrop of our work, and present the methodology followed by data-recording and analysis. Subsequently, we turn to some examples of work carried out so far to present a number of case-studies as examples of GIS based war heritage research.

Since the year 2006 a large team of researchers directed by Alfredo González-Ruibal, Incipit-CSIC) has been developing a wide range of research focused on the archaeology of the Spanish Civil War. Within this project, several different war and early postwar scenarios have been studied: frontline trenches, stable fortifications, military camps, concentration camps, fortresses, settlements erected by the relatives of political prisoners, fallen soldiers and mass graves. Meticulous archaeological methods-identical to those used on much older contexts- have been deployed in all of these sites.

The project's activities have been developed all over the Spanish territory, and amount to 12 archaeological interventions in 8 different locales: the forced labor camp of Bustarviejo (2006), the University City and the Casa de Campo park (2008, 2016) in Madrid, the Cueto de Castiltejón (2011) in La Fatarella, Tarragona, the Concentration Camp of Castuera, Badajoz (2011), the area surrounding the advance on the Alto Tajuña river around Abánades, Guadalajara (2010-2014) and the scenery of the Battle of Belchite, Zaragoza (2014-2015).

For all practical purposes, a battlefield constitutes an archaeological site, and it must therefore be studied using archaeological methods. Histories of violence leave strong material traces in any historical period, and, as sites, battlefields should, in all cases, be researched, preserved and disseminated. According to their age, battlefields are affected by such processes as natural degradation and anthropogenic alteration to different extents. But it is precisely in contemporary war contexts that materiality can reveal elements which have been neglected by written sources. Also, the modernisation and industrialisation of contemporary conflicts -as is the case in many other fields- have increased the diversity and volume of material culture that appears in battlefields. This has generated archaeological contexts of far greater entity and complexity, as well as larger and more complex war landscapes.

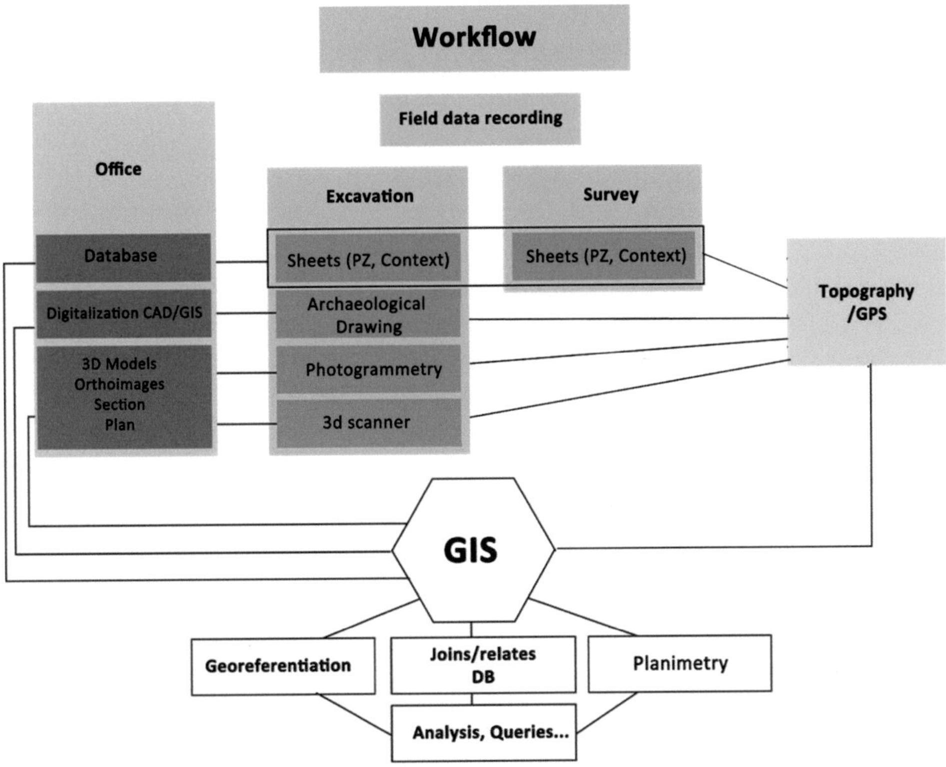

Figure 1. Workflow diagram used for data recording and management

The recording methods applied to this type of context are becoming increasingly standardized (Bellón *et al.* 2015; Cárdenas 2015), systematized and widespread. The use of GIS programs is also becoming more and more frequent beyond mere recording, and are now common tools of spatial analysis in conflict archaeology. Several works have highlighted the potential of these GIS tools for the study and analysis of war contexts in different periods, from Antiquity (Bellón *et al.* 2013; 2015; Cárdenas 2015; Rubio 2007; Rueda *et al.* 2015), to the Early Modern Ages (Rubio 2008; Rubio *et al.* 2011a) and, of course, the contemporary war landscapes of the Spanish Civil war themselves (Rubio and Hernández 2015; Rubio *et al.* 2011b).

Based on our own experience and building on some of these references, our project has allowed us to develop a practice-based and specific workflow model (figure 2) which has proven highly useful in our fieldwork. Both at excavations and archaeological surveys (including micro surveying) any element present on the archaeological record is thoroughly recorded using GPS and total station. The use of metal detectors is crucial in locating and recording metallic elements, as well as in providing a key safety feature in the event of our locating unexploded ordnance. The positional recording of materials includes a standardised code description of each piece and its incorporation into databases. These are then supplemented with specific field files for further description and inventorying.

These database-correlated points constitute the main information input about materials found in the field. They are then used as the basis for analyses and studies as well as for the elaboration of specific cartography. Referenced into XYZ-type coordinates, the points are then correlated to other vector elements (such as lines, or polygons which are referenced with a total station or GPS to record negative structures or units). Negative contexts and structures, are increasingly being systematically documented using close-range *Sfm* photogrammetry (Agisoft-Photoscan), which is then processed (via Blender) to obtain 3D models, specific orthophotos, or digital elevation models. In some cases 3D-scanners have also been used.

Other inputs and resources have been used as well as the field data. These, based on IDEE (the Spanish acronym for Spatial Data Infrastructure in Spain), as part of the European Union *Inspire* directive, come mainly in the form of WMS servers of the IGN (the Spanish National Geographic Institute) and its download centre. Within this public database, special use has been made of orthophotos from the PNOA (the Spanish acronym for the National Plan of Air Ortho-Photography) including both present-day and historical photographs, National Topographic (MTN25) raster maps, National Topographic Databases (such as BTN25) vector graphics, LIDAR data or 5-metre-grid Digital Terrain Models (DTM 05).

The main results (outputs) or products of this work have allowed us to develop specific 2-D and 3-D cartography (including photogrammetric topographic surveys- which allow virtual reconstructions and recreations) with databases associated to both artefacts and structures. Their relevance as GIS tools lies in the fact that they enable the management, understanding and analysis of research in conflict studies. Visibility studies, density studies, slope studies and cost-analyses have proven particularly useful for a better spatial understanding of the battlefields themselves, of the war action that took place on them, and therefore, of the way in which their protagonists experienced them.

Below is a series of examples taken from some of the sites to illustrate the application of this type of tools.

The Battle of the Ebro was the greatest and most decisive battle of the entire Spanish Civil War. In the summer of 1938, the Republican government, having suffered numerous defeats and its territory broken up into two separate parts, planned an ambitious counter-offensive to regain part of the territory which had been lost since March that year. Beginning on the 25th of July of 1938, the battle of the Ebro took place along a wide battlefield in the lower course of the Ebro river, between the provinces of Saragossa and Tarragona. Despite the Republicans' initial success and advances, the material superiority of the Franco army caused the campaign to fail, and the fighting to come to an end altogether by the 16th of November 1938.

These final confrontations took place precisely in the surroundings of the small municipality of La Fatarella, where part of the last Republican defensive line is still standing. In order to study this context, in September 2011 a joint team of researchers from DIDPATRI (University of Barcelona), Incipit (CSIC), BSC-CNS (Barcelona Supercomputing Centre) and the local association *Lo Riu* worked in several sites in the Raïmats area, excavating trench lines and forts. As a result, it was possible to interpret the Francoist assault and the last moments of the republican defence of the sector (González-Ruibal 2012a).

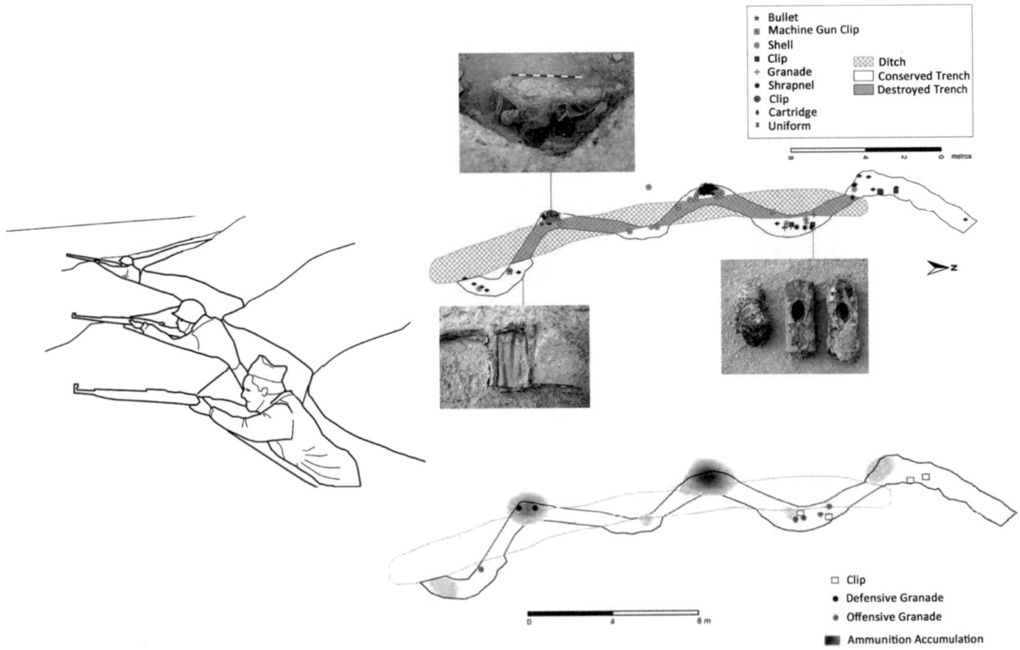

Figure 2. Trench at La Fatarella. Density of the remains and interpretation.

One of the focus areas of the research was the trench line of Hill 562: a tract of some 30 meters oriented north-south was exvcavated, of which six zigzag vertices were preserved. Although the trench had been partially destroyed by an agricultural ditch, the abundant material that was recovered (cartridges, shell casings, clips and grenades) and the careful recording of their pattern of distribution, permitted a detailed reconstruction of the events. By applying a point density analysis of the elements documented (figure 3), it has been possible to interpret the trench's three firing points: the three vertices pointing (westward) towards the enemy. The archaeological data also reveal that the trench was not very densely occupied, with only one soldier per shooting position. These analyses illustrate the early moments of the battle quite eloquently, as well as the Republican troops' final resistance. They were eventually forced out of their position by the Francoists. Attesting to these final episodes of resistance lay the body of a Republican soldier, who was exhumed where he had fallen in one of the firing positions.

This type of dispersion-based and density-based analysis of the materials documented has been applied to many of the contexts excavated or surveyed at other sites. In the case of Abánades (Guadalajara), the most continuously-excavated battlefield (2010 to 2014) allowed a reconstruction of the fighting in several of the excavated positions. (González-Ruibal 2010; 2012b; 2014; 2016b).

The offensive on the high course of the Tajuña river (*ofensiva del Alto Tajuña*) was a Republican manoeuvre devised to relieve the pressure from the Eastern Army, at the time undergoing the harsh Francoist offensive on Aragon. The western sector of the offensive, the area around Abánades, played a prominent role, with such positions as the Cerro del Castillo or the EnebráSocarrá witnessing heavy combat until the 24th of April, when the campaign came to an end.

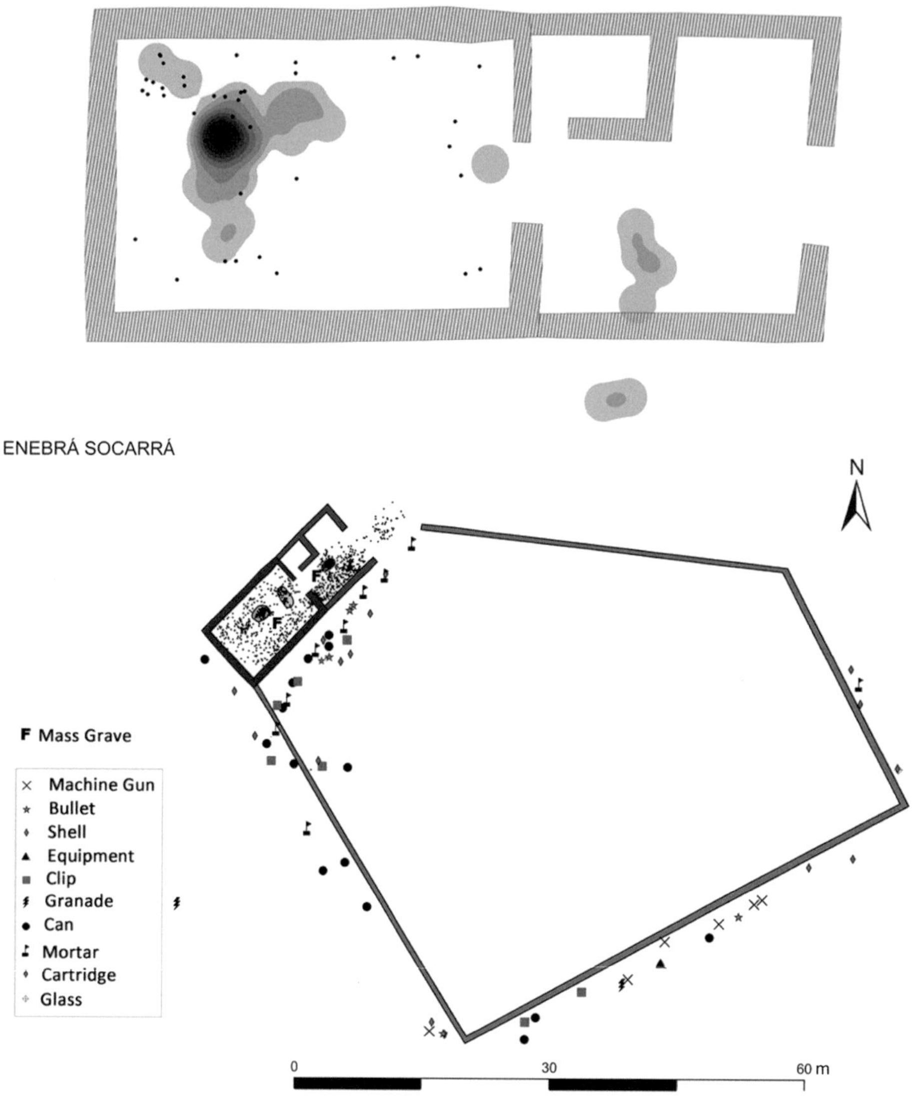

Figure 3. EnebráSocarrá. General distribution of materials and density of bullets and clips.

In the case of the EnebráSocarrá, a material dispersion analysis (figure 4) has allowed a precise interpretation of combats which do not appear in written sources. The Francoist troops, having fortified a former sheep pen, succeeded in defending their position, equipped mainly with 7.92-mm German Mauser rifles and an 81-mm mortar. 7.92 mm ammunition amounts to 77% of the total attributable to Francoist army rifles. The intense fire coming from them is attested by the 60 clips yielded by the sheep pen and the immediately surrounding areas. As for the attackers, the high number of shrapnel from artillery projectiles (145) collected in all the area, especially inside the sheep pen indicates the massive use of artillery and tank fire to take on the position.

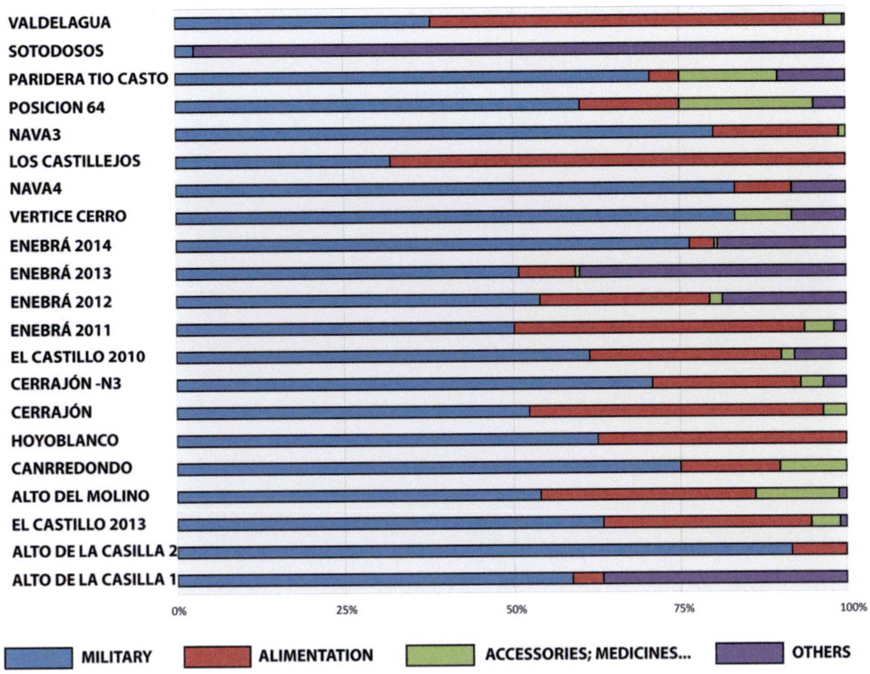

Figure 4. Statistics of the materials recorded at positions in Abánades.

Detailed inventorying of all the elements in each of the excavated positions has enabled statistical studies (Figure 4), both quantitative and qualitative, which allow us to obtain comparative data about the materials documented (military apparel, constructive elements, food, medicine and personal objects). These have allowed us to draw conclusions about the differences between sides, unit positioning (offensive or rear-guard), etc. by considering the continuity of occupation at each site.

In the case of the Cerro del Castillo in Abánades, the Francoist army's most advanced position in the sector, established after the offensive, we have been able to carry out studies of visibility, strategic territorial control and weapon efficiency (Figure 5).

This position, a hill fortified with bunkers and trenches, allowed long-distance control of troops, as well as granting access to the village of Abánades. The figure shows the visibility from two of the loopholes on the side of one of the south-slope bunkers on the hill. Using GIS tools (like visibility from the observer's point of view) which adapt the visibility analysis to the point of view of shooters placed at these loopholes, we have been able to ascertain how, in this case, the loopholes-observation and shooting points- covered the SW and S routes of access to the village. This sector was of particular strategic importance, as it granted access to the bridge over the Tajuña river. The bridge was a key element in controlling the movement of troops in the area, and it is precisely the overlapping point of both visibility areas, which indicates the effectiveness of the defensive system put in place.

The application of 500-metre and 1000-metre buffers, which represent the mean effective distance of rifles and automatic weapons respectively, has allowed us to represent graphically how this point was covered by the fire range of the weapons defending the Cerro del Castillo position. This provesthat this type of analysis enables the graphic, metric and objective representation of the way in which different positions were established according to their spatial characteristics and their military interest. Their positions thus reveal criteria beyond the mere subjective perception of their being built on elevated or protected places.

During 2014 and 2015 this team has carried out two archaeological campaigns in the surroundings of Belchite, a small town in Zaragoza province, as part of the *International Brigades Archaeology Project* (IBAP) (González-Ruibal *et al.* 2015). This project seeks to both study the material remains of the international brigades in the Spanish Civil War and develop an archaeological approach to the famous Battle of Belchite, fought in the summer of 1937.

Most of the archaeological works took place in the surroundings of Belchite town. During the 2014 field seasonarchaeological workwas carried out at one of the most important and best-preserved vestiges in the area: the sheepfold- *paridera-* of El Saso. Both defensive structures and part of the fortified sheep fold which served as the epicentre of life in the military compound. In the case of the latter, two main processes could be analysed. First, the evidence of its use by the military garrison. The study of artefact dispersion (food remains, glass, coins or writing material), enabled a study of the areas where soldiers spent their day-to-day and the reconstruction of their diet (lamb stew, dry fish and roasted barley tea in lieu of coffee). We were

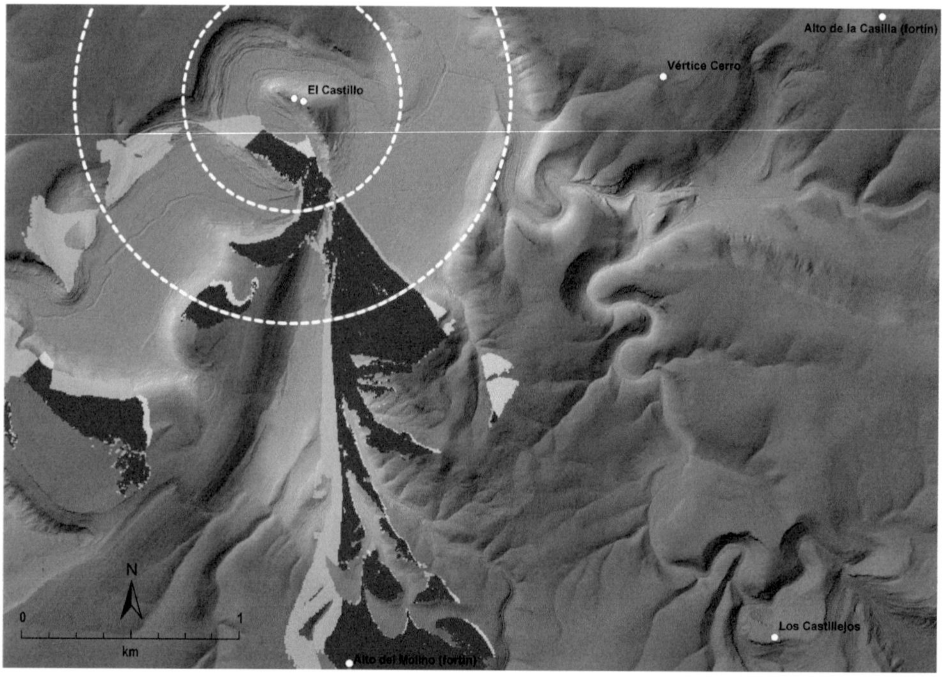

Figure 5. Ranges and visibility between positions from the fortified position of El Castillo.

also able to document the structure's process of destruction. Inside the sheep fold we found a great number of shrapnel fragments, clearly more densely concentrated towards the centre of the excavated area. Density analyses of these elements allowed us to locate the centre of this concentration, which, significantly, coincided with a destroyed pillar. It was possible, through the analysis of shrapnel dispersion, to interpret the pattern as the impact of an artillery grenade which had impacted and destroyed the pillar sustaining the roof, thus causing its collapse.

During the 2015 campaign we were able to excavate one of the forts defending the southern sector of the village. Again, dispersion analysis of the materials recorded indicated the existence of an activity are anext to the access, where a large fire may have existed for cooking or heating purposes. Also, the presence of specific materials, especially pistol shell casings, enabled a reconstruction of military action, indicating that the fort may have been taken by assault and that the fighting took place at a short distance inside the structure itself. Bullet holes documented at several points on the inside of the fortification also attest to close-quarters combat.

This fort was part of a strategic defensive system which included another three twin buildings and an important second-line stronghold which visually controlled both the defensive line and the transit along one of the main roads leading to Belchite. The defensive system's outline contemplated coordinating fire from these fortified posts, supporting each other in a crossfire. However, practice did not meet the expectations of the theoretical planning, as this fortified system fell rapidly at the start of the fighting in 1937.

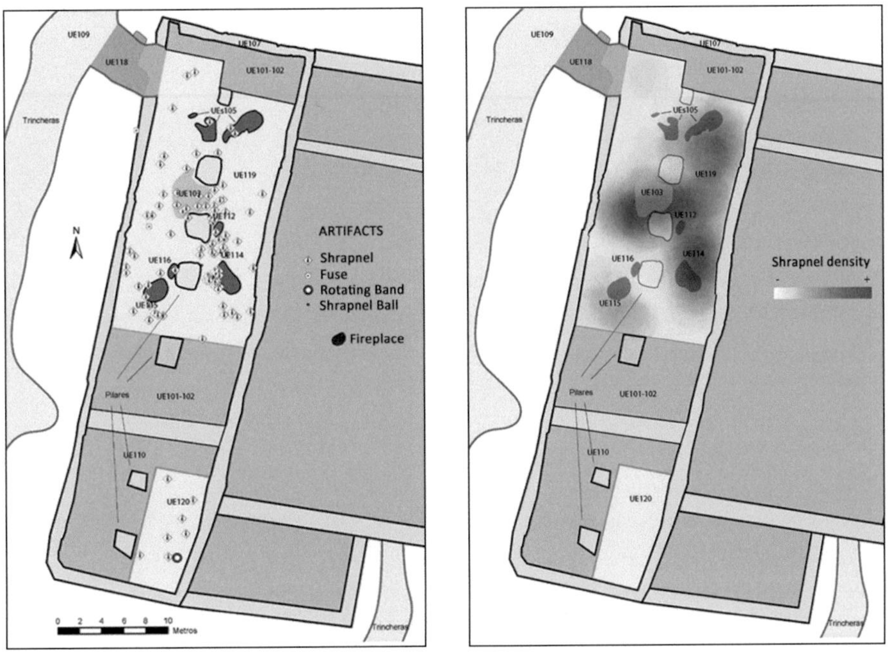

Figure 6. Artefact distribution and shrapnel density inside the lambing shed at El Saso (Belchite)

A site where we have been able to document the traces of the International Brigades is the surroundings of Mediana de Aragón. The Francoist attempt at a counterattack initiated on the 31st of August forced Republican commanders to send forces to the frontline. The containment of this advance was entrusted to the XI and XV International Brigades.

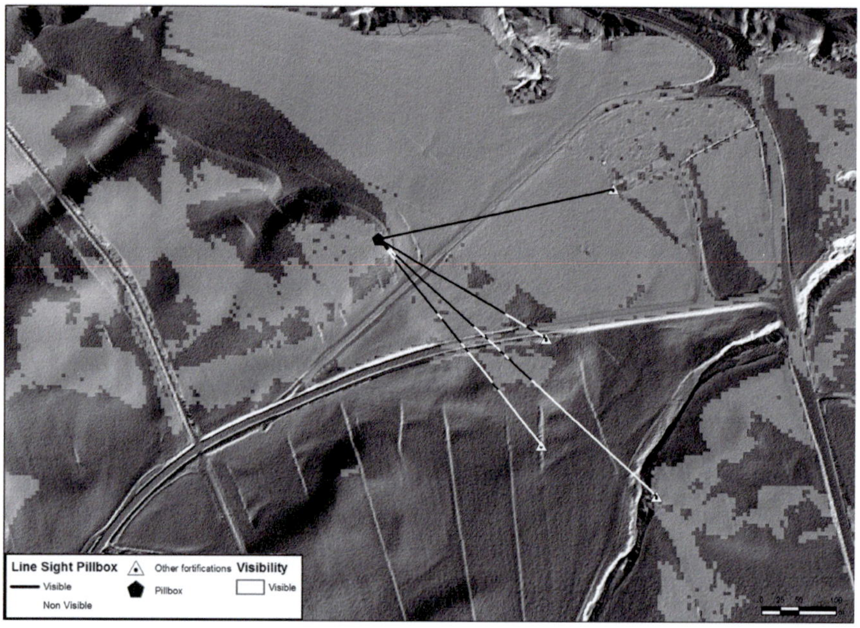

Figure 7. Visibility between fortifications at the seminar sector (Belchite)

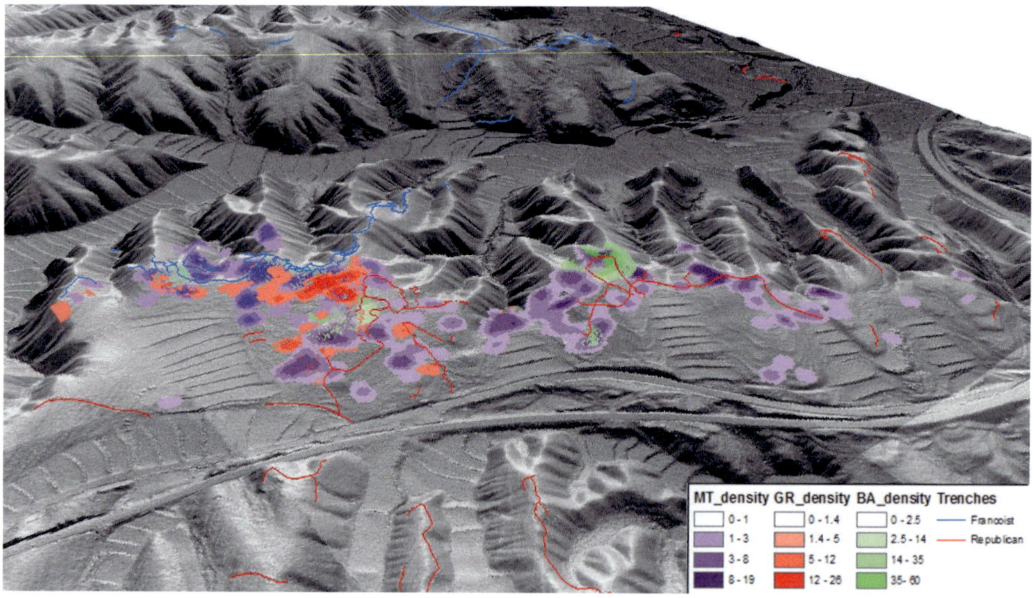

Figura 8. Shrapnel and grenade density at the "Parapet of death"

The scenario of this fighting at Medianais today a perfectly-preserved battlefield. Kilometres of trenches and positions on both sides are scattered along hectares of an abrupt and complicated terrain. The place chosen for the 2015 campaign was the so-called "Parapet of death", so named because it was the Francoist army's furthermost post, where the hardest fighting took place. This sector was surveyed intensively along the frontlines, allowing us to obtain useful information about how the combats developed.

During the survey, over 1800 elements of all kinds were been found, including large amounts of war material (especially remains of grenades, shrapnel and bullets), which indicate the characteristics and crudeness of the fighting.

Spatial and dispersion analyses have allowed us to identify and delimitate the positions of both sides: the Francoist forces occupied the higher and more visually advantageous parts, while the Republicans had established their last defence line before the road between Belchite and Mediana, which is connected through deep communication trenches with the firing positions closer to the enemy. In some cases, the distance between both trench lines was no greater than 40 metres. It is precisely at these closer points where higher concentrations of shrapnel (most of it from mortar rounds) and hand-grenade fragments were documented, eloquently represented by the data from the density analyses for these types (Figure 10) as opposed to surprisingly scarce ammunition, indicating that combats were very intensive and crudely pragmatic. Traces of hand grenades appear where combats were particularly violent.

References

Bellón Ruiz, J.P., Gómez Cabeza, F., Ruiz Rodriguez, A., Molinos Molinos, M., Rueda Galán, C., Lechuga Chica, M.A. and Pérez Cano, F. 2015, Una metodología arqueológica para el estudio de campos de batalla, in: J.P. Bellón, A, Ruiz, M. Molinos, C. Rueda and F. Gómez (eds.). *La Segunda Guerra Púnica en la Península Ibérica. Baecula, arqueología de una batalla*, 233-260.

Bellón, J.P., Rueda, C., Osanna, M. and Ruiz, A. 2013, *Numistro: de loco ad pugnameligendo*. Primeros resultados del análisis arqueológico de una batalla de la Segunda Guerra Púnica en Lucania. *Siris*, 13, 91-115.

Bleed, P. and Scott, D. D. 2011, Contexts for conflict: conceptual tools for interpreting archaeological reflections of warfare, *Journal of Conflict Archaeology* 6, 42-64.

Buchli, V. and Lucas, G. (eds.). 2001. *Archaeologies of the Contemporary Past.* London and New York: Routledge.

Cárdenas Anguita, I. 2015. Análisis SIG de un escenario arqueológico de batalla, in J.P. Bellón, A, Ruiz, M. Molinos, C. Rueda and F. Gómez (eds.). *La Segunda Guerra Púnica en la Península Ibérica. Baecula, arqueología de una batalla,* 261-275.

Carman, J., 2013. A Heritage of Conflict, and conflict of Heritage, in: S. Bergerbrant. and S. Sabatini (eds.). *Counterpoint: Essays in archaeology and heritage studies in honour of Professor Kristian Kristiansen*, 747-751. BAR International Series 2508.

Desfossés, Y. and Jacques, A. 2002, Vers une definition et une reconnaissance de l'archéologie de la Première Guerre mondiale, *Actes des colloquies 'La Bataille en Picardie, combattre de l'Antiquité au XXème siècle'*, Amiens. 203-220.

Espenshade, C. T., Jolley, R. L., & Legg, J. B. (2002). The value and treatment of Civil War military sites. *North American Archaeologist* 23(1), 39-67.

Hernández Cardona, F.X., 2007. Presentación de la monografía "Campos de Batalla, espacios de guerra", *Iber, Didactica de las Ciencias Sociales, Geografia e Historia* 51. 5-6.

Falquina Aparicio, A., Rolland Calvo, J., Marín Suárez, C., Compañy, G. and González-Ruibal, A. 2010. De estos cueros sacaré buenos látigos. Tecnologías de represión en el Destacamento Penal franquista de Bustarviejo (Madrid). *Ebre 38. Revista Internacional de la Guerra Civil (1936-1939)* 5. 247-271.

González-Ruibal, A. 2007. Making things public: Archaeologies of the Spanish Civil War. *Public Archaeology* 6(4). 203-226.

González-Ruibal, A. 2010. *Arqueología de la Guerra Civil Española en el Frente de Guadalajara. Informe de las excavaciones arqueológicas en los restos de la Guerra Civil en El Castillo de Abánades. Campaña de 2010.* Incipit-CSIC. Technical Report. Available at: http://digital.csic.es/handle/10261/29654

González-Ruibal, A. 2011. Digging Franco's Trenches: An Archaeological Investigation of a Nationalist Position from the Spanish Civil War. *Journal of Conflict Archaeology* 6(2). 97-123.

González-Ruibal, A. 2012a. *El último día de la batalla del Ebro. Informe de las excavaciones arqueológicas en los restos de la guerra civil de Raïmats, La Fatarella (Tarragona).Incipit-CSIC.* Technical Report. Available at: http://hdl.handle.net/10261/47780

González-Ruibal, A. 2012b. *Informe de las excavaciones arqueológicas en los restos de la Guerra Civil de Alto del Molino Abánades (Guadalajara). Campaña de 2011. Incipit-CSIC. Technical Report.* Available at:http://digital.csic.es/handle/10261/49097

González-Ruibal, A. 2013. *Arqueología de la Batalla Olvidada. Informe de las excavaciones en los restos de la Guerra Civil de Abánades (Guadalajara). Campaña de 2012. Incipit-CSIC. Technical Report.* Available at: http://digital.csic.es/handle/10261/81034

González-Ruibal, A. 2014. *Arqueología de la Guerra Civil en el término municipal de Abánades (Guadalajara). Campaña de 2013: El Castillo y La EnebráSocarrá. Incipit-CSIC. TechnicalReport.* Available at: http://digital.csic.es/handle/10261/101689

González-Ruibal, A. 2016a. *Volver a las Trincheras.* Madrid: Alianza.

González-Ruibal, A. 2016b. *Arqueología de la Guerra Civil en el Valle del Tajuña (Guadalajara). Campaña de 2014.* Incipit-CSIC. TechnicalReportAvailable at: http://digital.csic.es/handle/10261/129847

González-Ruibal, A., Ayán Vila, X., Caesar, R. 2014. Ethics, Archaeology, and Civil Conflict: The Case of Spain, in A. Gonzalez Ruibal and G. Moshenska (eds.). *Ethics and the Archaeology of Violence,* Volume 2 of the series Ethical Archaeologies: The Politics of Social Justice, 113-136.

González-Ruibal, A., Franco Fernández, A., Falquina Aparicio, A., Fernández Blancafort, I., Laíño Piñeiro, A., and Martín Hidalgo, P. 2010. Excavaciones arqueológicas en el Frente de Guadalajara: Una posición franquista en Abánades (1937-1939). *Ebre 38. RevistaInternacional de la Guerra Civil (1936-1939)* 5: 219-244. Barcelona.

González-Ruibal, A. and Marín Suárez, C. 2010. Guerra en la Universidad: Arqueología del conflicto en la Ciudad Universitaria de Madrid. *Ebre 38: revista internacional de la Guerra Civil, 1936-1939* 4: 123-143. Barcelona.

González-Ruibal, A., Rodríguez Simón, P. and Franco Fernandez, M.A. 2015. *Arqueología de la Batalla de Belchite. Campaña de 2014.* Incipit-CSIC. Technical Report. Available at: http://digital.csic.es/handle/10261/114184

Moshenska, G., 2008. Ethics and ethical critique in the archaeology of modern conflict. *Norwegian Archaeological Review* 41 (2), 159-175.

Quesada Sanz, F. 2008. La "Arqueología de los campos de batalla". Notas para un estado de la cuestión y una guía de investigación. *Saldvie* 8, 21-35.

Rubio Campillo, X., 2008. An archaeological study of Talamanca battlefield, *Journal of Conflict Archaeology* 4(1-2), 23-38.

Rubio Campillo, X., 2007. Campos de batalla de la antigüedad: el caso de Ilerda, *Iber: Didáctica de las Ciencias Sociales, Geografía e Historia* 51, 35-51.

Rubio Campillo, X., Cecilia Conesa, F. and Yubero Gómez, M. 2011a. Aplicación de nuevas técnicas de investigación en laarqueología del conflicto: la batalla de Talamanca (1714), in OrJIA (coord.). *II Jornadas de Jóvenes en Investigación Arqueológica*, 201-209.

Rubio Campillo, X. and Hernández Cardona, F.X. 2015. Combined arms warfare in the Spanish Civil War: the assault on the Republican defence line at Fatarella Ridge, *Journal of Conflict Archaeology* 10-1, 52-69

Rubio Campillo, X., Rojo Ariza, M.C., Yubero Gómez, M. and Cardona Gómez, G. 2011b.L'us de Sistemesd'Informació Geogràfica en l'arqueologia del conflicto, in: R. Arnabat Mata and F. X. Hernández Cardona (coord.). *Estratègies de recerca i transferència del coneixementhistòric-arqueològic. El cas de l'aviació republicana (1938-1939)*, 159-176. Llibres de Matrícula, Calafell.

Rueda Galán, C., Bellón Ruiz, J.P. and Valderrama Zafra, J. 2015. Baecula. S.I.G. aplicados al estudio de un campo de batalla de la Segunda Guerra Púnica, in: A. Maximiano and E. Cerrillo-Cuenca (eds). *Arqueología y tecnologías de información espacial: una perspectiva ibero-americana.* Archaeopress Open Access.

Saunders, N. J. 2007. *Killing Time: Archaeology and the First World War.* Sutton: Stroud.

Schofield, J. 2005. *Combat archaeology. Material culture and modern conflict.* London: Duckworth.

Scott, D. D., Fox, J., R. A., Connor, M. A. & Harmon, D. 1989. *Archaeological Perspectives on the Battle of the Little Bighorn.* Norman: University of Oklahoma Press.

Scott, D. D. and Mcfeaters, A. P. 2011. The archaeology of historic battlefields: a history and theoretical development in conflict archaeology, *Journal of Archaeological Research* 19, 103-32.

Notes on contributor

Antonio Franco Fernández (Muros 1975, MA 2001, University of Santiago de Compostela -U.S.C.-, Spain). He is currently in charge of the Council Department of Archaeology for the town of Muros (A Coruña, Spain) and he is associate researcher for Incipit-CSIC, with over 15 years of experience directing or codirecting excavations in Spain, Panama and Somalia, for the U.S.C., the Incipit-CSIC, the Smithsonian Tropical Research Institute or the Fundación El Caño. He is a trained topographer and surveyor and a GIS expert.

Section 4

Archaeology and the public

Disseminating to a wider audience

Geographic Information Systems: an effective tool for the management of the Cultural Heritage of Cantabria

Gustavo Sanz Palomera[1]

Introduction

The development of Geographic Information Systems has resulted in them becoming an effective tool that is often used by public administration for the management of their responsibilities. The use of software tools such as ArcGIS by different departments in the Government of Cantabria enables collaboration between them and greater fluidity in the transmission of information in their sphere of interest.

Starting with the premise that what is not known cannot be protected (García and Alazaga 2008, 120-130), through its Archaeological Section, the Cultural Heritage Service in the Cantabrian Directorate-General of Culture has fostered the use of Geographic Information Systems to improve traditional databases and archaeological and cultural inventories. They have thus become an essential tool for management and protection, not only of fragile archaeological heritage but of all cultural heritage (Espiago and Baena 1997, 7-67), traditionally consisting of buildings or constructions.

However, it is both archaeological and built heritage that have the greatest impact on the territory as legislation in Cantabria authorises the administration to establish a protection area or buffer zone around the different sites to safeguard their unique nature and characteristics, whether listed as a Property of Cultural or Local Interest or as an Archaeological Site.

New technologies enabled a traditional inventory system, consisting of index cards with information about the main characteristics of the sites and their locations, to be converted into a digital database in Access, which is easier to handle and use than the paper forms, in 2006. When this database was integrated into a

1 Consejería de Educación, Cultura y Deporte del Gobierno de Cantabria. Calle Vargas, 53, 39010 Santander (Spain) sanz_g@cantabria.es.

Geographic Information System, using ArcGIS, it became possible to manage large volumes of data very quickly (Martínez *et al.* 2008, 176). Additionally, it is easy to visualise the archaeological sites or Cultural and Local Interest Properties and their protection areas, and any impact they might suffer in the case of projects planned by public administrations or private individuals.

Law 11/1998, 13 October, on the Cultural Heritage of Cantabria

Law 11/1998, 13 October, regulating the cultural heritage of Cantabria, was the result of the transfer of responsibilities in cultural matters from the Government of Spain to the Autonomous Community of Cantabria. Like similar laws in the different autonomous communities in Spain, it is based on Law 16/1985, 25 June, relating to Spanish Historical Heritage, and the Spanish Constitution, enacted on 6 December 1978.

In its Article 49.1, the law states that "properties forming the Cultural Heritage of Cantabria can be listed as:

a. Monument
b. Historical ensemble
c. Cultural place
d. Archaeological zone
e. Natural place".

To protect these properties and their cultural, historical or artistic value, in Article 50, the law establishes that they should be assigned a protection area and in its Section 1 it describes this further: "the protection area of a property listed as of Cultural Interest or Local Interest, is understood to be the place near to the property, whether or not it is in a built-up zone, allowing its appropriate perception and comprehension, considering both the time when it was built and its historical evolution, supporting it environmentally and culturally, and whose alteration might affect its view or its values".

Section 50.2 explains that "the protection area can include buildings or groups of buildings, plots of land or fields – in all cases with the ground underneath it – urban and rural areas, geographical features and natural and scenic elements; whether or not they are very near or distant from the property, or form a continuous or discontinuous area".

In connection with this, Article 18.c specifies that the documentation for listing the property as of Cultural or Local Interest must include the area affected by the designation. Therefore, "the perimeter of the protection area will be fixed precisely, noting the geographical features and natural characteristics that form the area, stressing those that enhance its protection, view and study".

Once the documentation for the designation as a Cultural Interest or Local Interest Property and its protection area has been accepted, any urban activity must be authorised by the Department of Culture and Sport, as laid down in Article 52.1: "Any urban activity inside the protection area of a Property of Cultural Interest or Local Interest, including changes of use, as long as the town planning regulation protecting it has not been passed, will be approved by the Department of Culture and Sport, which will be authorised to determine the criteria and conditions of the activity, according to the general resolutions in this Law and those in the documentation of the designation, if applicable".

The importance that the Law about Cultural Heritage in Cantabria allots to the naturally fragile archaeological heritage can be seen in Section 1 of Article 89, where it establishes that "the properties forming the Archaeological Heritage of Cantabria are given the following types of protection:

a. Archaeological Site. A place with material or latent remains of human activity or of its natural context.
b. Archaeological Zone. A series of archaeological sites displaying unity in terms of their chronology, typology, location or relation with other values of a cultural or natural kind.
c. Archaeological Park. Site, group of sites or archaeological zone with significant elements allowing its social use as a visitable area for purposes of education and enjoyment.
d. Area of Archaeological Protection. Place where material evidence, historical antecedents or other indications show the existence of archaeological or paleontological remains.

However, what is more important for their protection is that the Cantabrian Cultural Heritage Law equates archaeological sites with listed Properties of Cultural Interest, as Section 2 in Article 89 states that "all archaeological sites in the Regional Archaeological Inventory will be protected in exactly the same way as Cultural Interest Properties, even though their designation has not been formally initiated". In connection with this, just as Properties of Cultural and Local Interest should be assigned a protection area, Law 11/1998 in Article 89, Section 3 states that "all archaeological sites and zones will be assigned a protection area from which they will be inseparable, with special attention paid to their natural context".

Therefore, the Cultural Heritage Law gives archaeological sites in the Archaeological Inventory of Cantabria (INVAC) the highest level of protection. However, INVAC does not only contain known or supposed archaeological and paleontological sites, but also areas of archaeological protection, as established in Article 90, Section 2, which states that "the designation of areas of archaeological protection will be published in the Cantabrian Official Gazette, listed in a registry created for that purpose and included in the Regional Archaeological Inventory".

These areas of archaeological protection are the equivalent of the protection areas established around Cultural Interest Properties and, according to Article 90.1, "they will be determined by the Councillor of Culture and Sport, after hearing interested parties and the Municipality affected, and in accordance with a report issued by the authorised technical commission". When the archaeological sites are of particular interest, because they possess rock art[2] or are of great scientific importance, they require greater protection than is provided by their inclusion in the Archaeological Inventory of Cantabria. The Department of Education, Culture

2 Law 16/1985, 25 June, on Spanish Historical Heritage states in its Article 40.2 that "caves, rock-shelters and places with rock art" not only belong to the archaeological heritage forming part of Spanish historical heritage but that "according to this law, they are designated as Cultural Interest Properties". Therefore, all the protection regulations applied to such properties must also be applied to them.

and Sport can initiate, by itself or on request, the process for the declaration of a site or several sites that share the same character or chrono-cultural ascription as a Property of Cultural Interest, together with its protection area, as established in Article 16.1 of Law 11/1998.

The protection of built heritage. The determination of protection areas

The Department of Education, Culture and Sport can initiate, by itself or on request, the process for the declaration of a property or several properties that share the same character or chrono-cultural ascription as a Property of Cultural Interest, together with its protection area, as established in Article 16.1 of Law 11/1998. Alternatively, it can be listed as a Property of Local Interest, together with its protection area, according to Article 28.1 in the same law.

As seen above, Law 11/1998, Article 50.1 defines the consideration of the "surroundings of a property listed as of Cultural or Local Interest". The procedure to determine the protection area of a Property of Cultural or Local Interest, catalogued as a Monument, Historical Ensemble or Cultural Place is:

First the property or properties are identified on a map. However, those listed before Law 11/1998 came into effect were not required to possess a protection area. Those which are to be listed on the initiative of the Department of Culture or on request are assessed in terms of their position using orthophotos or the land registry.

Following the precepts of Articles 50.2 and 18.c in Law 11/1998, whenever possible highly visible landmarks are used, such as geographical features (such as peaks, summits, rivers or streams) or man-made structures (tracks, roads, buildings, etc.), even though these may be altered over time due to urban changes to the area.

Tools in Geographic Information Systems help the administration design the protection areas as they also enable a study of the visibility of the properties within the area being protected from various viewpoints near the properties. Above all, the most important point is that the perimeter of areas should be easily identified by everyone.

Finally the latest land registry information is used, despite the differences between the municipal boundaries established by the administration of the autonomous community and those found in the land registry, which is managed by the national administration. Land registry information is vital in this case because Law 11/1998 in its Articles 17 and 28 established that all the property owners affected by the possible definition of the protection area around a Cultural or Local Interest Property must be informed so that they can present any allegations they deem necessary.

When the area has been provisionally designed, technicians visit the site of the Property to verify whether this preliminary design fulfils the function of guaranteeing its protection and visibility. To do this, photographs are taken from several positions and the degree of visibility from each one is assessed. If necessary, any defects noted during these visits are corrected. Then the corrected area is submitted for the approval of the corresponding Technical Commission; the Built Heritage Commission in the case of cultural buildings and places. The Directorate-

General of Culture can then open the case and request the binding reports of two consultant institutions, before submitting it to public information and informing all the property owners affected by the designation, as explained above.

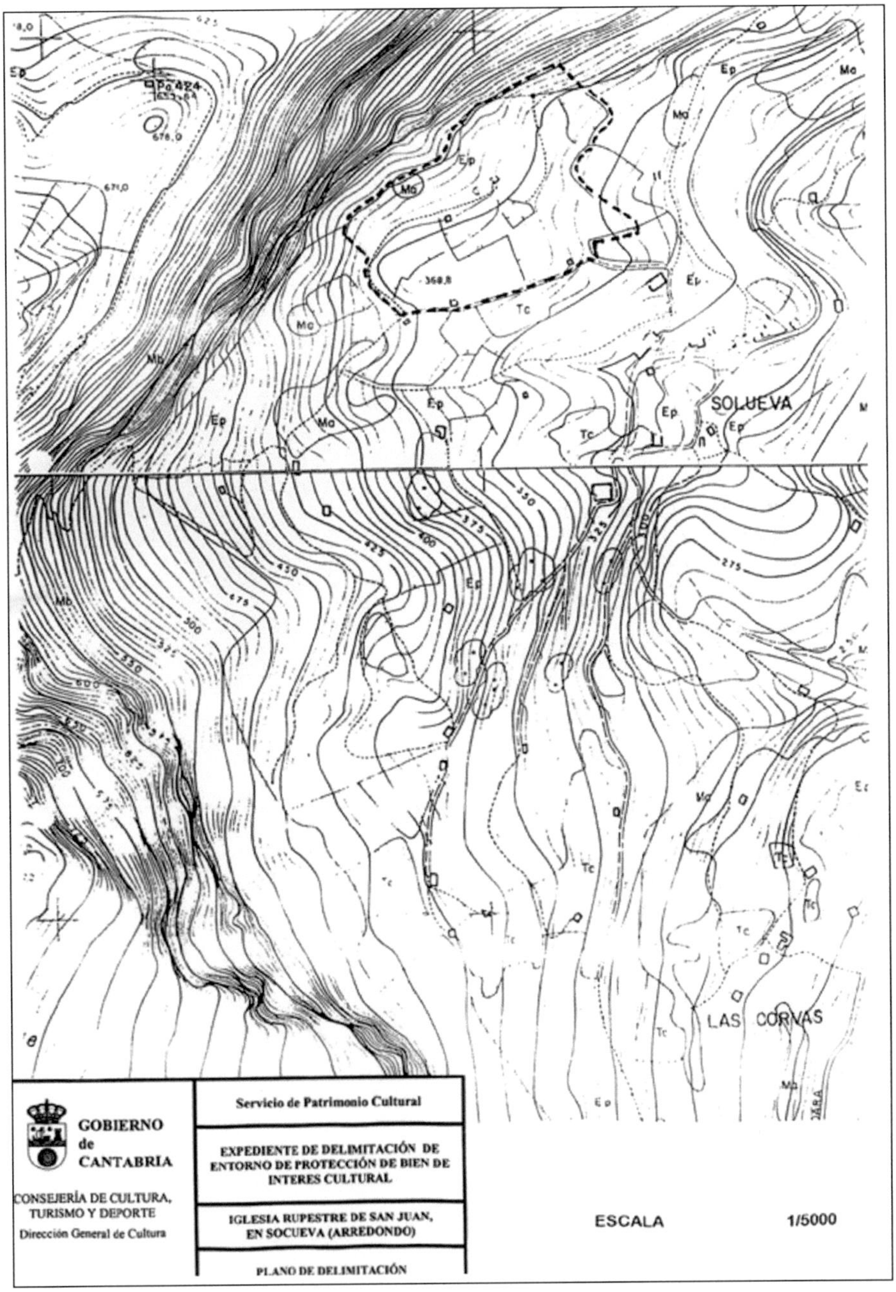

Figure 1. Map of the protection area of the Cultural Interest Property "The underground church of San Juan de Socueva" in 2005 (Directorate-General of Culture, Government of Cantabria).

If any allegations are made that the Commission considers to be fair, the protection area is redesigned and the opinion of the consultant institutions is requested once more; it is re-submitted to public information and the owners are informed before the definitive declaration by the Government. If no allegations

Figure 2. Aerial image of the redesigned protection area of the Cultural Interest Property "The underground church of San Juan de Socueva" in 2012 (Directorate-General of Culture, Government of Cantabria).

are made, the declaration can go ahead so that the Property and its area are listed in the General Registry of Cultural Interest Properties of Cantabria or the General Catalogue of Local Interest Properties, as appropriate.

The advantages of the use of new technologies by the Administration can be illustrated by the following examples. For instance, the use of a GIS revealed that the protection area of a Cultural Interest Property, the underground church of San Juan de Socueva (Arredondo), did not protect the Property as, according to the plan that accompanied the declaration, the church lay outside the area. Therefore this had to be designed again.

Geographic Information Systems make it easier for the public to understand aerial photographs so that they can recognise the sites more easily and GIS also make it possible for maps and plans to be published with true scales. In many cases, this avoids the protests and allegations made by the general public. This occurred, for example in the process of the designation of the protection area around the Cultural Interest Property, the Botín Historical Garden in Puente San Miguel, where the boundary was established 4m outside the physical limits of the garden.

Although in the document of the designation, this type of map or plan is shown true to scale, and this is also sent to all parties interested in the declaration, the adaptation to the measurements of its publication in the Cantabrian Official Gazette means that the size and scale of the plans is often distorted. However, this does not affect the general public's ability to understand them, as has been observed in recent years.

Figure 3. Aerial image showing the protection area around the Cultural Interest Property, the Botín Historical Garden (Directorate-General of Culture, Government of Cantabria).

The protection of archaeological heritage. The design of protection areas

One of the responsibilities of the Cultural Heritage Service is to improve the protection of archaeological properties located in the area of the Autonomous Community of Cantabria. The analysis of the information in the INVAC using GIS was able to locate the archaeological sites which, because of their characteristics and/or position, were most liable to suffer the impact of human activity and had to be protected with all the guarantees furnished by Law 11/1998, because they contain rock art or possess great scientific significance.

In connection with this, once the large task of updating and verifying the INVAC had finished, the Archaeological Section has proposed the design and declaration of several archaeological zones with their respective protection areas. These include the karst complex of La Garma (Ribamontán al Monte and Ribamontán al Mar), the megalithic ensemble of Collado de Sejos-Cuquillo (Polaciones and Mancomunidad de Campoo-Cabuérniga), the Forts of El Brusco and El Gromo (Santoña, Noja and Argoños), the Fort of El Rastrillar (Laredo), the schematic art site of Braña de los Pastores (Cabezón de la Sal), the archaeological sites of Cueto de Morín (Villaescusa) and El Puyo (Miera) and the military structures of Monte Picota (Piélagos).

In the case of the fortification of El Gromo, dated in the period of the Napoleonic Wars, human activity placing a large trough for livestock in the middle of the fort had begun to destroy the weak earthen ramparts, as cattle grazed and walked freely around the whole site, which went unseen because it was surrounded by a plantation of eucalyptus trees. The farm had been established breaking the regulations not only of the Farming Department but also of the Directorate-General of the Environment as the fortification was inside Santoña Salt-marshes Natural Park. And, most certainly without considering archaeological and cultural heritage.

Figure 4. Appearance of El Gromo earthen fortifications, before its declaration as a Cultural Interest Property (Directorate-General of Culture, Government of Cantabria).

Figure 5. Aerial image showing the discontinuous protection area of the Cultural Interest Property with the category of Archaeological Zone "Fortifications of El Brusco and El Gromo" (Directorate-General of Culture, Government of Cantabria).

The procedure for the design of the protection area of a Cultural Interest Property with the category of archaeological zone, for its declaration,[3] is usually more complex than for the protection area of a Cultural Interest Property with the category of Monument, Historical Ensemble or Cultural Place. This is due to the nature of the property to be listed and because they are usually located in geographically more complex places, such as mountainous or forested zones that mean it can be extremely difficult to locate the property. In such cases, the administration collaborates with companies that use highly precise topographic tools, such as differential GPS, which gives an extremely precise location of the boundaries of the areas.

In the case of caves with rock art that form part of complex underground systems, geologists and geographers carry out a study of the whole catchment area of the streams in the cave and design a protection area that at times might seem too large or out of proportion, but which is the most suitable to avoid any future construction project affecting the drainage of the cave system and consequently the conservation of the prehistoric art on the cave walls.

The procedure is rather similar to the one described for built heritage in the previous section: initial assessment using orthophotos or the land registry, GIS-aided design of the protection areas (including visibility analyses). Again, whenever possible, visible landmarks are used, either as geographical features or man-made. Finally the latest land registry information is used to inform property owners.

On-site visits are done before submitting for approval the proposed area ofprotection by the corresponding Technical Commission; the Archaeological and Paleontological Commission in the case of archaeological sites and zones. The Directorate-General of Culture can then open the case and request the binding reports of two consultant institutions, before submitting it to public information and informing all the property owners affected by the designation, as explained above.

Again, after any possible allegations are reviewed, the designation can proceed so that the Property and its area are listed in the General Registry of Cultural Interest Properties of Cantabria.

A particular example: the documentation for the northern route of the Pilgimage Road to Santiago de Compostela being listed as World Heritage

There is no doubt that the preparation of the documentation for the designation of the northern route of the Way of St James as World Heritage is where Geographic Information Systems have shown their full potential. The case was managed by the Directorate-General for the Protection of Historical Heritage of the Government of Asturias, supported by the Ministry of Culture and Sport as well as by the Autonomous Communities of Galicia, Basque Country, Cantabria and La Rioja.

The use of orthophotos, proposed by the Cantabrian Cultural Heritage Service, enabled the design of the new pedestrian routes, attempting to make them safer and more accessible for pilgrims, avoiding the use of communication routes with heavy

3 Articles 16 to 23 of Law 11/1998, 13 October, on the cultural heritage of Cantabria.

Figure 6. Plan of the associated property "Collegiate of Santillana del Mar" included in the documentation for the designation of the Northern route of the Pilgrimage Road to Santiago de Compostela as World Heritage (Directorate-General of Culture, Government of Cantabria).

traffic. This helped the proposal being accepted by the assessors of the UNESCO Committee and, most importantly, its understanding by individuals who might see their interests affected by the designation. By combining the various alternatives of graphic representation on the maps and plans that formed part of the appendix of the documentation, it was possible to produce a joint design that facilitated the exchange of data and information between the different administrations involved in preparing the documentation.

Whereas previous documentation for UNESCO only contained land registry information, roads and tracks and the protected properties and their protection areas, the new tools available to society and the public administration were able to make it clearer, not only for the UNESCO experts who assessed the candidature but also for anyone: which properties were being protected, what their protection areas were and where were they really located. This was possible thanks to the symbiosis of traditional documentation (land registry information, coordinates and surveying) together with more recent documentation (satellite orthophotos).

Advantages of GIS for the Heritage Administration in the management and protection of cultural heritage

The form of the new model of record for the Archaeological Inventory of Cantabria (INVAC) was designed in Access so that it would be compatible with, and could be integrated in, a Geographic Information System. In this way, the utility of what would have been a simple database (Fernández and García 2003, 1-2) would be increased exponentially by combining it with the representation in a system of coordinates that located the sites on the Earth's surface (Alonso *et al.* 2007, 31). This same principle is applied to listed Properties forming part of the cultural heritage of Cantabria (Cultural Interest Properties, Local Interest Properties and Inventoried Properties) as the tools in Geographic Information Systems allow the administration to carry out many more procedures than the design of protection areas for listed Properties or those in the process of being designated.

Among the many advantages of GIS (Fernández and García 2003, 3-15) for the authorities in charge of protecting cultural heritage, the following may be highlighted:

- Speed in localising the cultural properties: an aerial view is an straightforward way to access substantial information in a simple and agile way (orthophotos, land registry information, topographic data, typology of the cultural properties) and notice any errors that might appear in the information about each Property.
- Accuracy in the location of the Properties and their protection areas: however, higher precisions are only used in certain cases, such as designing the areas around caves with prehistoric art, where geological studies are necessary in order to determine the drainage basin of the water courses in the whole cave system.
- Documentation and overall access to the information about the Properties: thanks to the GIS tools, we can access all the information about an archaeological site or Property contained in a database (its coordinates in the ETRS89 datum, history of the Property and any research carried out, year it was listed, its protection area, bibliographic information, links to websites mentioning the Property and its designation in the Cantabria Official Gazette, etc.).

Assessment of the impact of projects and preparation of reports: this is probably the most powerful tool that a GIS provides for public administrations to manage cultural heritage and above all archaeological heritage. With these programmes, the administration can locate and diagnose, and sometimes prevent (Benítez and

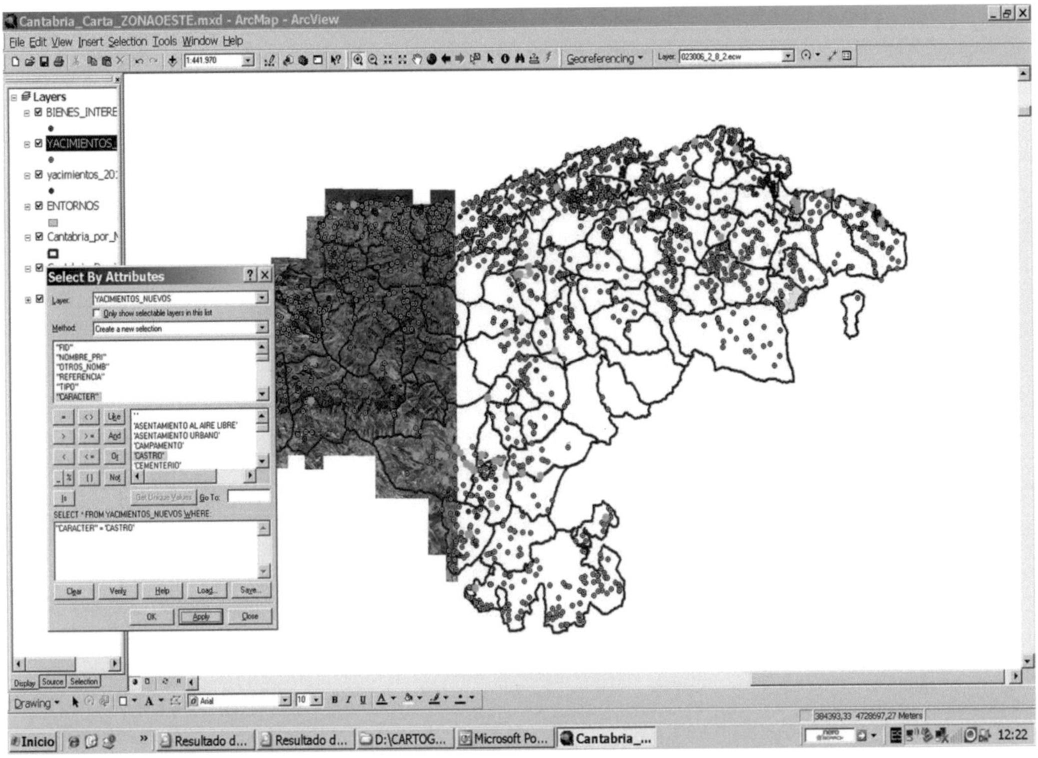

Figure 7. Screenshot of a query showing sites of the same typology (Directorate-General of Culture, Government of Cantabria).

Alonso 2011, 112) possible problems and risks that a project or plan might entail for properties or sites and their protection areas, in particular, and on the terrain in general.

In connection with this, the assessment of the effects of any project or plan will result in drafting a report with the preventive or corrective measures required to avoid the impact that project or plan might cause to properties belonging to the cultural heritage of Cantabria.

Preparation of archaeological heritage management and protection plans: GIS allows areas with greatest archaeological potential to be delimited and, after designating the Archaeological Zones and their protection areas, they can be used to draft special plans such as the ones for Altamira and La Garma Karst Complex, currently in progress.

Design of research strategies according to specific needs for the management and protection of archaeological heritage: these are promulgated by the Department of Education, Culture and Sport for the Regional Archaeological Plan,[4] which determines the lines of research whose funding is considered a priority, from both the territorial and chrono-cultural points of view, in order to fill the gaps

4 Published in the Cantabrian Official Gazette, No. 109, on 8 June 2010.

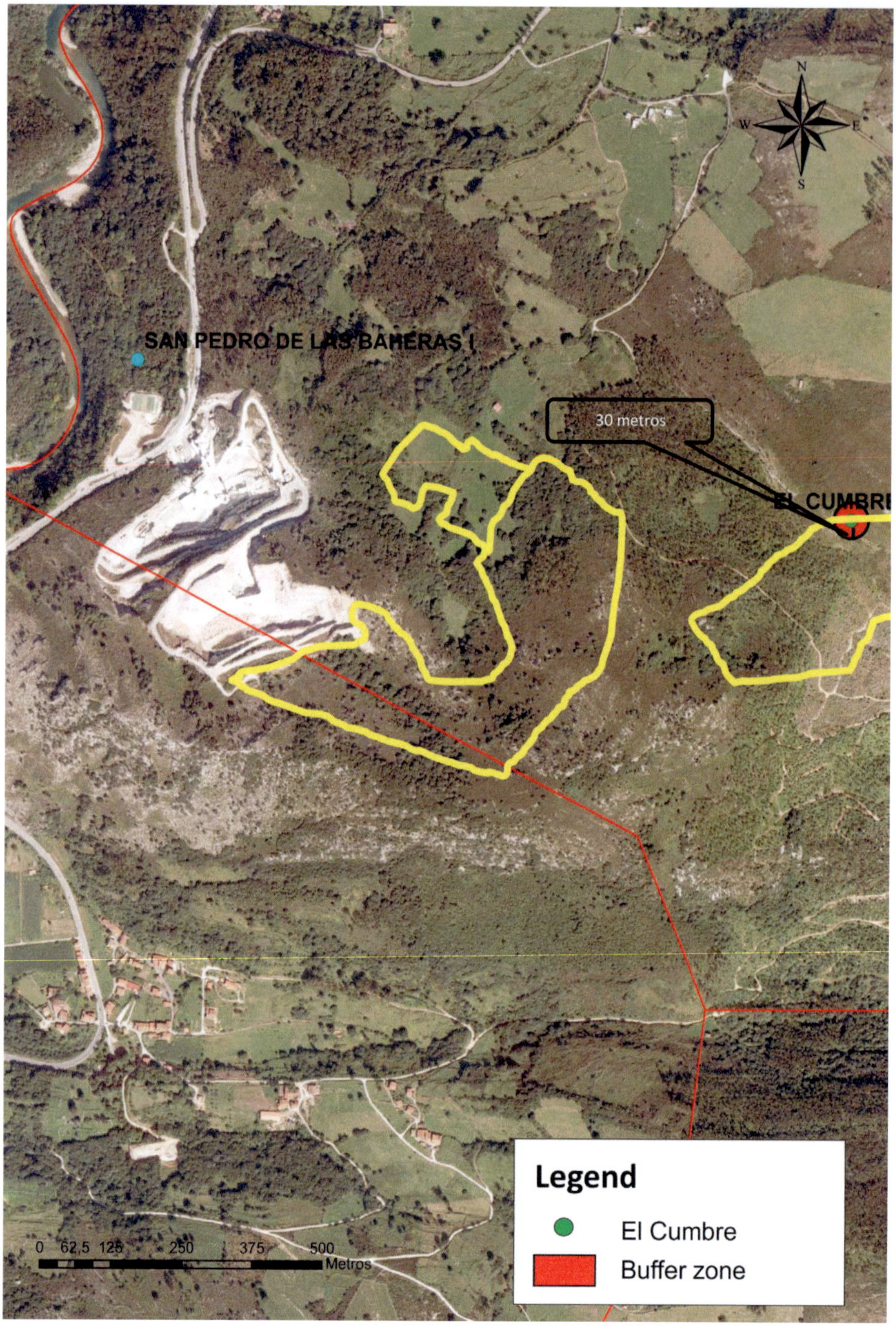

Figure 8: Example of a report analysing the impact of a mine on an archaeological site listed in the Archaeological Inventory of Cantabria, produced using a Geographic Information System (Directorate-General of Culture, Government of Cantabria).

in previous research. Certain historical periods and places in the autonomous community of Cantabria have been neglected by archaeological research, as can be seen in the map showing the different types of sites that are currently listed in the Archaeological Inventory of Cantabria INVAC).

- Facilitate the exchange of information between different departments and public administrations: some of the most serious problems suffered by the administration are its slowness and lack of communication between the different departments. GIS allow information about cultural heritage (location and nature of designated properties and their protection areas, and archaeological sites) to be selected and sent to other governmental departments. In this way they will be able to verify, beforehand, the possible effects that their projects might cause to cultural properties and above all to archaeological sites, due to their great fragility. They can then inform the Directorate-General of Culture, which will determine the appropriate preventive and corrective measures.
- Collaboration with national security forces: data with the position of all the cultural properties and archaeological sites listed in the INVAC and INVASUC (Underwater Archaeological Inventory of Cantabria) have been submitted to the Civil Guard and Legal Police.
- Facilitate the understanding of the general public: this is perhaps the most important aspect of GIS, as an orthophoto can be understood by all individuals who are capable of recognising roofs, trees, streams etc., which can be more difficult to identify on maps or land registry plans. In this way, the number of complaints and allegations made to the designation of a Cultural Interest Property or a Local Interest Property and their protection areas has decreased greatly since the declarations of Cultural and Local Interest Properties have been published in the Cantabria Official Gazette using GIS.
- Design of dissemination strategies (cultural routes and ways): GIS can be used to design cultural routes and ways that allow the general public to discover all the wealth of this heritage which is sometimes unknown and neglected.
- 3D Analysis of the territory: the use of 3D visualization techniques can help the general public understand the impact that some plans or projects might cause to archaeological sites and Cultural Heritage Properties in Cantabria. To date, the use of this tool is still quite limited.

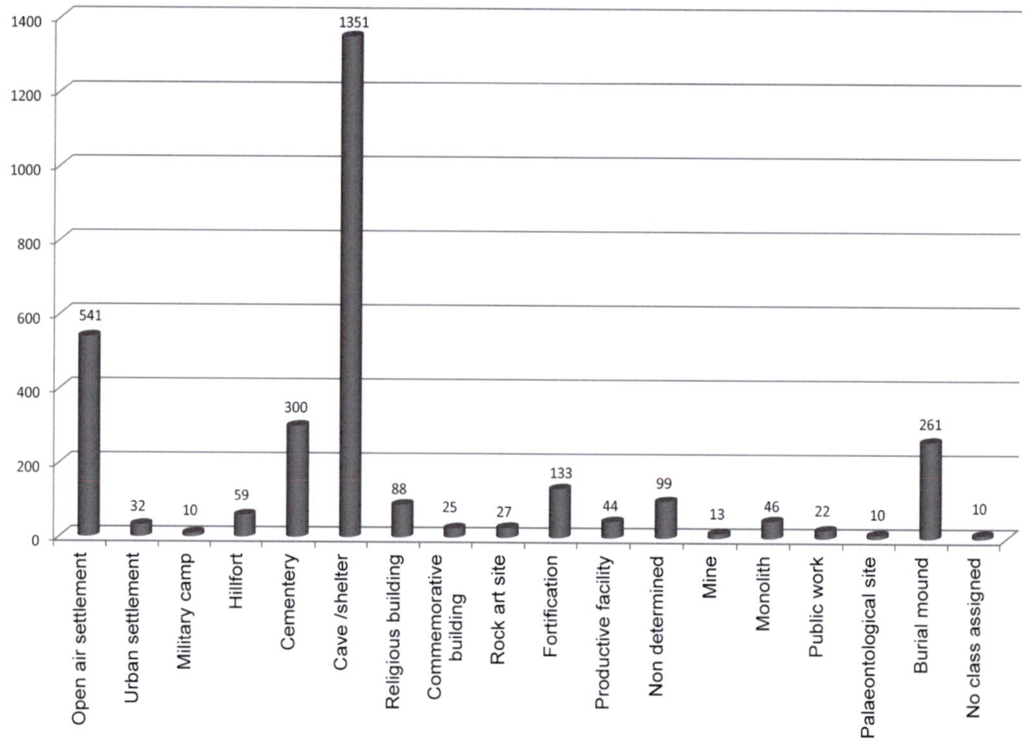

Figure 9. Typology of the sites listed in the Archaeological Inventory of Cantabria (Directorate-General of Culture, Government of Cantabria).

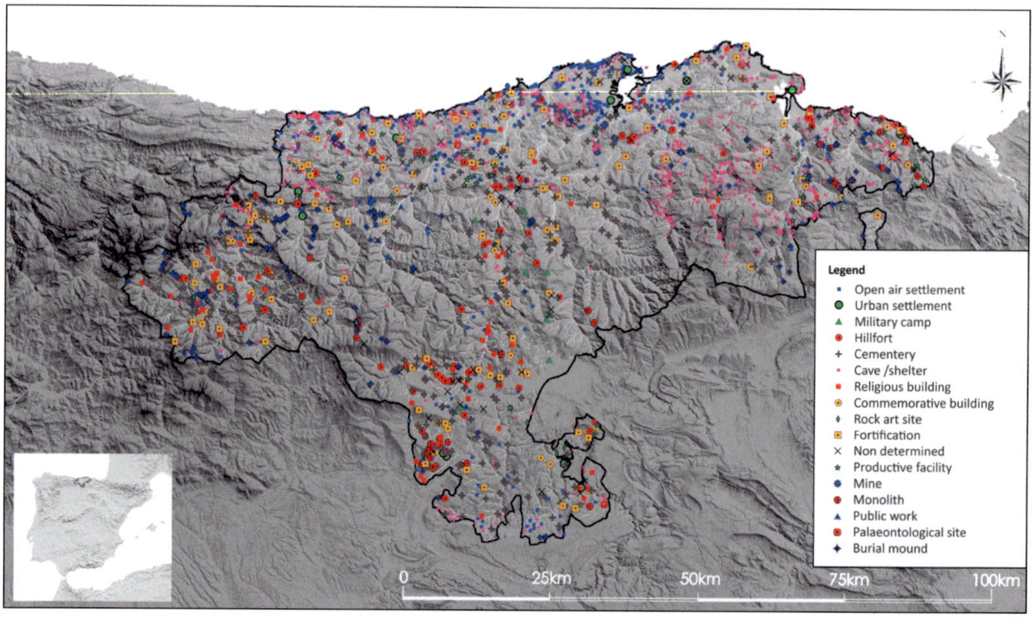

Figure 10. Map showing the position of all the sites currently listed in the Archaeological Inventory of Cantabria (Directorate-General of Culture, Government of Cantabria).

Conclusions

Possession of a good knowledge of our cultural heritage is the only way to protect it. Its nature and characteristics, together with the passing of time and the pressure exercised by humans, through urban development, forestry or industrial work, make it necessary to protect this heritage with all the tools admitted by the Heritage Law, even though sometimes archaeological heritage is only assumed to exist.

The importance of new technologies within the management tasks entrusted to the administration, in this case to the Cultural Heritage Service in the Cantabrian Directorate-General of Culture, must be stressed. The use of Geographical Information Systems facilitates the exchange of information with other government departments, such as Rural Services, Town Planning and Urban Environmental Assessment, Industry, Urbanism and the Environment, or with local or national administrations. This makes the administration's response to the general public's demands more agile.

Additionally, reports and the design of protection areas around Cultural and Local Interest Properties are easier to understand by the public, as the boundaries of protection areas are identified better on the orthophotos that are used with GIS.

Finally, new technologies are able to detect errors in old documentation, improve the visualisation of data in the hands of the administration and ensure the reliability of the data. All these aspects contribute towards improving the protection and management of the Cultural and Archaeological-Paleontological Heritage of the autonomous community of Cantabria.

References

Alonso Villalobos, C. *et al.* 2007. SIGNauta: un sistema para la información y gestión del patrimonio arqueológico subacuático de Andalucía. *PH: Boletín del Instituto Andaluz del Patrimonio Histórico* 63, 26-41.

Benítez López, D. and Alonso Villalobos, C. 2011. Aplicabilidad de los SIG para la gestión del patrimonio arqueológico subacuático andaluz: SIGNauta. *PH: Boletín del Instituto Andaluz del Patrimonio Histórico* 77, 110-112.

Bosque Sendra, J. 1992. Sistemas de Información Geográfica. Madrid.

Casado Soto, J.L. 2000. La carta arqueológica de Cantabria (CARSUCAN) y otras actuaciones en el Patrimonio Marítimo regional, in: Ontañón, R. (ed.). *Actuaciones arqueológicas en Cantabria. 1984-1999*. Santander, 57-67.

Casado Soto, J.L. 2009. Arqueología subacuática en Cantabria. El laboratorio para investigaciones arqueológicas subacuáticas (LIAS), in: Nieto, X. and Cau, M. A. (eds.). *Arqueologia Nàutica Mediterrània. Monografies del CASC* 8, 101-106.

Espiago González, J. and Baena, J. 1997. Los Sistemas de Información Geográfica como tecnología informática aplicada a la arqueología y a la gestión del patrimonio, in: Baena J. *et al.* (eds.). *Los SIG y el Análisis Espacial en Arqueología*. Madrid, 7-67.

Fernández Cacho, S. and García Sanjuan, L. 2003. Los SIG en la tutela del patrimonio arqueológico de Andalucía. ATLAS. Territorios y Paisajes en la Prehistoria Reciente de Andalucía, Departamento de Prehistoria y Arqueología, Universidad de Sevilla (online), https://grupo.us.es/atlas/documentos/articulos/desarrollo_local/desarrollo_local.pdf
(search: 14/ 1/ 2016).

García Rivera, C. and Alzaga García, M. 2008. La Carta Arqueológica Subacuática de Andalucía como instrumento para la tutela de un patrimonio emergente. *Mainake* 30, 129-143.

Fernández-Posse, M.D. and Álvaro, E. 1993. Bases para un inventario de yacimientos arqueológicos, in: *Inventarios y cartas arqueológicas: Homenaje a Blas Taracena*: 65-72.

Gómez Vega, B. 2000. Pasado y presente de la Arqueología subacuática en Cantabria. *PH*: *Boletín del Instituto Andaluz del Patrimonio Histórico* 33, 173-176.

Martín de la Cruz, J.C. and Lucena Martín, A.M. 2003 (eds.). Actas del I Encuentro Internacional de Informática Aplicada a la investigación y la Gestión Arqueológicas. Córdoba.

Martínez Casas, I. *et al.* 2008. Los Sistemas de Información Geográfica en su aplicación al ámbito arqueológico. Panorama actual, in: Iglesias Gil, J. M. (ed.). *Actas de los XVIII Cursos Monográficos sobre el Patrimonio Histórico (Reinosa, julio 2007). Cursos sobre el Patrimonio Histórico* 12, 173-189.

Sanz Palomera, G. in press. El inventario arqueológico de Cantabria (INVAC). Una eficaz herramienta para la gestión del patrimonio arqueológico de Cantabria. *Sautuola* XX.

Sources

General Catalogue of Local Interest Properties in Cantabria.

Law 16/1985, 25 June, on Spanish Historical Heritage.

Law 11/1998, 13 October, on the Cultural Heritage of Cantabria.

Decree 36/2001, 2 May, of the partial development of the Cantabrian Cultural Heritage Law.

Archaeological Inventory of Cantabria (INVAC).

Underwater Archaeological Inventory of Cantabria (INVASUC).

General Registry of Cultural Interest Properties in Cantabria.

Notes on the contributor

Gustavo Sanz Palomera completed his degree course at the Complutense University of Madrid and obtained a Ph.D. in History at the same university in 2005. His publications include *La Annona y la Política agraria de los emperadores Antoninos para el Occidente del Imperio* in British Archaeological Reports International Series, and numerous contributions in books and papers in specialised journals and conference proceedings, dealing with various aspects of the economy, the army, society, religion and politics in the times of the Roman Empire. Since 2007 he has been working as an archaeologist in the Cultural Heritage Service of the Government of Cantabria. He is also the co-director of the archaeological excavation of a shipwreck in San Vicente de la Barquera (Cantabria).

A map for Gondar. Cartographic system for the touristic development of the Amhara Region (Ethiopia)

Cristina Charro Lobato,[1] *Eduardo Martín Agúndez*[2]
and Agustín Cabria Ramos[3]

Introduction and objectives

The proposed project focuses on the Amhara Region and the city of Gondar, relying on the tourism sector of Ethiopia as an agent capable of promoting economic development and, therefore, capable of improving the quality of life of its inhabitants. We recognize the importance of tourism in Africa as a catalyst for socio-economic development, as it is the sector in which Africa will see growth above the global average over the next decade. This tourism will be based on recognition of the historical, cultural and natural heritage, and should be aimed at motivating local people to acquire a commitment to preserve their heritage and, at the same time, boost their own identity. Heritage-based tourism, besides being a driving force of cultural promotion, is also an economy booster, as it comprises a variety of activities such as cultural visits, recreation and associated educational activities, which entail accommodation and gastronomy, besides shopping or consumption of various products. These activities not only stimulate the economy, but strongly promote the culture, is associated with leisure and, in several occasions, with new experiences through various alternatives (Lara and López-Guzmán 2004).

The project takes on a specific component of technological innovation, providing new resources for tourists (digital maps, GPS routes, etc.) which will facilitate their access to information on the usual tourist destinations in the area,

1 (UCM, Developmap) Departamento de Prehistoria, Facultad de Geografía e Historia, Universidad Complutense de Madrid (UCM) mccharro@ghis.ucm.es.

2 (IGN, Developmap) Instituto Geográfico Nacional (IGN)- Ministerio de Fomento; emartin@fomento.es.

3 (IGN, Developmap) Centro Nacional de Información Geográfica (CNIG). Instituto Geográfico Nacional (IGN)- Ministerio de Fomento; agustin.cabria@cnig.es

as well as promoting new ones. The inclusion of these new touristic areas will lead to a fairer and more sustainable development of the region (Dunn *et al.* 1997), as they are commonly rather isolated. As a starting point, the main objective was to develop a digital geodatabase in order to produce thematic touristic maps of the city of Gondar and its surroundings using Geographical Information Systems (GIS). The second objective was to create a geoportal to publish cartography through Spatial Data Infrastructures (SDI), to make it available from anywhere, together with a set of cultural routes designed to be implemented in GPS devices. These routes were included in the paper map, the one objective currently achieved.

The situation of Ethiopian heritage-based tourism

At present, Ethiopia only gets a minimal quantity of the overall tourism generated in the Horn of Africa (Table 1); besides, the average stay of tourists in the city of Gondar and its surroundings tends to be short. It is a non-stable tourism, where stays rarely exceed one night. However, in recent years the annual increase of visitors has grown significantly, reaching 100.000 visits from both local and foreign tourists in 2011 (Feseha 2010), although it is still a small number, considering that this is a very interesting region in terms of cultural heritage and other activities.

Focusing on the city of Gondar, it received a total of 30.000 foreign visitors during the year 2011, with German, French and Spanish among the first in the ranking. These figures, provided by the Office of Tourism of the City of Gondar, include 36 different countries of origin (Table 2).

At a national level, World Heritage Sites have a strong demand, and are subject to various tourism initiatives -in addition to other activities: preservation, promotion, distribution, etc. (UNESCO 2016). When compared at a continental level, Ethiopia has nine UNESCO World Heritage Sites, besides three more on the provisional list, similar to Morocco, Tunisia and South Africa. However in these countries the number of visitors can be up to 25 times greater (Table 1).

COUNTRY	WHS	VISITORS (in millions)
Ethiopia	8	0.377 (2008)
Morocco	8	8.3 (2009)
Tunisia	8	6.9 (2009)
South Africa	8	9.592 (2009)
Algeria	7	1.9 (2009)
Egypt	7	1.,914 (2009)
Tanzania	7	0.75 (2009)
D.R. of the Congo	5	0.30 (2004)
Senegal	5	0.875 (2007)
Zimbabwe	5	1.956 (2009)
Kenya	4	1.783 (2007)
Mali	4	0.180 (2008)

Table 1. Number of visitors in several African countries, ranked by their number of World Heritage Properties. Self-produced based on data from the World Tourism Organization (UNWTO).

RANKING	ORIGIN	FASIL GHEBI VISITORS (Jan-Jun 2011)
1	Germany	1913
2	France	1262
3	Spain	1237
4	UK	1061
5	USA	1035
6	Italy	947
7	Netherlands	748
8	Belgium	413
9	Poland	252
10	Austria	243
...	...	
	Total	**12973**

Table 2. Total amount of tourists visiting the Royal Compound in Gondar, between January and June 2011.

Comparing Ethiopia with a country as visited as Egypt, which has seven World Heritage Sites, the difference is almost 32-fold. The only countries with a lower number of visitors are the Democratic Republic of Congo and Mali, with 5 and 4 World Heritage sites, respectively.

Tu sum up, Ethiopia, although being one of the major owners of WH Properties in Africa, is among the less visited. This is due, partially, to how information is spread by some of these countries, mainly from private sector initiatives with relatively difficult accessibility of information. This is mainly due to the scarce promotion of tourist attractions via the Internet, the tool most widely used by the leading countries of origin of tourists. In fact, this is one of the conclusions reached in several International Tourism Fairs, meeting points for tourism professionals, devoted to tourism development in Africa.

A map for Gondar

The idea of creating a touristic map for Gondar was based on several assumptions. Firstly, the awareness of its strong potential as a tourist destination, due not only to what it has to offer, but to its safe context and the existing communication network. Secondly, the realization that the full range of touristic possibilities offered by the area, beyond the Royal Compound and Simien Mountains National Park, was little known. Finally, through initial contact with potential local partners, we found that there was a strong interest in the project and the willingness to collaborate in providing the information available. So with small financial support provided by some Spanish entities (see Acknowledgments section), we decided to create the Gondar City Tourist Map, a first edition paper map.

Archaeological and cartographic backgrounds

As mentioned, there is a deficit of touristic publications within the Ethiopian public sector. Not even through the World Heritage Centre website are there links to thematic maps, which must be obtained by other means. In northern Ethiopia there are several WHS -Gondar, Axum and Lalibela-but there is no associated cartography to assist in locating and visiting them. It is possible to find maps of Axum and Lalibela generated by the Zamani Project (http://www.zamaniproject.org/), an African cultural heritage enhancement project that has tried to generate useful information for various fields, primarily research and conservation. This was different from our purpose, but very useful as it provided a basis for initiatives such as the one presented here.

In Ethiopia our team knew of just one previous project with similar characteristics, the one developed by the German team of the Institute for Geo Research (GTZ) that made a touristic map of Addis Ababa city with the collaboration of the Ethiopian Mapping Agency and the German Embassy. A proof of its success was that the 2008 edition of the map had been sold out by that very same year, even though its scope was relatively small, around the metropolitan area, with a legend showing the main places of interest.The project presented here wanted to go beyond the Gondar area, by including other places with great cultural value. In this way, Gondar would be able to become a well-known city not only due to its World Heritage castles complex, but also due to the heritage in its surroundings, of great historical interest.

There had been some other partial cartographic initiatives in the city of Gondar and its surroundings, some relating to archaeological monuments. Among them, we must highlight the maps of the Gondar castles complex that Sandro Angelini (1971) and Flemming Aalund (1985) made for UNESCO within the framework of conservation projects. Also, for research and conservation purposes, maps were made within the project devoted to the archaeological study of the presence of Jesuit Missions in the Lake Tana Region, led by professor Victor Fernández Martínez (see Fernández 2013; Fernández *et al.* 2012 for an overview), in which one of us collaborated (del Arco and Martín 2009). Finally, there was the huge project by Hiob Ludolf Center for Ethiopian Studies of creating an *Encyclopaedia Aethiopica*, recently finished (Uhlig *et al.* 2014). Some other initiatives, purely cartographic, were the touristic map by Demilie Arega (unspecified year) of *Lake Tana and its Environs*, the *Simien Mountains Ethiopia 1:100.000* map edited by the University of Bern and the Swiss Centre for Development (2010), the unfinished project of a map of the city of Gondar by the local company Girma Semu Consult, and the most recent one, a draft map of Gondar made by the University of Gondar in 2012. The latter was a local initiative following a workshop our team had held there in 2011 to get some feedback on the project, which demonstrated the effectiveness and profound uptake of the idea.

Natural and Cultural Heritage information content

We wanted this work to become the basis for tourism development in this area, based on two main concepts: archaeological parks and cultural routes. An archaeological park is the space that comprises an archaeological area and its landscape. Archaeological routes are itineraries which link different points under a coherent theme; so the objective was to create an image of a coordinated identity of different activities. Both features are appropriate for both cultural and touristic leisure, and are the centre of geographic areas of great interest, which makes it necessary to contextualize these areas in their environment. This can be done through thematic cartography, together with explanatory texts. These were offered as a formula for the development of thematic cultural tourism, as well as representing a basis to encourage measures for the protection and preservation of the local heritage, the development of new places open to visits, and activities which tend to offer a wider knowledge of the culture. With this objective in mind, we started gathering information about various cultural monuments and natural locations, both in the Amhara Region and Gondar and its nearby area, as a basecamp to access the touristic attractions of this area.

The Amhara Region, in the north of the country, is limited geographically to the south by Lake Tana and to the north by the Simien Mountains National Park. In this area there are a significant number of sites of high touristic potential. Some of the most important will be briefly mentioned, such as Simien Mountains National Park (WHC – 1978); Alatish National Park; Lake Tana – the source of the Blue Nile- and a few kilometres to the south are the famous Tis Isat Falls. All over the region we can find orthodox churches of great cultural interest (Phillipson 2009), as much for their unusual building structures as for the paintings inside them. There are also scattered remains of monuments made by Spanish and Portuguese Jesuits during the Seventeenth Century, composed mainly of residences, churches, pools, a palace and a cathedral (Danqaz, Debsan and Gorgora Nova Jesuit Missions). There is also a small architectural legacy of the brief period of Italian occupation (1936-1941), such as a monument to the dictator Mussolini on the north shore of Lake Tana (Mussolini's Lighthouse, inaugurated in 1938). It should be added that some of the main roads built in that period are preserved, and are now being restored. In the surrounding area there are several interesting cities that are broadening their services to tourists, allowing extended stays in the area. We can highlight the cities of Bahir Dar (capital of the Amhara Region), located 175 km from Gondar; Lalibela at 360 km, a World Heritage Site (WHC-1978); and Axum at 357 km, also a World Heritage Site (WHC-1980), all linked by road and with frequent flight connections from Gondar.

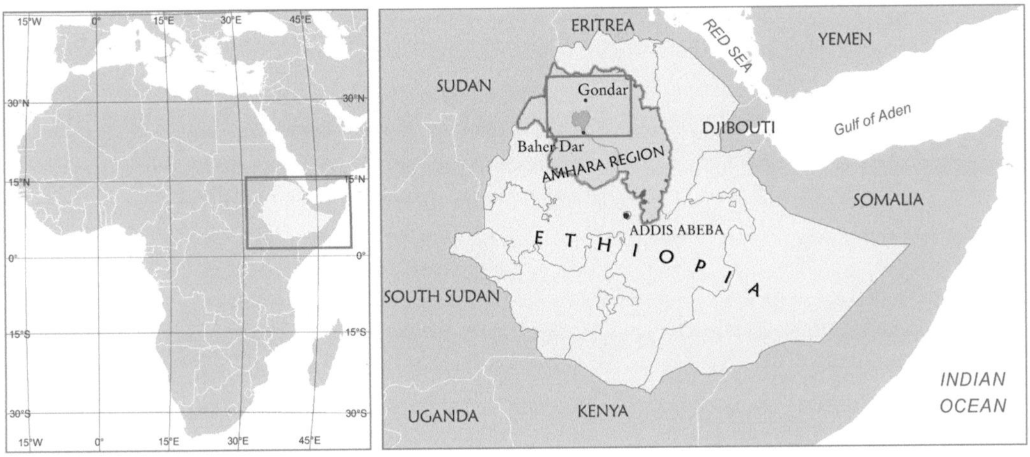

Figure 1. Location maps of the area of study. International context map (left), and regional context map (right); the blue box marks the area of study (Authors).

There are many buildings of historical interest worth visiting in Gondar (Ethiopia's capital until 1855), among them outstanding examples of the so-called Gondarine Ethiopian architecture (Royal Compound (WHC – 1979) of Fasil Ghebi and Qusqwam castles, together with Fasiladas Bath -a large rectangular pool where the feast of Timkat takes place annually), public buildings built during the Italian occupation (area of Piazza) and religious monuments. Among the latter, the best known is the church of Dabra Berhan Selassie, which was built in the 17th Century, part of the World Heritage compound due to its murals reflecting Ethiopian hagiography and religious history, as well as the ceiling paintings showing angel heads with wings, considered representative of Gondarine art and one of the most recognizable symbols of the city. Near Gondar City there are some

Figure 2. Mosaic of various attractions of the Amhara Region (authors).

monuments from the first half of the 17th Century built during the reigns of the emperor Susenyos and his son Fasilides on their royal camps, such as Guzara Castle.

This information was included in the map, as brief insights to the monuments and routes that can be visited in both the Region and the City.

Geographic information content

A large part of our work involved concentrating on finding sources of geographic information, with the premise of locating resources at the lowest possible cost. The search was conducted via various existing geo-information portals on the Internet, for example, the USGS (United States Geological Survey), OpenStreet Maps, and other more general search engines. We sought the largest amount of resources to perform the tasks for future phases of work so that, in turn, they would be useful layers for GIS information and its derivative products. The resources finally used are briefly listed below.

Medium resolution imagery – Landsat: Landsat-7 with its sensor ETM+, with a spatial resolution of 30 meters in the visible spectrum bands (NASA 2016a), making it possible to obtain images of 15 meters thanks to the panchromatic band, through techniques of data fusion. This resolution was very suitable for showing the region on a small scale. In addition, the resource is freely accessible on the USGS website (USGS 2016). Four different images covered the area, merged in a single mosaic.

ASTER Digital Elevation Model (ASTER DEM): the Digital Elevation Model (DEM) of the ASTER project (Advanced Spaceborne Thermal Emission and Reflection Radiometer) is an initiative of the Ministry of Economy, Trade and Industry of Japan and NASA (USA) (Japan Space Systems 2012). It provides information on the relief in a regular grid, and a geometric resolution of 30 meters, which allows the representation of the topography of our working area. As for Landsat, all ASTER project information is available for free on the USGS website. The information collected by the sensor is divided into 1° quadrants. Our work area was represented by a total of 12 fragments, merged into a single entity.

MOD44B Vegetation Continuous Fields-MODIS: forest areas of the region are an item to add to the geographic base, useful to supplement the information displayed, as our working area is unique in having quite a few unpopulated areas. Adding this layer provided a more balanced appearance in the final composition, and a more accurate representation of the landscape. To this end it was decided to use the MOD44B processed images acquired by NASA (2016b) representing the percentage of vegetation cover held by each pixel of the image, with a spatial resolution of 250 meters.

DIVA-GIS Vector Bases: DIVA-GIS initiative (DIVA-GIS 2016) is a free GIS software that offers on its website a vector basis of all countries in the world with information on administrative boundaries, roads, railroads, altitude, terrain cover and population density. It was therefore a useful source of geographic information for our GIS, used on certain scales.

QuickBird imagery: high-resolution satellite orthophotos covering the city, a particularly valuable and very useful resource for generating geographic information. The transferred images covered an area of 10.072 Has of the urban area of Gondar, provided an excellent geographic information base for subsequent office work, and constituted a fundamental layer in the GIS. The approximate value of this information was €2.000, confirming the acceptance and commitment to the project by the Ethiopian counterpart.

Official cartography -Vector Bases from land registry: CAD layers where plots and buildings were represented, provided by the land registry office of the city of Gondar, dependent on the City Hall.

Official cartography – Ethiopian Mapping Agency- 250K Series: the need for this cartographic base arose from the lack of information in the outlying areas of the city of Gondar, so to provide our mapping with a set of geographical features to frame the other thematic information, administrative boundaries, hydrography and roads communication were represented, drawn from official EMA maps. The 250K series (1:250,000 scale) was chosen because it facilitated the vectorization of the mentioned layers, only 8 map sheets being needed on this scale, compared with 100 sheets for the 50K series (1:50,000 scale). Moreover, it proved to be an ideal complement to the DIVA base, which overlapped with Landsat images, providing an overview of the region on a small scale.

Fieldwork: entities of interest were recorded to capture their geographical position in coordinates, in addition to collecting the data necessary for their thematic classification and taking pictures of each, all collected on a geodatabase. The recording was set initially for the area of influence of Gondar, reaching out 150 kilometers. Finally, due to the meager financial support obtained, the recording system of these routes was just tested, which will serve for future validation of fieldwork campaigns. A team of three people walked the city with a handheld GPS device, camera, notebook and maps of the city. Points were captured at the same time as they were drawn on

CODE	Type	Type (English)	CODE	Type	Type (English)
ALC	AlquilerCoche	*Car rental*	IPU	InstitucionPublica	*Public institution*
BUS	EstacionBus	*Bus station*	BAN	BancoCambio	*Money change*
TAX	Taxi	*Taxi*	INF	Informacion	*Information*
AER	Aeropuerto	*Airport*	SOU	Souvenir	*Souvenir shop*
OAR	OficinaAero	*Aerial office*	MON	Monumento	*Monument*
HOT	Hotel	*Hotel*	MUS	Museo	*Museum*
PEN	Pension	*Hostel*	IGL	Iglesia	*Church*
CAF	Cafeteria	*Café*	MEZ	Mezquita	*Mosque*
BAR	Bar	*Bar*	SIN	Sinagoga	*Sinagoge*
RES	Restaurante	*Restaurant*	CIN	Cine	*Cinema*
SUP	Supermercado	*Supermarket*	PAR	Parque	*Park*
MER	Mercado	*Market*	HOS	Hospital	*Hospital*
COR	Correos	*Post office*	CLI	Clinica	*Clinic*
LOC	Locutorio	*Phone center*	FAR	Farmacia	*Pharmacy*
INT	Internet	*Internet center*	GAS	Gasolinera	*Gas station*
AYU	Ayuntamiento	*City hall*			

Table 3. Classification of tourism entities to register (Authors).

the maps, to verify their location, ensuring that the final accuracy of the registered coordinates did not adversely influence the work. To facilitate the planning of daily work, the urban area was divided onto 20 sheets, also used to review the names of streets and districts and their boundaries. A thematic classification of features to record assessed the recording and later representation in the tourist cartography, reflected in Table 3. Ultimately, the entire city of Gondar was recorded, capturing the position of more than 400 entities with their own photographic record.

Bibliographical research: To collect part of the thematic layers on heritage and tourist features of interest, in addition to the fieldwork, we analyzed publications over the heritage of the Amhara Region. Helped by descriptive reviews on how to get to various monuments and once located on satellite and mapping images, more than 10 sites the team was unaware of were located, and formed part of the representation of a small-scale map of the Amhara Region. In this way, we ensured the accuracy of their locations, and facilitated future reviewing, checking and updating in future fieldwork campaigns.

GIS creation/implementation

The thematic information obtained, along with the geographic databases and heritage information, was integrated into a Geographic Information System, enabling the visual hierarchy, processing and analysis of information, and the generation of various geographic products. For this task ESRI ArcGIS was chosen both because it has a great graphic potential for generating quality end products, and for being the one used by Ethiopian institutions.

The spatial reference system was chosen from between the official ADINDAN datum (ellipsoid Local Clarke 1880 – projection UTM 37N), and the global system (WGS84 ellipsoid – UTM 37N). The latter was selected because most of the geographical information was available in that system. Bases or geographical information layers referring to the ADINDAN datum were converted to WGS84 using the parameters of the following geocentric Cartesian translatory: $\Delta X = -165$ m, $\Delta Y = -11$ m, $\Delta Z = 206$ m.

The ASTER GDEM – EGM96 global geoid was used for altimetry.

With regard to geographical extension, two space frames were selected, making it possible to work with different purposes regarding the information to represent. For each frame there were different cartographic databases. First, the urban representation of the city of Gondar, where touristic entities were located on a cartographic base spatial reference, marking limits for representing both the center and the peripheral areas. The geographic information layers were: a) QuickBird Image, b) 20 m Contour lines – obtained from ASTER DEM through the tool 3DAnalyst > Surface Analysis > Contour- with later filtering and manual redraw; c) Parcel and Building layers, filtered from a CAD file; d) Forest, River and River Area layers, Streets, Bridges and Green Areas layers -vector polygon shapefiles- were obtained by hand digitizing over the Quickbird image.

The second spatial framework represented the surroundings of Gondar in a radius of 150 km, suited for day trips. The geographic information layers were: a) Hillshade, obtained using the tool 3DAnalyst > Surface Analysis > Hillshade; b) Boundaries_0 and Boundaries_1, containing national and regional administrative areas from DIVA-GIS; c) National Park, digitized after geo-referencing the Simien Mountains National Park map; d) Forest, representing vegetal coverage, obtained from MOD44B images, by reclassifying its radiometric resolution 2^8 – 256 levels/ pixel- to show areas with absence of forest (0-14.99 value), forest (15-100 value), water (200 value) and no data (256 value); e) River lines, Water area, Roads, Urban area and City were obtained by digitizing the previously geo-referenced EMA 1:250.000 cartography.

Gondar City Tourist Map

Integration of all geographic layers in the same geodesic system was followed by the representation of all the available geographic information using an attractive visual composition. In addition, text, images and sketches were inserted to enrich the physical description of places, improving the cartographic product. This was a very time consuming task, combining cartographic representation techniques with editorial skills. For this purpose, ESRI ArcGIS was used again, along with Adobe Corel DRAW to design symbols.

Map format

A study on common dimensions of tourist paper maps was performed. The selected format is above average -but within regular dimensions- as it was decided to use the maximum available space on the paper, allowing more information to be added. The folded design makes the various map contents easy to see, without needing to open the entire worksheet. The dimensions are: folded: 22x12 cm, unfolded: 70x100 cm, advertising: 12x12 cm. The paper was of excellent quality (90 g/m2), currently used by the Spanish National Geographic Institute (IGN) in their National Topographic Maps.

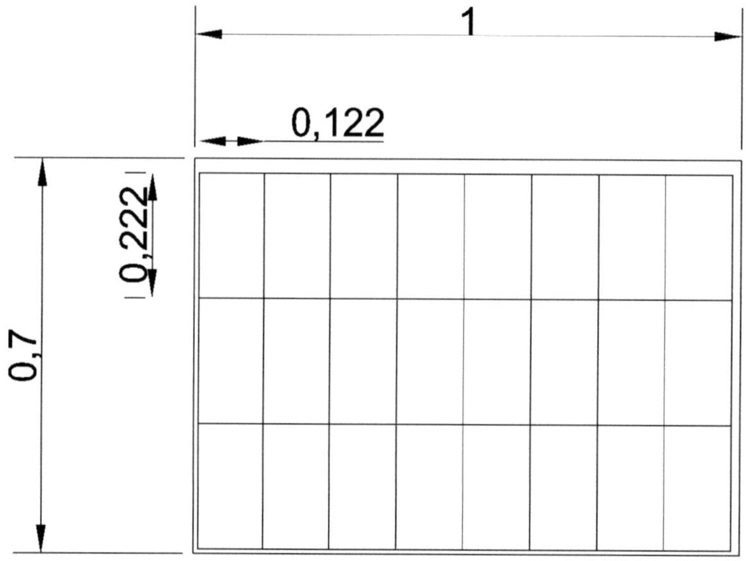

Figure 3. Schematic of folds and dimensions of the sheet (Authors).

The following data frames were implemented:

- Gondar City map, 1:6,000 scale.
- Map of surroundings of Gondar, 1:30,000 scale.
- Amhara Region map, 1:500,000 scale.
- Fasil Ghebi Compound Map, 1:2,000 scale.

Thematic information

Thematic information of the city of Gondar came from places and sites registered with handheld GPS devices during the 2012 campaign, after the position adjustment of some points -due to some accuracy errors- and filtering of entities in areas with an overabundance of points, to improve readability. In addition, some thematic types were not used to lower the information display load.

The process of aggregating thematic information in the Amhara Region was different, since these entities were extracted from publications and were located on satellite images.

Symbolization

One of the most laborious phases in map design is the creation of each of the symbols representing the different types of thematic layers. It must be borne in mind that the symbol should evoke its meaning instantly, without necessarily having to consult the legend. In addition, it must be easily legible and have a size and color that does not excessively stand out from the cartographic background. A total of 36 original symbols were created, except for a few more universal ones, such as airports, bus stations, company logos and World Heritage Sites. All symbols have the same format (see Figure 4). It was decided to create a specific set of symbols that helped to quickly interpret information regarding tourist routes of the Amhara Region collated on the map.

Labelling

One of the most important aspects to consider in any map is labelling, as it requires control for each of the texts and compliance with certain rules. Different labelling was used for hydrography, localities, geographic points and peaks, important places and place names. Names recorded in this map were a sensitive issue, because there are different ways of writing the same place name, depending on the source consulted. For instance, we can find written the name "Gondar" as "Gonder" or "Gondor". It was chosen to keep the toponyms as in the EMA cartography, but not

Figure 4. Shape and dimensions of the cartographic symbols of thematic entities (Authors).

every element of our project was registered on them. Therefore, the team contacted Dr. Andreu Martínez D'Alos-Moner, member of the project *Encyclopaedia Aethiopica*, for consistency regarding the names used.

Structure and final composition

In order to give an added value to the cartographic product, the following features have been included (marked in Figures 5 and 6):

1. Front cover with description of the inside
2. Back cover with credits and advertising space
3. Global and continental location maps
4. Technical specifications of the map
5. Geographical description and historical overview
6. Route maps
7. Photographs of the sites included in the map
8. Summary of the proposed routes
9. Text describing the routes
10. Legend of symbols used
11. Map of the outskirts of Gondar
12. Map of the Fasil Ghebi compound
13. Sketch of the minibus network

Figure 5. Amhara Region side of the map, 1:500,000 (Authors).

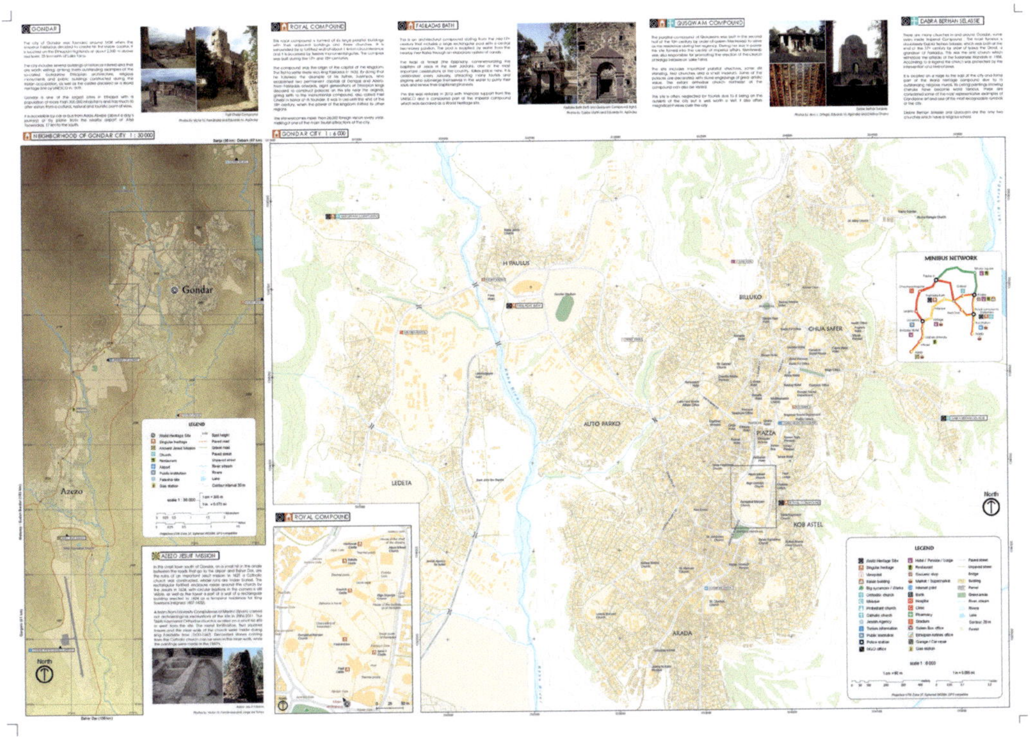

Figure 6. Gondar side of the map, 1:30,000 and 1:6,000 scales (Authors).

The map was divided into two different parts, one for each side of the sheet. On the first side we find the information regarding the Amhara Region, while the second shows everything about the city of Gondar and the surrounding areas.

The final presentation of the product was determined by the folds already described, so that the element of the front and back covers occupied the entire cover.

Conclusions and future prospects

This Gondar City Tourist Map was an important step in promoting touristic quality in Ethiopia, and make it known to the rest of the world. It was also a proof of the quality of the work performed, as was recognized by The British Cartographic Society with the *National Geographic Society New Mapmaker Award* in 2013, and it is available in several countries in Europe through different stores (Land&Karte, Stanfords, Librería Desnivel, etc.).

We considered it relevant to present here this Amhara Project for two reasons. First, because it is an experience of cooperation that goes beyond the purely academic sphere in which, until then, the project members had participated. Second, because it demonstrates, once again, the usefulness of GITs not only as tools for information management and analysis, but for social action and divulgation. There are many different possible platforms and formats, so it has become one of the most powerful tools we can rely on in areas in which heritage features play a central role.

In this sense, we could not assess its impact, for several reasons. The first is that, despite having run with costs of printing and distribution, signed collaboration agreements and held an official ceremony, the maps are still in boxes, have not been distributed within Ethiopia. Secondly, we have not reached the objectives proposed in origin, except very partially: in addition to the paper map was the design of a geo-portal, currently under development, which cannot succeed due to lack of funding. The latter also affects other scheduled tasks, such as training of local staff for maintenance and final disposal of information for the Ethiopian exploitation of the map. In a similar situation is the Complutense University archaeological project, which has failed on museological conservation of monuments. However, University of Gondar has made a small draft map of the city, by their own local initiative, which emerged from this project, which we consider a success from the point of view of awareness and local commitment.

Nevertheless, it was an experiment in designing a model that could be extrapolated to any country, region or area -with sufficient heritage potential to boost the tourism sector- in terms of methods, techniques and resources. The model takes into account the constraints of each location of action to establish an optimal methodology. Furthermore, it applies techniques based on Geographic Information Technologies so the model could be implemented with a minimum cost, also through the use of both freely available geographic data and open sources.

Acknowledgements

The authors want to thank the Spanish Agency for International Cooperation and Development (*Sistema cartográfico para el fomento turístico de la Región Amhara (Etiopía) – Acción preparatoria* 10-CAP2-1562) which funded the starting phase of the project, the University of Alcala (*Sistema cartográfico para el fomento turístico de la Región Amhara (Etiopía) – Recopilación de información UAH* – 11/2011), whose economic support financed part of the fieldwork, and the material support provided by IGN, as well as Mar Zamora for her involvement during the first phase of the project. Gondar City Council, North Gondar Administrative Zone Culture & Tourism Department, The Office of Culture and Tourism of the Amhara Region and other authorities of Gondar welcomed us and provided various administrative support and available information. The University of Gondar showed their willingness in collaborating and holding the workshop *Heritage, Tourism and Cartography*. The Ethiopian Mapping Agency through Ato Sultan Mohammed attended us and found the project as an opportunity to promote this Region. Ángel Chica from Spanish Technical Cooperation Office (OTC) played an active role as a source of information and local contact, as did Mercedes Pérez from the Spanish Embassy in Addis Abeba. Chalachew Nigatu had a big role as local guide, as well as Dawit Tibebu, for their kind help in contacting the local authorities. Special thanks to Víctor M. Fernández, Jorge de Torres and Andreu Martínez for their company, comments, help and support before, during and after the whole project. There have been people who have worked selflessly in the production of the map, especially the members of the association Developmap Cartography for Development, as well as Carlos Martín and Ana Ortega, whose support were decisive in many different requests. Finally, we would like to express our gratitude to all the local people we met during our fieldwork, for they are the ones that really do know their city.

References

Aalund, F. 1985. *Ethiopia, Operational action for the safeguarding of the immovable cultural heritage and its reintegration into morden life, Master Plan for the Preservation and Presentation of Cultural Heritage.* UNESCO.

Angelini, S. 1971. *Ethiopia, The Historic Route, A work-plan for the development of the sites and monuments.* UNESCO.

del Arco, V. and Martín, E. 2009. *Registro cartográfico a escala 1:500 y cartografía del detalle a escala 1:250 de los restos arqueológicos de los asentamientos jesuitas del s. XVII en Etiopía.* Madrid: Universidad Politécnica de Madrid.

DIVA-GIS (2016). [Internet]. Available at:<http://www.diva-gis.org/>[last accessed on 30/06/2016].

Dunn, C.E., Atkins, P.J., Townsend, J.G. 1997. GIS for development: a contradiction in terms? *Area* 29 2, 151-159.

Fernández, V.M. 2013. Enlivening the dying ruins. History and archaeology of the Jesuit Missions in Ethiopia, 1557-1632. *Culture and History, Digital Journal* 2/ 2.

Fernández, V.M., Martínez, A., Torres, J. de and Cañete, C. 2012. Archaeology of the Jesuit Missions in the lake Tana Region: Review of the work in progress. *Aethiopica* 15, 72-91.

Japan Space Systems 2012. ASTER GDEM. [Internet]. Available at: http://www.jspacesystems.or.jp/library/archives/ersdac/GDEM/E/index.html [last accessed on 30/06/2016].

Lara de Vicente, F. and López-Guzmán, T.J. 2004. El turismo como motor de desarrollo económico en ciudades Patrimonio de la Humanidad, *PASOS Revista de Turismo y Patrimonio Cultural* 2(2), 243-256.

Mulugeta Feseha 2010. *Participatory Tourism: The future of Ethiopia: community based ecotourism development: from research to implementation, model from Adwa, northern Ethiopia.* Addis Ababa: Mulugeta Feseha.

NASA 2016a. The Landsat Program. [Internet]. Available at:<http://landsat.gsfc.nasa.gov/>[last accessed on 30/06/2016].

NASA 2016b. MODIS [Internet]. Available at:<http://modis.gsfc.nasa.gov/data/dataprod/mod44.php>[last accessed on 30/06/2016].

Phillipson, D.W. 2009. *Ancient Churches of Ethiopia. Fourth-Fourteenth Centuries.* New Haven: Yale University Press.

Uhlig, S. *et al.* (eds.) 2014. *Encyclopaedia Aethiopica.* Wiesbaden: Harrassowitz Verlag. http://www1.uni-hamburg.de/EAE/

UNESCO 2016. [Internet]. Available at:<http://www.unesco.org/new/en/unesco/> and <http://whc.unesco.org/ >[last accessed on 30/06/2016].

USGS 2016. Earth Explorer [Internet]. Available at:<http://earthexplorer.usgs.gov/>[last accessed on 30/06/2016].

Notes on contributors

Cristina Charro Lobato is an Archaeologist and Social and Cultural Anthropologist, specialized in Geographic Information Technologies. She works as Digital Cartographer for a multinational GPS device company, while working on her PhD on Iron Age Iberian Landscape Archaeology. She has conducted research projects in Spain (Iron Age) and Ethiopia, and participated in archaeological projects in Spain, Portugal, France and Ethiopia.

Eduardo Martín Agúndez is a Cartography and Geodesy Engineer, Geographic Engineer staff of the Spanish National Geographic Institute. He has extensive experience in the design and implementation of Geographic Information Systems, and has participated in different archaeological projects always providing a cartographic vision. These include collaborations with the Complutense University of Madrid in archaeological projects in Ethiopia and with the *Encyclopaedia Aethiopica* at the University of Hamburg.

Agustín Cabria Ramos is a Geographic Engineer in the Spanish National Geographic Institute, an institution of the Government of Spain, with responsibility for Astronomy, Geodesy, Cartography, Geophysics, Seismicity and Volcanism, among others. He currently works at the National Center for Geographic Information in outreach efforts related to geosciences products. He taught as an associate professor at the University of Alcalá (Madrid) for 14 years, where he participated in projects of thematic mapping of Ethiopia. He has great interest in Archaeology, and has done volunteer social work in Mexico and Guatemala.

Concluding discussion

P. M. van Leusen[1]

Introduction

This book celebrates 10 years of extracurricular 'geomatics in archaeology' teaching at the Mérida Institute of Archaeology and the Institute of Heritage Sciences in Santiago de Compostela. Since the start of the courses in 2006, much has improved especially in the availability, range and quality of digital geodata – but much has also remained the same: every year, fresh students of archaeology sit down at their introductory courses to learn about the practice and theory of spatial data. The course coordinators and editors of this volume rightly stress that, whilst GIS is the tool, Geomatics – the gathering, storing, processing, and delivering of geographic information or spatially referenced information – is the learning goal. From my own very similar experience teaching GIS-in-Archaeology courses for the past 15 years, I can add that GIS technology has remained essentially stable and mature since the turn of the millennium, and that having GIS and geomatics skills still provides archaeologists (who have tended to be 'early adopters') with an advantage in the job market.

The editors note, in their introductory paper, the twin problems facing educators in this type of course: on the one hand, teaching the students to handle a very complex piece of software and associated geodatabase; on the other, teaching them to understand its limitations in the face of spatial archaeological data and theory. As the editors explain, the volume consists mostly of excerpts from master's or doctorate theses, ongoing academic research projects and, in a few cases, work by professional archaeologists working for commercial or government agencies. As is to be expected, the quality of these contributions is very variable, in part due to differences in the ease with which the authors express themselves in English. A sensible thematic order has been imposed on the volume by its four sections, on (1) Documenting and characterizing the archaeological record, (2) Methods and procedures for spatial analysis, (3) Case studies, and (4) Dissemination. I will organize my discussion according to these four themes, and then attempt to draw some general conclusions.

---

1 University of Groningen.

1. Documenting and characterizing the archaeological record

The very first paper in this section, on patterning in assemblages from surface survey, demonstrates that 'characterizing' the archaeological record really requires complex spatial analysis (for which see section 2). This introduces the problem of how to 'sell' the products of such analyses to heritage managers and to the general public (section 4). Kernel Density Estimation is used by Sevillano Perea to visualize complex derivatives of the original data, which can no longer be grasped intuitively. In order for such analyses to be accepted by other stakeholders in the field of Cultural heritage Management, researchers will have to demonstrate that their geomatics approaches are robust in the face of 'dirty' archaeological data and in the presence of (too) many potential explanatory variables.

In the concluding section of the second paper, by Garcia Sanchez, an interesting point is being made of whether surface survey is destroying the spatial patterns in the survey record by removing essential components of the site assemblages. This issue has been debated since the 1970s, and deserves to be studied experimentally, but here I want to point out that the rural archaeological record is being destroyed anyway by tillage and other activities, we seem to be unable to stop it, and therefore 'preservation by record' is the next best thing we can do. So the real question is, can we produce a good archaeological record in the field or do we need to collect, clean, study and store the artefacts where any expert can access them at any time?

The third and fourth papers in this section stress the importance of digital filtering techniques for pattern recognition when dealing with 3D point clouds generated either from laser scanning or structure-from-motion (photogrammetry). These are two of the main new data sources that geomatics courses in Archaeology now need to incorporate. Costa Garcia and Fonte rightly stress the importance of field verification in the documentation process: any identification of archaeological features in RS data 'at the desktop' should be regarded as preliminary until verified.

2. Methods and procedures for spatial analysis

The three contributions in this section nicely demonstrate that working on methods and procedures for spatial analysis in Archaeology requires a strong *methodological* framework. This is not explicitly present in any of the papers, and indicates an aspect that should be strengthened in training. The first paper (by Paniego Diaz, a historian) discusses and compares some uncontroversial types of line-of-sight analysis, but the relative utility of these methods depends on the *purpose* that has been defined for it – and this is lacking. By comparison the third paper, by Garcia Garcia, is much clearer in defining the methodological goal ('a method to measure wind protection as a criterion for past settlement location choice') before proposing a procedure to achieve this.

The second paper of this section, by Señoran Martín, shows good intentions in attempting to generate specific hypotheses about movement within megalithic landscapes but fails on two methodological counts: first, that his theorizing about new concepts of mobility does not give rise to any *testable* predictions, and secondly, that the use of spatial statistics is not getting him where he needs to be. The latter point perhaps requires some elaboration: setting yourself the goal of proving that some archaeological distribution is spatially non-random is not interesting – human behavior is non-random – so what we really want to know is *in what way* they are non-random! For example, does settlement behavior follow a Poisson curve with respect to distance-to-a-resource, and if so, which function does it approximate? This is advanced stuff, more suitable for the higher-level courses, but at a more basic level we may note that establishing a statistical correlation between the locations of megaliths and the modeled least cost network is not enough for this author to be able to infer a causal relationship between the two. After all, many other factors could (and probably did) also affect mobility. These are general failings in archaeological geomatics, so should not be taken as criticism of this particular work, but still they must be addressed here: if theory, methodology, and methods are not properly connected, the student of geomatics is left empty-handed!

3. Case studies analyzing spatial patterns

The purpose of a case study is to investigate and illustrate, for a particular case, some general problem and/or its solution. In the context of geomatics in Archaeology, case studies normally explore a specific GIS approach or procedure to see if it can be made to support a pre-existing hypothesis (sometimes not made explicit by the researcher). Case studies, moreover, have a problematic relationship with theory: only if the latter gives rise to some practical question or testable hypothesis is it useful to include an extensive theoretical discussion. It is much more important to include a good discussion of *methodology* (that is, the justification of the geomatics approach chosen by the researcher), since this determines the value of the case study to the reader.

The 'case studies' section of this volume contains seven contributions, several of which illustrate the muddle that results from not formulating an explicit hypothesis. In the very first paper, the cost distance analysis performed does not appear to contribute significantly to the authors' stated goal, which is to increase our understanding of mobility in relation to lithic raw material procurement. In the fourth paper (GIS analysis on Roman caves) no clear spatial hypotheses have been formulated either, and the authors' conclusions appear unsupported by the evidence provided. In fact these two cases could more properly be called 'exploratory data analysis' (EDA), where the goal is simply to investigate various types of potential relationships between variables and to see if any interesting patterns can be produced. In this light we may also understand the third paper, by Liceras-Garrido *et al.*, as an exploratory location analysis of Iron Age sites.

We can contrast this to the contribution by Costa García on Roman military sites, which is clearly more mature in describing research hypotheses and approaches and in explaining the significance of the results obtained. This author also is a positive exception in that he is aware, and informs the reader, of the biases that plague his dataset. I regard this as a crucial element in archaeological geomatics teaching: almost inevitably, archaeologists must work with not just imperfect or incomplete data sets but with severely biased ones, because they are the end result of a series of depositional, postdepositional and discovery processes that each skew the evidence in a particular direction. Thus, every case study has to argue that, *despite* these biases, the data are still suitable for the intended geomatic analysis!

The contribution by Sebastián Lopéz on rock art illustrates this same point in a different way. She is attempting to find support for her ideas about the relationship between cave paintings and transhumance in the Mediterranean uplands, on the premise that rock art is the first archaeologically visible sign of this economic practice. Deciding whether there is a *significant* relationship (in terms of cost distance or viewshed control) between caves with rock art and transhumance routes requires one to define the 'background' against which the selected dataset is being compared. For example, how many more caves, with or without rock art, might exist in the study area? Do these caves perhaps have a similar relationship to the transhumance routes?

Taking a much more strongly theory-driven approach, López Lillo argues that the apparent even distribution of habitations within his Andean study area, along with the empty area around the central mound of La Bolsa, are indicative of a type of society that is not adequately covered by the Sahlins-Service classification. Instead, he proposes an alternative that I am unable to reproduce, since it is protected by an impenetrable forest of Clastrean theory, but my point here is that his geomatics approach is well justified and integrated in his theoretical discourse. The reader need not necessarily agree with the author's assumptions or conclusions, but at least a fruitful discussion of his argument is now possible.

Finally, warfare is one field of human behavior in which we might expect rational location and allocation decisions will be made. The application of geomatics to battlefield archaeology, as in the final paper of this section by Franco Fernandez and Rodrígues Simón, allows for various kinds of useful GIS studies especially in the area of line-of-sight analysis (but recall that early versions of GRASS GIS also had functions to calculate artillery trajectories); their paper also illustrates how 3D visualizations can help explain a complex situation to a nonspecialist public.

4. Dissemination

This section is called 'archaeology and the public', but the first of the two contributions (by Sanz Palomera) mainly discusses the use of GIS to create new protected areas and prepare impact assessment studies. Although the author stresses that the introduction of GIS has facilitated the public understanding of heritage management information, his main concern is in its advantages as an information management tool – increasing the speed and accuracy with which heritage information is recorded, shared among stakeholders, and queried. The efficiency and effectiveness of heritage management departments is improved because having a shared spatial database allows more facile exchange between departments, less time consumed in impact assessment and preparation of management plans, more time available for field checking of the accuracy and precision of the data, and higher-quality dissemination products. The final contribution to the volume, by Charro Lobato *et al.*, gives an interesting and honest account of what geomatics is involved in something as 'simple' as making a map for heritage tourism in Ethiopia. In a sense, this was one of the most instructive papers for me personally, because it is not about analytical geomatics at all (which is after all what we focus on in academic geomatics courses) but rather about creating a public information product that needs to succeed in a particular socio-economic and political context.

Conclusion

What can we learn from this volume celebrating ten year of geomatics teaching in one of Spain's foremost centers for spatial archaeology? As will be clear from my discussion of the contributions in section 2 and 3, I believe the instructors should consider strengthening the methodological component of their basic and advanced courses. There is a painful but necessary lesson here: if no clear research question is formulated, then no clear methodology can be devised – and then it will remain unclear what, if anything, is the significance of the results. Students achieve methodological strength not so much from the 'how to' component of the instruction – learning geomatics techniques and approaches – as from discussing the possibilities and limitations of archaeological data, theory and methodology on the basis of case studies. In that sense, this volume can play an important role as a 'reader' for future versions of the courses. I recognize, of course, that instructors try to achieve their learning goals in the face of the twin pressures of limited time available and students' desire to focus on the technical skills rather than the methodological issues...

A final point that I want to make here concerns the advantages and disadvantages of publishing these contributions in English. It is important for some of these young Spanish researchers to participate and compete in the international English debate, but this can only work if their English is of sufficient quality that it can be understood, and understood correctly, by readers from other language areas. The editors of the current volume have laudably engaged a professional editor to correct the submitted texts, but this costly measure only serves to stress the fact that the forced use of English in the Master curricula and scientific publications of other countries is indeed unfair, placing non-native readers and speakers of English at a disadvantage. Unfortunately the alternative, choosing to publish in your own language – the road often chosen in France, for example – leads to international isolation. A true Devil's dilemma!

The subtitle of this volume talks about harvesting the results of 10 years of geomatics training. To continue this metaphor, any harvest of this type contains adolescent plants that have not fully grown yet, senescent ones that have suffered from insects and blights, as well as healthy examples that provide nourishment to the reader. If we liken the editors of this volume (and the instructors of the Archaeology and Geomatics courses) to farmers, then we may suggest that they should now add fertilizer to those patches of the field in which the plants are malnourished, and apply pesticide where they are under attack.